The world is a poor affair if it does not contain matter for investigation for the whole world in every age. Nature does not reveal all her secrets at once. We imagine we are initiated in her mysteries: we are as yet, but hanging around her outer courts. *Seneca.*

He who sees things grow from the beginning will have the best view of them. *Aristotle.*

JAMES HUTTON
(From a portrait in the possession of Sir Victor Warrinder, Bart.)

The Birth
and Development
of the
Geological Sciences

FRANK DAWSON ADAMS

———

Dover Publications, Inc., New York

Published in Canada by General Publishing Com-
pany, Ltd., 30 Lesmill Road, Don Mills, Toronto,
Ontario.
Published in the United Kingdom by Constable
and Company, Ltd., 10 Orange Street, London
WC 2.

This Dover edition, first published in 1954, is an
unabridged republication of the first edition published by Williams & Wilkins in 1938.

International Standard Book Number: 0-486-26372-X

Library of Congress Catalog Card Number: 55-1083

Manufactured in the United States of America
Dover Publications, Inc.
31 East 2nd Street
Mineola, New York 11501

CONTENTS

Chapter I

INTRODUCTION

Every student of geology when first reading that great classic of the science, Lyell's *Principles of Geology* cannot fail to have his attention arrested by the introductory chapters in which an account is given of the strange explanations which were put forward by ancient writers to account for certain occurrences or phenomena which they observed in nature. Outstanding among these was the presence of fossil shells, skeletons of fish, or the remains of other marine animals embedded in the rocks of the earth's crust on the summits of mountains high above the level of the sea. In the opinion of most ancient writers these were not the remains of animals at all, but had been brought into being by quite other forces than those of life. Some held that they had been developed by the action of the stars and gave their reasons for this opinion. Others were of the opinion that they owed their origin to some process allied to fermentation, which had been set up in the rocks in which they were embedded while these were still in a soft and plastic state. Others described them as "Freaks" or "Jokes" of nature, quaint and amusing productions of the creative power when at play; still others as the discarded remains of preliminary and experimental attempts at creation on the part of the great architect of the universe, preparatory to the actual creation of the world in which we live, the account of which is given in the book of Genesis and the result of which was pronounced by the Divine Creator to be "good." Yet others taught that these fossils had been developed through the influence of the powers of evil for the express purpose of deceiving mankind and giving rise to doubts concerning the truth of the teachings of the Bible, while champions of the Bible on the other hand arose, combating this opinion with all their might, and teaching that these fossils *were* the actual remains of living creatures, having been brought into their present positions on the mountain tops by the waters of the Mosaic Deluge, and that those preserved in the rocks confirmed instead of opposing the record of Holy Writ.

It is difficult for us now to understand how these old writers, very many of them men of keen intellect and outstanding ability, endowed with all the learning of their times, could put forward explanations so trivial, quaint and even absurd to account for these objects whose origin now seems so self-evident.

In order to gain some clear conception of the reasons which led these ancient writers to entertain such strange opinions, the author, as opportunity offered, went back to the original texts of some of the more distinguished of those writers and by perusal of their works was led on to search out those of others who are less well known and thus to the exploration of many by-ways and corners of that little known field of ancient and especially of medieval literature which deals with the subject of the mineral kingdom.   This revealed a vista of thought, indeed a whole world of quaint and peculiar interest, differing from that in which we live, not only in its content of knowledge but in its methods of approaching the study of nature and its whole outlook on the physical universe.   In the belief that a consideration of this might be of value to others who are interested in geology but who, either through lack of leisure or through failure of means of access to the requisite sources of information, were not in a position to carry out such studies for themselves, the results of these researches have here been set down, and so this book came to be written.   Since the development of any science is based largely on the labors and discoveries of a succession of outstanding men, who, as it were, erect the structural frame work of the whole—the contributory bricks to fill in and complete the building being supplied by a host of subordinate workers—especial attention has been paid to the work and personality of such prominent leaders in the history of geology.

In following out the speculations concerning the facts and processes of inorganic nature, from Classical Times down through the Middle Ages and the Renaissance until they took on new forms in the light of modern times, it was found that the attention of the ancient writers was not spread over the whole range of what now constitutes geological science, but was concentrated chiefly on certain definite subjects which especially attracted their attention.   Chief among these are the following—The origin, nature and properties of gems and other minerals; The origin or "generation" of stones; The origin and development of ores and metals; The nature and significance of "Figured Stones" in the rocks of the earth's crust; The causes which led to the existence of mountains and mountain ranges; The genesis of springs and rivers; The causes and significance of earthquakes and the inferences which can be drawn from these concerning the character of the earth's interior.

In the present work, a history of the development of human knowledge along each of these lines is presented in successive chapters. Other chapters deal with geological science in Classical Times and with the development of the science of mineralogy from the time of the early Greeks down to the beginning of the nineteenth century.

And since it is difficult to understand the views which prevailed on many geological subjects during medieval times without some knowledge of the "mental climate" of this interesting period, a chapter on the Conception of the Universe in the Middle Ages has been included. A number of quaint ideas and curious stories which were met with in exploring some of the recondite by-paths of this ancient literature have been gathered together in still another chapter.

Volcanoes and volcanic action have attracted much attention throughout the ages—in those countries where volcanoes exist—but no attempt is made to give a consecutive account of the evolution of opinion on the subject. To do so would entail much repetition since extensive references to this subject are of necessity made in treating of the other subjects referred to above.

And finally a somewhat extended account is presented of the birth of *Historical Geology* which took place with the rise and fall of the Neptunian Theory toward the close of the eighteenth century. At this time it was discovered that the rocky strata of the earth contained a succession of records which, if deciphered, would present a history of the earth from its beginning down to the present day; and furthermore that the fossils found in these strata were the illustrations in this book of nature, showing the animals and plants which lived upon the earth's surface in the successive periods of its history. The more accurate methods for measuring geological time, which have been recently discovered, add immensely to the interest of this history, in that they have shown that these records embrace a period of some 1,500,000,000 years.

With Historical Geology established on a firm basis in the first half of the nineteenth century, the geological sciences developed by leaps and bounds, geological surveys and geological societies were established in all civilized countries and geology met with world-wide recognition and widespread economic application. The present work does not attempt to embrace the history of these latter developments; this would require not one but many volumes. In it the attempt is made to trace the history of the geological sciences from the earliest period in which we have any written records in Europe, that of the early Greeks, down to about the year 1825 A.D., when Historical Geology began its rapid development.

In a few of the chapters where the subject seems to require it, an outline of the later history of the particular subject of which the chapter treats, is briefly sketched. And, finally, having traced the growth of knowledge and the development of opinion down through the successive centuries, in each of the subjects which are discussed, at the conclusion of each chapter or in some instances in the concluding chapter of the book, a backward glance is taken over the long and difficult road which was followed, with a view to ascertaining if possible what it was that so often turned the seekers aside from the highroad of knowledge into by-paths of misunderstanding or of error, in which they wandered in some cases for ages. An understanding of these causes may possibly be of some advantage to those continuing the same lines of investigation now, by enabling them to avoid similar dangers.

But the question very naturally presents itself in each case as to whether it is quite certain that the goal of a true and final understanding of the subject has at length been reached, or whether, on the other hand, while it can be seen that the earlier views were wrong and that a great advance has been made toward a true understanding of the subject, this goal has not yet been attained and in some cases is not yet in sight.

BIBLIOGRAPHY AND SOURCES

The *sources* from which a knowledge of the early history of geology is derived may be arranged under the following general heads, references to individual works being given in the several chapters:

(a) *Early Grecian works.* Aristotle's *Meteorologica* and the book of Theophrastus *On Stones* are the most important sources in this group. Neither of these is a formal treatise. Aristotle's work has rather the character of a series of lecture notes, while that of Theophrastus seems to be a fragment of a larger work of which the rest has been lost. It deals chiefly with the practical uses of the various minerals or stones which are mentioned.

(b) Later works of Classical Times. Here Pliny's *Natural History* is by far the most important source of information. It is an immense storehouse of facts and fancies drawn from the works of many other writers, some of which are now lost while the names only of their authors have survived.

(c) A succession of Encyclopedias, the earliest of which was the

Etymologies of Isidore of Seville (A.D. 570–636)—one of the very few works of importance which were written when Europe was still in the Dark Ages—and in the Middle Ages, the encyclopedias of Neckam, Bartholomew the Englishman, Vincent of Beauvais, Berengarius and many others, each drawing extensively on the works of his predecessors.

(d) A group of popular books on Natural History of which Conrad von Megenberg's *Book of Nature* may be taken as an example. Many of these were "Herbals," of which the *Hortus Sanitatis* is the most important. While these latter deal primarily with plants, many of them treat of all three kingdoms of nature.

(e) The Medieval Lapidaries, commencing with that of Marbodus and including those of Camillus Leonardus, de Boodt and many other important writers.

(f) A number of more specialized works dealing with certain geological subjects, which commenced to appear with the dawn of the Renaissance. Among these the works of Agricola, Gesner, Aldrovandus, Caesius, Becher and Boyle may be especially mentioned.

(g) Books which appeared in ever increasing numbers from the opening of the eighteenth century until the close of the period embraced in the present work. These are treatises on special subjects, geological descriptions of certain countries, books of travel, and some general works which are the precursors of our modern text books and manuals of geology. One of the very earliest, if not the first of such text books (embracing mineralogy, petrography and palaeontology but not including physical geology) is the work of J. S. Schrötter, entitled *Vollständige Einleitung in die Kenntniss und Geschichte der Steine und Versteinerungen*, in three quarto volumes which was published at Altenburg (1774–78).

It may be mentioned that one of the best guide books for the student who wishes to explore the mazes of the ancient literature of the geological sciences, is the work of D. Giacinto Gimma, entitled *Della Storia Naturale delle Gemme, delle Pietre è di tutti i Minerali, ovvero della Fisica Sotteranea*, in two quarto volumes, which was published at Naples in 1730.

The following general histories of geology have appeared in modern times and are especially worthy of mention:

Sir Archibald Geikie, *The Founders of Geology*, London, 1897, being the Williams Memorial Lectures delivered at Johns Hopkins University in 1896. A second edition of this work, revised and enlarged, was published in 1905.

Karl Alfred von Zittel, *Geschichte der Geologie und Palaeontologie*,

Munich and Leipzig, 1899. A somewhat abridged translation of this excellent work, by Maria M. Ogilvie-Gordon, was published in London in 1901.

George Perkins Merrill, *Contributions to the History of American Geology*, Washington, 1906.

George Perkins Merrill, *Contributions to a History of American State Geological and Natural History Surveys*, Washington, 1920.

In preparing the present history the writer has in all cases gone back for his material to the original authorities. As it is often necessary to consult a work repeatedly he endeavored so far as possible to secure the texts themselves for his own library, following the example of that great lover of books, "Richard de Bury by divine mercy Bishop of Durham . . . while we filled various offices to the victorious Prince and splendidly triumphant King of England, Edward the Third from the Conquest, whose reign may the Almighty long and peacefully continue," who in the year 1344, setting forth for the guidance of others the manner in which he brought together his great collection of books says: "Besides all the opportunities mentioned above, we secured the acquaintance of stationers and booksellers, not only within our own country, but of those spread over the realms of France, Germany and Italy, for no dearness of price ought to hinder a man from the buying of books, if he has the money that is demanded for them, unless it be to withstand the malice of the seller or to await a more favorable opportunity of buying."[1] During the course of twenty years or more, a diligent search for these was made whenever they were to be found and a library of somewhat over one thousand volumes in various languages was brought together, many of these being books of extreme rarity. When it was impossible to secure the works themselves recourse was had to one or other of the great libraries where these were to be found and notes or abstracts of the volumes sought were thus obtained.

It is believed that few works exist, which are of real importance in connection with the present studies which have not been seen. In examining these the truth of Bacon's[2] remark has been brought home, "Some books are to be tasted, others swallowed, and some few to be chewed and digested."

Finally, the author desires to acknowledge his indebtedness to authori-

[1] *The Love of Books, the Philobiblon* of Richard de Bury, newly translated into English by E. C. Thomas (The King's Classics) London, 1902, pages 1, 17, 54.
[2] Essays. *On Studies.*

ties of the Redpath and Osler Libraries of McGill University, the Library of the Geological Society of London, the Vatican library, the Library of Congress in Washington and especially to those of the Library of the British Museum in London for permission to make use of the resources of these great collections. He also desires to acknowledge his indebtedness to Dr. W. W. Francis, Librarian of the Osler Library, for valuable advice and assistance.

CHAPTER II

THE GEOLOGICAL SCIENCES IN CLASSICAL TIMES

A search for the beginning of the geological sciences in Europe takes us back to the earliest literature of that continent, that of the early Greeks. Among these peoples intellectual pursuits were held in high esteem; literature, philosophy and the arts flourished. The rest of Europe has left no records of any similar interests or achievements in those early times comparable to those of the Grecian peoples. While, however, the soil of Greece was a favorable one for the growth and development of intellectual ideas, there has been a wide diversity of opinion on the question whether these had their origin in Greece or were derived from other and more remote sources.

1. THE EARLIER GRECIAN WRITERS

The early "Sophi" or "Wise men of Greece," who later came to be known as "philosophers," engaged in many daring speculations concerning the origin of the universe. The earliest of these philosophers was Thales, who held that all things had their origin in water. According to Anaximenes they were derived from air, while Heraclitus considered fire as the essential principle of the universe.

The literature of Greece was voluminous, and it embraced a wide range of subjects in poetry, drama, history, philosophy and other branches of human learning. But a large part of it has been lost, having been swept away in the disasters which overtook that country or through the vicissitudes of time. Nothing remains of the works of many writers whose names are renowned. The works of others have survived in mere shreds and patches from which can be gleaned but a very imperfect idea of the teachings which their authors intended to present.

Agricola,[1] the Father of Mineralogy, an accomplished classical scholar whose mind always reverted to the teachings of the Greeks and Romans, devoted much time to the study of the early history of the sciences. Writing at the time of the Renaissance he gives the names of no less than twenty-seven authors among the Greeks who were known to have written concerning "Stones," but whose works at the time when he wrote in 1547 had disappeared completely and whose names alone remained. These are Democritus, Socrates the Rhodian, Xenocrates, Sudines, Callistratus, Isamenias, Horus, Satyrus, Archelaus, Megasthenes, Bocchus, Nicanor, Iacchus, Juba, Zachalias, Agatharchides of Samos, Thrasyllus of Mendes, Heraclitus of Sicyon, Nicias of Malea, Dorotheus the Chaldean, Theophilus, Dercyllus, Dionysius the African, Diogenes, Orpheus, Epiphanius and Didymus of Alexandria. A few verses attributed to Orpheus have survived, to which reference will be made later.

The works of some of these writers were still extant at the opening century of the Christian era, for Pliny in his Natural History refers to certain statements to be found in them. Most of these, however, have to do with animals or plants, as for instance those of Juba, spoken of by Holland in his translation of Pliny as "King Jube," Sudines, Democritus and Horus. Pliny records that

Horus, King of the Assyrians, devised first this receit against drunkennesse—the ashes of swallows bills incorporate with myrrh will secure any man from drunkennesse and cause him to beare his drink well in case the wine he drinketh be spiced therewith.

Zachalias the Babylonian (Pliny, Book 37, Chap. 10) in the books which he wrote to King Mithradates, referred to the occult and magical powers resident in gems.

---

[1] *De Natura Fossilium* (Dedicatory Letter), Basel, 1547.

The information which Pliny obtained from these vanished authors would indicate indeed that the contributions which they made to sciences of the mineral kingdom were of such trivial importance that their loss is one to which we may reconcile ourselves, even with thanks. There were, however, certain other Greek writers whose works have come down to us, many of them in very fragmentary form, in which there are certain references to geological questions, more particularly in the field of dynamical geology, which are of interest. Schwarcz has reviewed, during many years of incessant labor, the whole body of ancient Greek literature prior to the time of the expedition of Alexander the Great (334 B.C.) and has brought these scattered references together and discussed them in his book.[2]

His exhaustive study shows that while some of the Greeks in these early times observed certain striking geological phenomena and speculated in a general way with reference to their cause and origin, there have come down to us from the Greeks but very few contributions of any importance to geological science. In Webster's quaint phraseology, "From the Grecians who were accounted great masters of all kinds of learning, there hath floated down to us but little scantlings of this kind of knowledge."[3]

It may be of interest here to set down in their chronological order the names of those writers among the early Greeks who have contributed something of geological significance.

Thales of Miletus (c. 636–546 B.C.) thought that the alluvial deposits at the mouths of rivers showed that water could change into earth.

Anaximander of Miletus (610–547 B.C.), the immediate successor to Thales.

Pythagoras of Samos (540–510 B.C.) founded at Croton in Southern Italy the famous Italian school. No writings of his have survived. His reputed teachings as set forth by Ovid are referred to below.

Xenophanes of Colophon (fl. 540–510 B.C.).

---

[2] Julius Schwarcz, The failure of geological attempts made by the Greeks from the earliest ages down to the epoch of Alexander, London, 1868, p. xviii.
See also:
Ernst von Lasaulx, *Die Geologie der Griechen und Römer*, Abhand. der Bayer. akad. d. Wissenschaft, Munich, 1852. (Deals almost exclusively with geography.)
Harold O. Lenz, *Mineralogie der alten Griechen und Romer*, Gotha, 1861.
Wm. Whewell, *History of the Inductive Sciences*, London, 1857.
[3] *Metallographa, or an History of Metals*, London, 1671, p. 26.

Anaxagoras of Glazomenae (500–428 B.C.). Quoted on earthquakes by Aristotle.

Xanthus of Sardos (fl. 480 B.C.).

Herodotus of Halicarnassus (484? B.C.) "The Father of History." The date of his death is uncertain.

Empedocles of Agrigentum (fl. 444 B.C.).

Democritus of Abdera (460–357 B.C.). Quoted on earthquakes by Aristotle.

Xenocrates of Chalcedon (396–314 B.C.).

Aristotle of Stagira in Macedonia (384–322 B.C.).

Theophrastus of Eresus on Lesbos (c. 370–287 B.C.).

Eudoxus of Cnidus (fl. 366 B.C.).

The references scattered through this great mass of literature do not present a formal treatment of any geological question but rather incidental references to geological phenomena. Schwarcz groups them under several headings among these:

(a) The belief in a Central Fire within the Earth, held by the most learned Pythagoreans, although it is not known from whom they derived the theory in question.

(b) Instances of the alteration of the relative positions of Land and Water, as Myrsilus states that Antissa was once an island, and later writers state that it was once connected with Mount Ida in the same manner that Pithecussae and Prochyta were united with Mount Misenum, Sicily with Rhegium or Ossa with Mount Olympus. Pindar in his *Olympiacs* narrates how the island of Rhodes emerged from the depths of the sea. Similar occurrences are cited by many of these ancient writers. By far the most complete and striking account of the replacement of portions of the sea by land is the well-known description by Herodotus, one of the most observant and least speculative of these wise men of Greece, of the origin of lower Egypt, which he gives for the purpose of showing that this portion of Egypt occupies what was formerly a gulf of the sea.

(c) The Remains of Animals or Plants enclosed in the rocks of the Earth's Crust.

These references may be reviewed in their chronological order. The earliest is that by Xenophanes of Colophon, the founder of the Eleatic school of philosophy, who flourished about the year 540 B.C., and was one of the earliest, if not the first person in Europe, so far as is known, who observed geological phenomena and attempted to explain them,

(v. Lasaulx, part III, p. 518).   His actual statement has not come down to us but its content is given in one of Origen's works, according to which Xenophanes says that in certain places where earth is mingled with the water of the sea, the former is dissolved by the latter, the proof of this being seen in the fact that far inland and even on the tops of high mountains sea shells are found.   In the quarries of Syracuse imprints of fishes and the remains of seals are to be seen; on Pharos impressions of laurel leaves deep down in the rocks; and on Malta the reproductions of all manner of things living in the sea.

Xanthos of Sardis, who lived in 500 B.C., in the fragment of his work which has survived notes that sea shells were to be found in the rocks of the earth's crust far from the seacoast in Armenia, Phrygia and Lydia, and concludes from this that the sea must once have covered the districts where they occur.   Herodotus, who lived about the same time, describes similar occurrences in Egypt and gives it as his opinion that the nummulites found in such abundance in the Eocene limestones about the pyramids were probably the indurated remains of lentils left over from the stores of food brought together for the use of the builders of the pyramids.

The mathematician Eudoxus of Cnidus, who flourished about 366 B.C. mentions fossil fishes in the rocks of Paphlagonia.

Eratosthenes (276–196 B.C.).

Aristotle (384–322), the greatest of all the Greek naturalists, says in his treatise, *De Respiratione*, that "A great many fishes live in the earth motionless and are found when excavations are made," words which would seem to indicate that he believed that these fishes had at one time been alive within the earth and had during the course of time died and become fossils.   Such an opinion is not one which Aristotle would be expected to entertain, but Theophrastus, his favorite pupil and his successor in the presidency of the Lyceum, writes (in his work *On Fishes*) of such fossils found in the rocks near Heraclea in Pontus and also in Paphlagonia, that they had either been developed from fish spawn left behind in the earth or else had wandered from the rivers or from the sea into underground passages in search of food and had been turned into stone.   It would seem, therefore, that if Aristotle had a true conception of the origin of these fossil fish he did not communicate it to his pupil.

The occurrence of fossil fishes is also mentioned by Polybius, Strabo, Livy and Juvenal.   Pomponius Mela regards the stories of their existence as fables. Theophrastus[4] also mentions that there is a fossil ivory and great stone-like bones which grow in the earth through some plastic force at work in nature.

There are two of these ancient Greeks, the fragments of whose works indicate that they had really engaged in palaeontological investigation. The first is Anaximander of Miletus (615–547 B.C.), one of the earliest philosophers of the Ionian school and the immediate successor of Thales, its founder; the other is Empedocles of Agrigentum, who flourished about 444 B.C. Besides his investigations in palaeontology, the latter engaged in studies of the action of the volcanic forces, and, according to tradition, came to his death by falling into the crater of Mount Etna.

Anaximander held that mankind had derived its origin from fishes, and Cuvier, referring to this, says in his *Histoire des Sciences naturelles*:

> Quoi qu'il en soit, Anaximander ayant admis l'eau comme le seconde principe de la nature, prétendait que les hommes avaient primitivement été poissons, puis reptiles, puis mammifères, et enfin ce qu'ils sont maintenant. Nous retrouverons ce système dans des temps très rapprochés des nôtres et même dans le dix-neuvième siècle.

In this statement, however, Cuvier would appear to have gone further than the statements of Anaximander warrant. Schwarcz, (p. 47) whose very wide acquaintance with the literature bearing on this question entitles his opinion to consideration, says, referring to this passage of Cuvier, that the great palaeontologist could hardly have read, in any classical author, of reptiles representing transitional forms between fish and human beings "After the latter had put aside their piscine type." Anaximander, however, was evidently making an advance towards serious palaeontological induction, if we are right in understanding him to say that fishes are the oldest ancestors of the present fauna, because he found their fossil remains in the lowest fossiliferous strata. The question has been raised whether Anaximander based these advanced opinions on his own personal observations in field and quarry, or whether, like Herodotus, he drew them from teachings of Egyptian scholars.

Empedocles, in a fragmentary poem entitled *Concerning Nature* mentions four successive stages through which the earth passed in emerging from the primitive chaos and advancing towards a state resembling more or less that which it now presents. During the evolution there appeared in the earth forms of heads, arms, legs, double-faced, double-

---

[4] See von Lasaulx, op. cit., 1852, p. 523.

breasted creatures, bovine animals with human heads, and many other strange forms, all imbued with a vital principle out of which later there originated various species of the animal kingdom.[5]   It would seem that he derived this idea from observing the fragments of vertebrate skeletons and the shells of various cephalopoda and other fossil forms which abounded in the rocks in the caves and caverns of the Sicilian land which was his home.   He taught that plants appeared before the animals, and his four successive stages in the development of living forms might be interpreted as a faint foreshadowing of what is now known to have been the actual course of the evolution of living beings.   Zittel[6] however dissents from this conclusion holding that these speculations, like many other musings of Empedocles, are not worthy of any such serious interpretation, but belong rather in the same class as his conclusion that a great race of giants lived in Sicily in former times, based on the discovery in that country of some large bones of certain extinct mammals.   It may here be noted that similar stories of the former existence of giants upon the earth, many of these tales being based on similar discoveries, are told by Herodotus, Solinus, Pliny, Pausanius and other ancient writers and indeed by some more recent ones as well.

Of the early Greek writers, however, two stand out prominently in the history of the geological sciences, namely, Aristotle and Theophrastus. The name of Aristotle[7] will always be associated with that of his teacher, Plato.

Plato was a descendant of Solon, and was an accomplished gymnast and wrestler, a facile and ready writer, well trained in music, mathematics and letters, a brilliant young Athenian aristocrat.   He was the founder of the celebrated Academy.   This was technically a religious institution, established for the worship of the Muses, with officers, a constitution and landed property.   But philosophy, mathematics, astronomy and various other sciences were taught, women as well as men being admitted.   It passed through a series of changes in its tendency and doctrine and was eventually, as a stronghold of Paganism, despoiled and abolished by Justinian in A.D. 529.   The Academy commenced its work in a house and garden bought for Plato by some of his friends in 387 B.C., and was situated about twenty minutes' walk from Athens, near a gymnasium sacred to the hero Academus.   The attendance, however, increased so

---

[5] See Schwarcz, op. cit., p. 54.

[6] *Geschichte der Geologie*, p. 6.

[7] Gilbert Murray: *History of Ancient Greek Literature*, London, 1902.

that it soon overflowed from the house and garden into the public gymnasium, but later adequate buildings were acquired.

Aristotle attended the Academy and Plato was his chief teacher in philosophy. The pupil, however, did not see eye to eye with his master on all questions, the relations between the two being illustrated by Aristotle's celebrated sentence in his *Ethics*, about Plato and Truth, "Both being dear, I am bound to prefer Truth." When Plato died, Aristotle left Athens and retired to Mytilene, on the east coast of the Island of Lesbos, where he married the Princess Pythias and spent two of the happiest years of his life. "Here it was" says D'Arcy Thompson,[8] "that he learned the bulk of his natural history, in which, wide and general as it is, the things of the sea have, from first to last, a notable predominance." In 345 B.C. Aristotle went to Pella in Mysia at the invitation of Philip of Macedon, and became tutor to his young son, Alexander (afterwards surnamed the Great):

Nothing is known of these lessons. One fears there was little in common between the would-be rival of Achilles and the great expounder of the contemplative life except the mere possession of transcendent abilities. Aristotle's real friend seems to have been Philip. . . . A year after Philip's death Aristotle returned to Athens, and Alexander marched against the Persian Empire. Aristotle had always disapproved of the plan of conquering the Earth.

In 335 B.C. Aristotle founded at Athens the Lyceum, the organization which Gilbert Murray considers to be "The greatest intellectual feat of the age." The school of philosophy was inaugurated in a building situated, like Plato's Academy, on the outskirts of Athens. It was near the grove of Apollo Lykeios, and had a "peripatos," or covered walk, in which Aristotle often taught, and on this account the students of Aristotle were often called the Peripatetics. Aristotle had no democratic sympathies and the turmoil of Athenian politics was unmeaning to him. He preferred the life of contemplation. He is described as a man of poor health and unattractive personal appearance but of great energy of mind.

The Lyceum was an institution intermediate in character between Plato's Academy and the great schools connected with the Alexandrian libraries, and it was probably, like them, helped by royal generosity. Here Aristotle himself directed the work of his "fellow students and

[8] D'Arcy Thompson (see *The Legacy of Greece*, Oxford, 1922).

built the gigantic structure of reasoned knowledge which has been the marvel of succeeding ages."

Fifty years after Aristotle's death the Lyceum had dwindled away to an insignificant institution, and the master's writings were little read. Other centers of learning had sprung up and divided the greatness of Athens among them.

Aristotle's writings were divided by the later Peripatetics into works for publication and lecture material. His reputation in antiquity was based entirely on the former class, especially on the semi-popular dialogues. It is a curious freak of history that with the exception of the *Constitution of Athens* not one work of this whole class is now preserved. From Aristotle we have no finished and personal works of art like the dialogues of Plato. We have only notes and memoranda of the school. This explains the allusive and elliptical style, the anecdotes and examples which are suggested but not stated; it also explains the repetitions, overlapping and occasional contradictions.[9]

Many of the writings which have come down to us are the joint work of Aristotle and his disciples. Aristotle's works cover an astonishingly wide range, treating of almost all subjects of human knowledge cultivated in his time. They embrace history, poetry, philosophy, logic, metaphysics, and physical sciences. Roger Bacon said that "although Aristotle did not arrive at the end of knowledge he set in order all parts of philosophy.[10]

Dante calls him "The master of those who know," and saw him in the first circle of the Inferno, where were the unbaptized who "Seeing that they were before Christianity, worshipped not God aright ... sitting amid a philosophic family, all regarding him and doing him honour."

Conrad Gesner writes of him as "Vir in omni humanae sapientiae doctrinaeque genere incomparabilis, quavis denique hominis laude superior."[11]

Throughout the Middle Ages Aristotle was regarded as the head and chief of all philosophers; one whose opinion on any subject was authoritative and final. On this account in medieval works he is often referred to simply as "The Philosopher," or "The Stagirite," from his birthplace, Stagira in Macedonia. Mahaffy says concerning him;

While there never was any man who had a greater effect in promoting the knowledge of his own and of preceding generations, it may be said in proof of this greatness

[9] Gilbert Murray, p. 375 et. seq.
[10] Ibid.
[11] *Bibliotheca Universalis*, Zurich, 1545, p. 73.

that he also retarded more than any other man ever did the course of scientific discovery. For he bound the learned men of the Middle Ages by the superstitious veneration for his words which they accepted as almost inspired. In the 13th century he was all but canonized as a saint in the Roman Catholic Church.[12]

In the opening chapter of his work entitled *Meteorologica*,[13] Aristotle outlines the method which he follows in his discussion of the physical universe. He had already treated of the "First cause of nature" the heavens and the elements, in his treatises, *De Generatione et Corruptione* and *De Caelo*, and he continues his discussion of this great subject in the present work.

What Aristotle says in the *Meteorologica* of his "Four Elements" and of his "Two Exhalations" will be considered in Chapter IV of the present work, which deals with the Generation of Stones. Meteorology, Aristotle says, concerns itself with events which "Take place in the region nearest the motion of the stars," that is, in that portion of the heavens which lie below the sphere of the moon. "Such are the milky way and comets and the movements of meteors." He treats of winds and waters in the heavens, of thunderbolts and earthquakes. "Of those things," he observes, "some puzzle us while others admit of explanation in some degree." "When the enquiry into these matters is concluded," he says, "animals and plants will be considered," and he then closes with the words, "When that has been done we may say that the whole of our original undertaking will have been carried out."

It may be noted here that Aristotle makes no mention of an intention, after having treated of plants and animals, to proceed to the description of the third or mineral kingdom of nature. He does, it is true, at the conclusion of Book III, and in Book IV of the *Meteorologica*, speak of the composition of homogeneous bodies, the metals and "fossils," under which latter term he includes stones and certain minerals such as sulphur, ochre and realgar, but this is merely to point out that these are comprised respectively of the "vapourous" and "smoky" exhalations supposed to be given off by the earth. No reference is made to their character or properties, and no description of them is given. It is true, also, that the work concludes with the following sentence, "After the homogeneous bodies have been explained we must consider the non-homogeneous too and lastly the bodies made up of these, such as man, plants and the rest."

[12] *A Survey of Greek Civilization*, London, 1897.
[13] The works of Aristotle translated into English, *Meteorologica*, Book I, Oxford, Clarendon Press, 1923.

Rocks would come under Aristotle's definition of non-homogeneous works, and it has been by some supposed that at the time he contemplated writing another work dealing with minerals and rocks, which work would complete his survey of the whole realm of nature. This opinion, however, is contradicted by the statement in the opening chapter of the work, to which reference has just been made. This book on minerals and rocks, if it was ever written, has been lost, although eagerly sought for throughout the centuries. Albertus Magnus, speaking of his efforts to find it, says, "que diligenter quesivi per diversas mundi regiones."[14]

In many of the medieval Latin versions of Aristotle's *Meteorologica*, however, there is to be found appended to the Fourth Book an *additional chapter*, in three paragraphs, entitled *De Mineralibus*. That this chapter is not a direct translation is evident from the fact that it contains Arabic proper names, and internal evidence of other sorts leads to the same conclusion. The question then arises whether this contains an Aristotelian nucleus, or whether it is the product of another pen. This question has been the subject of a long-continued controversy into which it is not necessary to enter here,[15] but which may now be considered as having been brought to a conclusion by the researches of Holmyard and Mandeville. These authors show that *De Mineralibus* was not written by Aristotle but by Avicenna. Part of it is a direct translation and the rest of it is a résumé of certain sections of a work by Avicenna written in Arabic and entitled *Kitab al-Shifa*, the *Book of the Remedy*, which he composed in response to the request of his friend Al-Juzjani that he should write a general commentary on Aristotle's works. He was too busy to write a formal commentary, but compromised by writing a plain exposition, free from any attempt at a refutation of adverse views. He had already written the first book of his great *Canon of Medicine*, and thereafter he worked at the *Shifa* and *Canon* simultaneously. He was at this time Vizier of the Buwayhid prince, Shams al-Daula, and was living at Hamadhân in Persia. The book can be dated very precisely. It was written between 1021 and 1023. The Latin rendering which by many writers was referred to as Aristotle's *De Mineralibus* and which

[14] Quoted in: *Roger Bacon and the state of science in the 13th century* by Robert Steele. Studies in the history and method of science. Oxford, Clarendon Press, 1921, vol. ii, p. 135.

[15] F. De Mély, *Les Lapidaires de l'antiquité et du Moyen Age*. Paris, 1902, Tome III. Premier Fascicule, Les Lapidaires Grecs, xxxiv.

Julius Ruska, *Das Steinbuch des Aristotles*. Heidelberg, 1912.

E. J. Holmyard and D. C. Mandeville, *Avicennae de Congelatione et Conglutinatione Lapidum*. Paris, Librairie orientaliste, 1927.

appeared separately as a treatise by Avicenna under the title *De Congelatione et Conglutatione Lapidum*, is divided into three chapters, the first of which bears the same title as the whole treatise, the second the title *De Causa Montium* and the third that of *De Quatuor speciebus Corporum Mineralium*. This treatise of Avicenna is a remarkable work, having in several respects a distinctly modern outlook. In certain points he differs from Aristotle, as for instance in his statement that meteoric stones originate in the heavens and fall thence upon the earth, while Aristotle states in his *Meteorologica* that such stones took their origin on the surface of the earth and were blown up into the heavens by a violent wind, subsequently falling to the earth again. He also disagrees with Aristotle, as well as with the alchemists, in that he holds that the metals cannot be changed or transmuted into one another, but that they are each composed of a separate and distinct kind of earth.

The only individual (or separate) treatise on minerals and rocks which has come down to us from the Greeks is that of Theophrastus, which bears the title *Concerning Stones*. It is a short work which makes about fourteen octavo printed pages. Theophrastus was a pupil of Aristotle and his successor as head of the Lyceum in Athens. He is known to have written some two hundred works, the great object of which was the development of the Aristotelian philosophy. Of his works very few have survived, his treatise *Concerning Stones* being one of these, and another being entitled *Concerning Plants*. The latter is an excellent work, one of the finest on this subject which have come down to us from ancient times.

In Theophrastus' work on stones two facts at once strike the attention of the reader, the first is that it differs entirely in its style and content from any other work upon this subject written in classical times—in fact, no other work at all resembling it appeared until later on in the Middle Ages. It remained for 1800 years a most valuable and authoritative work on minerals, referred to and quoted by all, and shows that even at this early time there existed, probably among miners, quarrymen and other men engaged in the technical trades and industries, a body of knowledge of mineralogy, especially on its practical side, which had been drawn upon by Theophrastus in the compilation of his work.

The second fact is that it is very short and bears all signs of being a fragment of some larger work.[16] Pliny had a copy of this larger work

---

[16] See F. Moore, *Ancient Mineralogy, an inquiry respecting Mineral Substances mentioned by the Ancients*, New York, 2nd ed., 1859.

before him when he was writing his *Natural History*, for he quotes from Theophrastus concerning certain minerals which are not mentioned in the text which has come down to us. Theophrastus, furthermore, mentions but relatively few minerals, and omits the consideration of the metals entirely, stating that he had already treated of them in a separate work. This latter, however, has been lost. In the more extended form of this treatise on stones, Theophrastus undoubtedly included many additional minerals, since we know from Pliny and others that the ancient Greeks were acquainted with many other minerals which Theophrastus does not mention. The work furthermore is written in an abrupt style, suggesting a series of lecture notes rather than a formal treatise, and apart from an introductory statement concerning the part played by the four Aristotelian elements in the composition of minerals, the treatment of the subject is empirical. The minerals are not classified further than a general separation into those which can be melted or changed in some manner by heat, and those which remain unaltered when submitted to the action of fire.

In his brief statement concerning the various minerals and rocks, he continually refers to the uses to which they may be put and to their various practical applications in the arts and industries. He refers to the principal gems—"male and female"—and when engraved, to their use in signet rings. He speaks of amber, Lapis Lyncurius, and lodestone as possessing in common the property of attracting certain substances to them. He then mentions pearls and coral which grow in the sea, the latter "red in colour and resembling roots." He speaks of the several varieties of marble, alabaster and gypsum, and of the quarries from which they are obtained, describes how marble is burnt to obtain quicklime, and enlarges on the method of producing what is now known as Plaster of Paris by burning gypsum. He also treats of many different kinds of earths, marls and ochres and their uses as paints and for other purposes, and describes how the raw ochres may be changed in color by burning. He refers to pyritous minerals in general and some of the mineral fuels, and considers a few of the metallic ores, among them the cinnabar from Colchis "which they say is produced there in rocks and on precipices from which they get it down by darts and arrows." He describes, also, the manufacture of white lead by the action of vinegar on metallic lead.

Three manuscripts of this interesting but fragmentary treatise of Theophrastus are known, all of them in the Vatican Library. It was first printed, together with the other surviving works of Theophrastus and of Aristotle, by Aldus of Venice in 1496.   The Greek text with a Latin translation and commentary was reprinted as an introduction to De Laet's *De Gemmis et Lapidibus* in the edition of 1647 published in Leyden, and also in another edition of the same work which appeared in the following year.   The Greek text with an English translation by Sir John Hill was published in London in 1744.   This English text was translated into French in 1754 and into German by A. H. Baumgärtner (Nuremberg, 1770), by Carl Schmieder (Freiberg, 1807) and in abstract by H. O. Lenz[17] (Gotha, 1861).   Editions by F. Wommer also appeared in Leipzig (1852) and in Paris (1866).   The Greek text is also reproduced by F. de Mély,[18] and there is a translation into German by Karl Mieleitner.[19]

2. THE LATER GRECIAN WRITERS

After the time of Theophrastus there are very few writers among the Greeks, with the exception of Strabo, who contribute anything worthy of notice to geological science.   Those who refer to it make reference chiefly to the economic uses of certain minerals in connection with agriculture, mining or building construction.   There are:

Agatharchides (or Agatharchus)—181–146 B.C.
Diodorus Siculus—flourished about 44 B.C.
Dioscorides—flourished about A.D. 60
Dionysius (Periegetes)—flourished about A.D. 81–96
Strabo—wrote 7 B.C.

Agatharchides, the celebrated historian and geographer, in his treatise on the Red Sea,[20] mentions rich gold deposits on the Nile at a locality which he does not define accurately.   Lenz thinks that he is referring to the gold deposits which are known to exist at Fazoke, on the Blue Nile near the western border of Abyssinia.   He says that the gold occurs in a series of branching veins.   In working these deposits the rock is first disintegrated by fires built against it, then mined out and crushed to the size of a fine meal.   This is then washed by running water on a slanting

---

[17] *Mineralogie der alten Griechen & Römer*, p. 16 et seq.

[18] *Les Lapidaires grecs*, Paris, 1898.

[19] *Geschichte der Mineralogie im Alterthum und in Mittelalter*, printed in Johnsen's *Fortschritte der Mineralogie, Kristallographie & Petrographie*, vol. VII.   Jena, 1922.

[20] *Periplus Rubri Maris*.   For excerpts of this work see Lenz: *Mineralogie de alten Griechen und Römer*, p. 29.

table, the operator moving the material about with his hands during the operation.   By this process, he says, the rocky material is washed away while the "gold dust," being heavier, remains upon the table.   This latter is then removed, weighed and placed in an earthen crucible, with a lump of lead, some salt, a little tin and some barley or bran.   The lid is fastened on and the crucible fired continuously for five days and nights. When it is opened after cooling, there remains, he says, only a mass of gold which weighs only a little less than the "gold dust" which was originally put into the crucible.   Agatharchides also states that native gold occurs on the Arabian shore of the Red Sea in lumps varying in size from an olive to a walnut, and that the tribes in whose territory this occurs fashion these into necklaces and armlets.   They also barter this gold with the neighboring tribes for other metals, receiving for it three times its weight in copper, one half its weight in iron, or one tenth of its weight in silver.   These exchanges represent the relative scarcity of these three metals in this part of Arabia at this time.   He also mentions the occurrence of topaz on an island in the Red Sea known as Snake Island.

Diodorus Siculus was a Greek historian who wrote under the title of *Bibliotheca Historica*, a history of the world from the creation to the year 60 B.C.   This was in forty books, of which, however, only fifteen have come down to us.   The work lacks order of arrangement and critical power, but is valuable as a collection of materials the sources of which are now lost.   Scattered through this history here and there are references to matters of geological interest and, more particularly, to the occurrences of ore deposits in certain countries, and the methods of mining and smelting by which the metals were obtained from the ores. He says that in Arabia, Egypt, Ethiopia and India the intense heat of the sun is the cause of the development not only of many great and magnificent animals, but of a very great variety of splendid gems. Among these is rock crystal which is pure water, converted into a solid by the heavenly fire; also smaragdite and aquamarine which are found in veins and which take their colour from the heavens, topaz which has the colour of the sun, and carbuncle in which there is a certain light which was shut up or enclosed within it, when the gem took upon itself a solid form.   He also refers to the amber which occurs beneath the waters of the Baltic Sea and on the adjacent coast and also to the asphalt in the district about Babylon and in the Dead Sea.   He also mentions the volcanoes and volcanic rocks of the Phlegrean Fields, Vesuvius, Etna and the Lipari Islands.   He records that when Hamilcar reached

Etna in 394 B. C. he found that the volcano had shortly before his arrival displayed intense activity, and that the whole coast at the foot of the mountain was so covered with the lava that his army had to make a wide detour around the volcano instead of following the coast, along which the attendant Carthaginian fleet was sailing. He speaks of the earthquakes which took place in various parts of Greece, and of certain cities on the coast of that country which were swallowed by the sea. His most notable references, however, are those which he makes to certain mining areas. He says that the inhabitants of Saba in Arabia had accumulated enormous wealth by trading in gold and silver, and that alluvial gold was abundant in certain streams in Gaul. His account of the rich silver deposits near the present city of Cartagena in Spain and the methods employed in mining them is of especial interest. He says that before the Romans took possession of these deposits, the Carthaginians extracted from them immense quantities of the precious metal. The Romans, after they took them from the Carthaginians, worked the mines by slave labor and Diodorus gives a lurid account of the hardship and cruelty to which these unfortunate workers were subjected by their Roman taskmasters. He also speaks of the tin deposits of "Belerion," by which term he refers to Cornwall. Here, he says, the natives win the tin ore from the stony earth, and then smelt it and refine the tin. They bring the metal in ingots to the island of Iktis, supposed by some to the Isle of Wight, where it is taken over by the merchants from Gaul, who convey it by pack horses on a thirty days' journey to Massilia (Marseilles). Also of the well-known iron deposits of Aethalia (Elba), which, he says, are worked on a large scale, the ore being smelted there and the iron sent by ship to all the great marts of the world.

Pedanius Dioscorides, of Anazarbus near Tarsus, who flourished about the middle of the first century A. D., wrote the outstanding work on Materia Medica of ancient times. Singer[21] says that it was "The most influential botanical treatise ever penned," and that it remained the authoritative handbook on pharmacology for more than fifteen centuries down to the close of the Middle Ages. All that is known of Dioscorides himself is contained in the brief dedicatory letter which prefaces this work, which is to the effect that, following the life of a soldier, he visited many lands, and that he gathered the material for his work in part from personal observation and in part from the work of

---

[21] *Studies in the History and Method of Science*, vol. 2, p. 60, Oxford, 1921.

others.   A great variety of editions of the work have been published in a number of different languages: in many of them, what is believed to have been the original text has been so extended by interpolations and additions that the genuine Dioscorides constitutes a remnant only or a core.   The only translation of the great work into English was made by John Goodyer,[22] the renowned botanist of Petersfield during the years 1652 and 1655, but it was not printed, probably because the translation failed to find a patron and at that time "To set foot forward in forreine ground without the countenance of some worthie personage" was too "fearefull" a risk for a publisher to undertake.   Goodyer's manuscript translation has for centuries remained at Magdalen College, Oxford. In 1934, however, it was sent to press, and the great work of this celebrated author is now available for study by all those who are interested in the subject of which it treats.   What is regarded as the original text consists of five books, of which the last deals with *Wines and Minerals.* The opening words of the fifth book runs as follows:

Having given an account, most loving Areius [a friend to whom Dioscorides dedicates his work] in the four books written before, of Spices and of Oils and of Ointments and of Trees and both of ye Fruits of them and of their Tears; and moreover of living Creatures and of Honey and of Milk and of Fats and of those which are called Corn and Pot-Herbs and a full discourse about Roots and Herbs and Juices and Seeds.   In this, being the last of the whole work, we will discourse about Wines and of things called Metallic, beginning with ye Tract concerning ye Vine.

Since what he says about wine does not concern us here, we will pass on to the *Things called Metallic.*

As Theophrastus treats of the practical uses of minerals and stones, Vitruvius of their employment in the construction and adornment of buildings, so Dioscorides in his treatise concerns himself solely with the uses to which minerals and mineral products may be applied in the practice of medicine.   He mentions one hundred minerals, mineral substances and technical preparations.   This portion of the work is compendious, embracing, in the English translation, thirty-six quarto pages only, the reference to each substance usually being brief.   It is not always possible from his summary statement of their characteristics to be certain to what mineral he is really making reference.   In the case

---

[22] *The Greek Herbal of Dioscorides*, illustrated by a Byzantine, A.D. 512, Englished by John Goodyer, A.D. 1655, edited and first printed A.D. 1933 by Robert Gunther, with 396 illustrations, Oxford, 1934.

of technical preparations he describes the way in which each is made and in all cases he refers to its uses and its various medicinal properties properties which, in many cases at least the substance does not possess. Some of the properties are purely magical, but in early times medicine and magic were intimately related to one another. He mentions Kadmeia, Molubdaine and many ores and preparations of lead, zinc, copper and iron, also stimmi (antimony sulphide), chrysocolla from Armenia, Macedonia and Cyprus, malachite, cinnabar, quicksilver, orpiment, realgar and sulphur, as well as ochres raw and burned and various clays, alums and salts, also Melilites (or honey stone) which derives its name from the fact that "Men carrying it and giving it to their own masters, find them well affected and taking a forgetfullness of evils which they have committed even when oft repeated." Also of selenite "Which some have called Aphroselenon because it is found in ye night-time, full in ye increase of ye moon. But it grows in Arabia, being white, transparent and light: they give ye dust for a drink to ye epileptical; but ye women use them for an amulet and it is thought that being bound to trees, it makes them bear fruit."

Dionysius, surnamed Periegetes (or "Describer of the Earth"), was a Greek poet who flourished, probably, in the time of Domitian about A. D. 81 to 96. He was the author of a poem in 1186 hexameters, the Latin translation of which is entitled *Descriptio Orbis Terrarum*. It is chiefly geographical and of little value. In it, however, are a few references to the occurrences of minerals in various countries. He says that in the Rhipaean mountains (the Urals) near the Frozen Sea, amber and diamonds are found, and in the mountains of Pallene in Greece, the asterius (Star Sapphire) and the lynchis (a term which he here probably uses for the ruby) occur. He also mentions the occurrence of rock crystal and jasper at Thermoden in Pontus on the shore of the Black Sea, and elsewhere, and beautiful rolled agates in the River Choaspis in Persia. The land of Ariana (in Central Persia), he says, is unfruitful but yields many beautiful gems, and the country rock is traversed everywhere by veins holding lapis lazuli. In India the inhabitants find in the sands of the rivers gold, aquamarines, diamonds, green jasper and bluish transparent topaz.

Strabo, the Greek geographer and historian, was born about 63 B. C. and died some time after A. D. 20. Of his historical works only fragments have survived, but his *Geography* in seventeen books is the most important work of antiquity on that subject. According to Ettore Pais the book was written about 7 B. C., some later events being inserted in a subsequent edition. Strabo travelled very extensively and his descriptions are largely based on personal observation. Scattered through this great treatise on Geography there are here and there references to geological subjects. Thus he notes the evidence, seen in certain districts, of the elevation or subsidence of land areas, in some cases of great extent.[23] These movements, he thinks, were accompanied by earthquakes and probably had their source in the central fires. This elevation of land accounts for the existence of marine shells in the rocks of the earth's crust which are now high above the level of the sea. The existence of these fires explains the origin of volcanoes, which he thinks may be regarded as safety valves, providing a means of escape for the winds or fiery vapours which when pent up within the earth often give rise to earthquakes. He describes Etna, Vesuvius and the Lipari Islands, and notes the evidence that great subterranean explosions took place in connection with the eruptions of the volcanoes constituting this latter group. He notes the transporting action of running water and the existence of alluvial deposits at the mouths of rivers and along their course.

He gives an account also of the character and mode of occurrence of many celebrated ore deposits in various countries. The Spanish deposits are richer in gold, silver and copper and iron than any to be found elsewhere in the world. The gold occurs both in the country rock and in alluvial deposits.

In the chapter on Ariana he quotes Onescritus to the effect that a river in Carmania brings down gold dust, that there are to be found mines of silver, copper and minium and that in the district are two mountains, in one of which arsenic is found, while salt is obtained from the other. When making reference to the mines in a mountain of Pontus called Sandaracurgium, he adds his testimony to that of other writers concerning the bad labor conditions which prevailed in the mines in those early times. The mines, he says, are worked by slaves who are short-lived, owing to the "Strong odour given off by the ores which are destructive to life" (Book XII, Chap. 3) or, in other words, the men died off rapidly on account of overwork and bad ventilation.

Strabo also mentions the great quarries near Luna (now the Carrara quarries) (Book V, Chap. 2) from which were obtained the enormous

---

[23] *Geography*, Book I, Chap. III.

blocks of marble used for building purposes in Rome and Italy, also those of Pentelicus and Hymettus (Book IX, Chap. 2) yielding an excellent quality of marble, and those of the island of Paros, (Book X, Chap. 5) from which was obtained the finest marble for statues and other works of art. He says that when this Parian marble is taken from the quarries the spaces left by its removal become filled up with new rock in the course of time, a similar replacement by natural growth taking place also in the limestone quarries of Rhodes, the iron mines of Elba and in the salt mines of India (Book V, Chap. 2.)

Strabo also makes reference to many localities in which gems are found, especially in the several countries bordering on the Indian Ocean: carbuncles of various colors and pearls from India, emeralds and other gems from the localities on the shores of the Red Sea, and pearls from the Persian Gulf. He also repeats the story of topaz which could not be readily seen in sunlight but which could be seen at night, the seekers for the gems finding them and covering each at night with a little earthen cup which next day served to mark the position of the gem, which was then extracted, also the deposits of sand on the coast of Syria between Tyre and Acre used for the manufacture of glass at Sidon. (Book XVI, Chap. 2.)

Before passing from the consideration of the contribution of the Greeks to geological science, attention should be drawn to certain passages in the Metamorphoses of Ovid, to which special reference is made by Lyell.[24] These are to be found in Book XV of the poem and are put into the mouth of Pythagoras. In these, Pythagoras, one of the earliest of Grecian philosophers, is supposed to give an account of many facts and processes of nature which would seem to indicate that much more was known by the early Greeks concerning these matters than can be gathered from any of their writings which have come down to us. He speaks of the elevation and depression of great land areas and the great changes which are brought about by these movements. Of earthquakes and volcanoes and their causes, of the erosion by running water and its results, and of other things. Some of the illustrations of these phenomena which he cites as examples, are taken from events which did not happen till long after the death of Pythagoras. Lyell recognizes this, but thinks that the passage probably embodies the substance of the teachings of Pythagoras. Ovid lived, however, more than five hundred

[24] *Principles of Geology*, 12th ed., vol. I, p. 18.

years after Pythagoras, and much additional knowledge had been accumulating during this long period, and it seems much more probable that the statements which Ovid set forth in his Metamorphoses represent the popular conception of the teachings of Pythagoras current in the Augustan age rather than any actual knowledge possessed by the great philosopher himself.

3. THE ALEXANDRINE LAPIDARIES

Finally, reference should be made to a small group of lapidaries in the Greek language which appeared in the early centuries of the Christian era, and which were written in Alexandria or at least show a very marked Alexandrian influence, on the magical virtues and occult powers attributed to the precious stones or gems. This was due to an infusion of eastern magic which at this time pervaded the schools of the greatest centres of learning in the western world. It was furthermore believed by the Gnostics, who influenced Alexandrine teaching very strongly, that the powers resident in these gems were greatly enhanced or intensified by engraving upon them certain magical signs or symbols, and so there arose a great demand for Gnostic amulets or charms, worn to ward off the influence of evil spirits, diseases and accidents.

These Alexandrine lapidaries have been brought together, collated and carefully studied by De Mély,[25] in whose admirable work the Greek texts are reproduced, together with a French translation of the more important of these. The authors of the lapidaries are for the most part unknown, and the works make no contribution to an actual knowledge of mineralogy, but they nevertheless possess a distinct interest and value in showing the standpoint from which minerals were regarded at this time and the value attached to them. Three of the most important of these lapidaries, which may be taken as representatives of the whole, are:

(1) The *Cyranides* attributed to Hermes Trismegistus.

(2) A treatise *On Rivers and Mountains* by the Pseudo-Plutarch.

(3) A poem entitled *Lithica* attributed to Orpheus.

The *Cyranides* is the longest and most important of the three, the Greek text occupying one hundred and twenty-one quarto pages. It is really a medical work which is stated in the text to have been written

[25] F. De Mély, *Les Lapidaires de l'Antiquité et du Moyen Age—Les Lapidaires grecs·* Paris, 1898, 1902.

under divine inspiration for the alleviation of the sufferings of humanity. The prologue says that there were two books, the first written by Cyrus, King of Persia, and the second by Harpocration of Alexandria, and that out of these two Hermes compiled the present work, known as the *Cyranides*. Its attribution to Hermes Trismegistus (if indeed such a person ever lived) is wholly fanciful. It is a Gnostic work and was written between A. D. 227 and 400. The work is divided into four parts or books, designated as the *First, Second, Third* and *Fourth Cyranide*. Each of these consists of twenty-four chapters, that being the number of letters in the Greek alphabet, each successive chapter being designated by one of these letters. The first *Cyranide* is based upon "Litteromancy,"[26] a branch of magic which teaches that occult powers result from the union of different objects whose names begin with the same letter of the alphabet, an idea on which designs and cabalistic formulas for amulets and phylacteries are based. In pursuance of this idea each chapter is headed by the name of a bird, a plant, a stone and a fish. These represent respectively the four elements: air, earth, fire and water. This magical introduction is followed by a series of medical prescriptions, some of them based upon, or drawn from, one or other of the four symbols above mentioned, but others derived from other sources. These symbols, moreover, engraved on the stone make it a charm against certain diseases, infections or accidents. The *Second, Third* and *Fourth Cyranides* are bestiaries, treating respectively of quadrupeds, birds and fishes and their respective therapeutic virtues.

Scattered through the work there is mention made of fifty-two stones, of which thirty-six are of mineral origin and sixteen of animal origin, almost all of which are mentioned in the medieval lapidaries. Of the twenty-four stones which represent the element of fire in the *first Cyranide*, fifteen are found in Pliny's *Natural History* and seventeen are found in the medieval lapidary of Leonardus. The following is the list of these twenty-four stones in the order of their occurrence in the twenty-four successive chapters of this *Cyranide*:

Aetites	Porphyry
Dendrite	Taite or Panchorus
Hephestite	Chrysite
Cinedios	Beryl
Nemesite	Evanthus

[26] De Mély, p. lxiii.

Thyrsite

Lyngurium

Xiphios

Rhinoceros

Hyetite (This may be the Hyenia of Pliny)

Psorite or Porus (This stone is mentioned by Theophrastus and is a variety of marble)

Gnathos

Emerald

Jasper

Medea

Onyx

Sapphire

Crapaudine or Batrachite

Ocytocios

Gnathos is said to be a hard stone resembling a jawbone in shape. Rhinoceros is stated to be found within the nose or horn of the rhinoceros and to resemble a horn in appearance. Ocytocios is merely a small Aetites.

In the *Cyranides* is found the first mention of certain stories which became famous in medieval times, as that of the pelican which shed its blood to resuscitate its young (not to feed them as the legend usually runs), that of the hunting of the unicorn which could be captured only by beautiful women—the creature being of a very amorous disposition, and of the salamander which cannot itself be burned and will extinguish any fire into which it is thrown. Also a host of other ancient "facts" most remarkable and interesting, but flatly contradicted by all modern experience.

The second of these works, that entitled *Treatise on Rivers and Mountains*, has been attributed to Plutarch, who was born about A. D. 46 and died at a very advanced age, although the exact date of his death is not known. It is included in some editions of Plutarch's works, and omitted in others, but it is now generally agreed that the treatise was not written by him[27] but by some unknown author who is generally referred to as "Pseudo-Plutarch," and that it dates from the first quarter of the third century. It is a strange work and at the first glance the reader is at a loss to know whether he has before him a treatise on geography or one on religion. Further examination, however, shows it to be a magical and medical treatise, resembling the *Cyranides* in many respects, and like these it is divided into twenty-four chapters, the cabalistic number corresponding to that of the letters in the Greek alphabet. The original arrangement of the book was evidently such that each chapter was devoted to some one plant or stone found in a certain river or near the foot of a mountain and to the mythological

[27] See De Mély, *Les Lapidaires grecs*, p. liv.

traditions associated with it. Woven about these is a wealth of magical and mystical references and allusions, with charms for protecting treasure, capturing tigers, putting demons to flight, keeping dogs from barking, curing leprosy, fevers, hemorrhages, and for a great variety of other more or less useful purposes. The rivers and mountains are merely mentioned as those where these plants or stones possessing magical powers are to be found or where certain mythological events took place, from which in many cases the river or mountain derives its name. As authorities for these stories, books by unknown authors not mentioned even in the bibliographies of Pliny's *Natural History* are often cited. It is, in fact, doubtful in most cases whether these authors or their books ever existed. As examples of the treatment, two passages may be selected, one referring to a river and the other to a mountain. The river Pactolus:—"There is found in the Pactolus a stone called argyrophylax which looks like silver. It is rather difficult to recognize it because it is intimately intermixed with the little spangles of gold which are found in the sands of the river. It has one very strange property. The rich Lydians place it under the threshold of their treasure houses, and thus protect their stores of gold. For whenever any robbers come near the place, the stone gives forth a sound like a trumpet and the would-be thieves, believing themselves to be pursued, flee and fall over precipices and thus come to a violent death." Again: "There is on Mount Cronius a stone which, on account of the peculiar phenomenon connected with it is called Cylinder. Whenever Jupiter casts forth his thunder or lightning, this stone, being frightened, rolls down from the top to the bottom of the mountain, as stated by Dercylle in the first book of his treatise entitled *On Stones*."

There are twenty-four stones mentioned in his *Treatise on Rivers and Mountains*, some of them, as for example beryl, asterites and sardonyx, being well known; others are well-known stones which here appear under new names, as "cylinder," just mentioned as occurring on Mount Cronius, which is the well known Ceraunia or thunder-stone mentioned by Pliny and found in most of the medieval lapidaries; the Linurge is the Amianthus and the Collote is the Chelidonius. Others bear new names and cannot be identified, if indeed they are not wholly mythical, such as Corybas, Sicyone and Mynda.

The third of these works is a poem attributed to Hermes and bearing the title *Lithica*. It consists of 770 lines and while containing no mineralogical information concerning them, sets forth in detail the magical virtues of 27 stones, among which are quartz, diamond, jasper, topaz, opal, chrysolite, magnet, coral, agate, hematite, sardius, and the emerald. This poem was probably written in the fourth century A.D., and seems to have drawn upon an earlier Grecian lapidary ascribed to Damigeron. Only fragments of the original Greek work have survived, although a Latin translation of the whole text has come down to us.   This work of Damigeron is also of interest in that it formed the principle source from which Marbodus derived the material for his celebrated medieval lapidary.

4. THE ROMANS

Having now briefly reviewed the contributions made by the Greeks to the geological sciences, let us pass on and consider what their successors, the Romans, knew and wrote concerning the third realm of nature.

The Roman did not turn naturally to the study of pure science, he regarded it rather as one of the vagaries of the Greek mind.   He was a man of affairs, concerned with law and order, good administration, service to the state, military power and the extension of the Roman Empire: as contributory to these ends, he was interested rather in applied science.

There is a passage in chapter 46 of Book 2 of his *Natural History* in which Pliny, speaking of the winds, refers to the widespread interest in science among the Greeks even in times of war, where there was no place of safety on the land and the seas also were unsafe on account of widespread piracy; whereas in the Roman Empire in his own time when there was peace, he says regretfully that science was neglected, and men sought after gain rather than learning.

Galen,[28] writing about one hundred years later, reiterates the same complaint, saying that in his time there were no real seekers after truth, but all were intent upon money, political power or pleasure.

The earliest and one of the very greatest of the Roman writers was Lucretius—Titus Lucretius Carus—to give his full name, who was born about 99 and died 55 B.C.   It is probable that he saw the light in Rome and judging from his name he beonged to a noble family, nothing certain however is known concerning him.   His work *De Rerum Natura* is the greatest didactic poem of all time.   It is in six books, and presents a picture of the whole physical universe based on the teachings of Epicurus.

[28] See Thorndyke, *A History of Magic and Experimental Science*, vol. 1, p. 127.

The most abstruse speculations are clearly explained in majestic verse, and the subject enlivened by digressions of power and interest. He first refers to man's life as crushed and broken by the teachings of the religion of the olden time "Which brought forth criminal and impious deeds," of which he sets forth as an example the sacrifice of Iphigenia at Aulis, and which had its strength in the "Religious fears and threatenings of the priests—because everlasting punishment is to be feared after death" if the enmity of the gods is aroused. Epicurus, he goes on to say:

Was the first that dared to uplift mortal eyes (against these teachings). . . . For neither fables of the gods could quell him, nor thunderbolts nor heaven with menacing roar. Therefore forth he marched far beyond the flaming ramparts of the world as he traversed the immeasurable universe in thought and imagination: whence victorious he returned bearing the prize, the knowledge of what can come into being and what cannot, in a word, how each thing has its powers defined and its deep set boundary mark. Wherefore Religion is now in her turn cast down and trampled underfoot, whilst we by victory are raised to heaven.[29]

His poem was written to free men's minds from the slavery of superstition and banish from them the fear of death, by inducing them to use their reason and to look about them on the world of nature, to consider its structure, contemplate its beauty and to see "The source from which each thing can be made and the manner in which everything is done, without the working of gods."[30] He deals only with the physical world, the ethics and logic of Epicurus lie outside his plan. Lucretius is in fact a materialist, for him the gods, if indeed they exist at all, dwell apart from men and have no concern with them. In his system the place which they occupied in the old religion of the Greeks, is taken by pure chance. But, as remarked by that distinguished scholar J. D. Duff, "The instincts of Humanity are in this matter opposed to Lucretius, men will not buy immunity from their fears by the sacrifice of all their hopes. And so the poem of Lucretius, while most interesting and attractive, and while it endeavors to irradiate human life with a spirit of cheerful contentment, does not altogether achieve its object."

Lucretius evidently did not find his task altogether free from difficulties since he remarks (Book I, 136–138) "Nor do I fail to understand that it is difficult to make clear the dark discoveries of the Greeks, in

[29] *De Rerum Natura*, I, 70–80.   Translated by W. H. D. Rouse.   (Loeb Classical Library.)
[30] *De Rerum Natura*, I, 158.

Latin verses, especially since we have often to employ new words because of the poverty of the language and the novelty of the matters," and (Book V, 335–337) "Again the nature and system of the world has been discovered but lately and I am now found the *first to describe it in our own mother tongue*". The statement that "The nature and system of the world has been discovered but lately" leads us to the recognition of the fact that while Lucretius was an Epicurean and based his work on the teachings of Epicurus, during the period of over two hundred years which had elapsed since the time of Epicurus much information that was unknown to the master had been discovered, so that Lucretius was able to present a much more complete picture of the universe than that which Epicurus could envisage. Lucretius discusses the four elements—fire, air, earth and water—their respective properties and their relations to and combination with one another. This leads to the consideration of the constitution of matter as composed of atoms of these elements, and the elucidation of his Atomic Theory. He then treats of the sun and moon, the tides and the succession of the seasons, the ocean and its relation to the continents. He also, however, touches or in some cases enlarges upon, certain matters which have a more immediate geological significance, such as the process of rock decay and disintegration (Book V, 306):

> Again, do you not see that stones even are conquered by time, that tall turrets do fall and rocks do crumble, that the gods' temples and their images wear out and crack, nor can their holy divinity carry forward the boundaries of fate, or strive against nature's laws?
>
> Again, do we not see the monuments of men fall to pieces, asking whether you believe that they in their turn must grow old? Do we not see lumps of rock roll down torn from the lofty mountains, too weak to bear and endure the mighty forces of time finite? For they could not fall thus suddenly torn off, if they had endured all through from time indefinite all the wrenching of the ages without breaking up.

He treats of the origin of springs and rivers, caverns and pools of water beneath the earth's surface, of the violent winds within the earth which at times rend it asunder and give rise to earthquakes,[31] of volcanic eruptions and their causes, the exhalations of subterranean vapors found in gold mines and their shocking effects on the health of the miners, as well as the discovery and mode of occurrence of metals. The range of subjects of which he treats and his method of discussing them, fanciful

---

[31] The views of Lucretius on the cause and origin of earthquakes are set forth in chapter XI.

as this often is, show that he possessed a much wider knowledge of the earth and the forces which are at work in modifying its surface than that displayed by those who preceded him.

The next Roman writer whose work presents itself for consideration is Vitruvius. He was a Roman architect, who served as a military engineer under Julius Caesar in the African war of 46 B.C. and was inspector of military machines under Augustus. The dates on which he was born and died have never been ascertained with certainty. In his old age, however, he wrote a work which has achieved great renown, entitled *De Architectura*. Sontheimer[32] is of the opinion that this work was written between 37 and 32 B.C. Other authorities give somewhat earlier or somewhat later dates. This work is not only a comprehensive treatise on architecture, but an encyclopedia of the technical knowledge of the day, more especially in connection with the building trades. While it contains very little of general geological interest, there are scattered through it many scraps of information concerning certain minerals and rocks, and the uses to which they are put as building materials, paints and for other technical purposes. It contains the earliest reference to hydraulic mortar and the first study of architectural acoustics. Many local varieties of clays, sands, gravels and stone are referred to and the peculiar value of each for certain purposes in connection with building consturction is indicated. He also treats of the best materials for use in making lime for mortar, and points out that a certain mortar which will harden under water can be obtained by the use of puzzulano, a peculiar earth which occurs in the vicinity of Vesuvius. This is, of course, the now well-known ash "puzzulano" which was ejected at times by the volcano. He notes that in the district about Vesuvius there must be fires below the surface of the earth, because steam and hot water issue from the ground in certain places, especially about Baia, and the evidence which the district affords that in former times these fires broke open passages to the surface, through which molten rock was poured out over the surrounding country. He mentions the occurrence of pumice stone in association with these lavas, a peculiar variety of rock which, he says, is known only to occur in one other locality near Etna. Vitruvius also describes the materials from which paints and colors are obtained, among them ochres of various kinds, and mentions that a much brighter red color can be obtained from cinnabar. This mineral, he

---

[32] *Vitruvius und seine Zeit, Dissertation*, Tübingen, 1908.

says, was first found near Ephesus, but in the time of Vitruvius it was mined in Spain and purified in Rome.   It was used not only as a color, but as a source from which mercury was obtained.   He mentions the use of this metal in the art of silvering and gilding, and also in the separation of gold by amalgamation.   He makes a mistake, however, in describing this process, since he says that after the mercury has taken up the gold, the resulting mass is placed in a stout cloth and squeezed, when the mercury passes out through the cloth and the pure gold remains.   What remains is really the gold amalgamated with the mercury, while the excess of free mercury only is separated.   He also refers to malachite from which green paints are made and which he says occurs with copper ores in Macedonia, also white lead made by the treatment of metallic lead with acetic acid, a process of which again he gives an inaccurate description.   He also mentions the occurrence of asphalt at Babylon, the Dead Sea and elsewhere, as well as beds of rock salt and springs of saline waters in various localities.   His views on the origin of springs and rivers is referred to in Chapter XII.

Of all the writers in classical times Pliny has made the most outstanding and important contribution to our knowledge of the mineral kingdom, in that he has preserved for us a great body of information drawn from the works of older writers which would inevitably have perished had he not assembled and recorded it.   Caius Plinius Secundus, called Pliny the Elder A.D. 23 to 79, was born of an illustrious and wealthy family, held many important positions and spent his life largely in the imperial service.   He travelled extensively in the countries bordering the Mediterranean, among them Spain, Greece and Egypt, served in Africa, commanded a cavalry troop in Germany and was admiral of the Roman fleet at Misenum at the time of his death.

The younger Pliny in a letter addressed to Baebius Macer, in referring to his uncle the elder Pliny, says:

You will wonder that a busy man could find time to write so many books, books, too, containing much abstruse matter.   You will wonder more when I tell you that for some time he was a pleader, that he died at the age of 56 and that meantime he was much hindered and distracted by important state business and by his intimacy with our emperors.   But his intellect was quick, his industry perfectly marvellous, his power of remaining awake remarkable.   From the 23rd of August he began to study at midnight, and through the winter he continued to rise at one, or at latest two in the morning, often at twelve.   Sleep he could always command.   Before daybreak he would go to the Emperor Vespasian who also worked at night, and thence to his official duties. On returning home he gave what time remained to study. After taking a light meal, as our forefathers used to do, he would often in summer, if he had leisure, recline in the sun and have a book read to him, on which he took notes or from which he made extracts. He read nothing without making extracts, for he used to say that he could get some good from the worst book. After reading in the sun, he generally had a cold bath, then a light meal and a very short nap, after which, as if beginning another day, he would study till dinner time. During dinner a book was read to him and he made notes upon it as it went on. I remember one of his friends once stopping the reader, who had pronounced a word incorrectly, and making him repeat it. My uncle said to him, "Did you not understand the word?" "Yes", he replied. "Why then did you stop him? We have lost more than ten lines by this interruption." So parsimonious was he of his time. He rose from supper in the summer by daylight, in winter before seven, as regularly as if constrained by law. Thus he lived in the midst of his work in the bustle of Rome. . . . When travelling, he always had a scribe at his side with a book and writing tablet, whose hands in winter were protected by gloves, so that the cold weather might not rob him of a single moment . . . in fact, he thought all time lost which was not given to study. It was by his intense application that he completed so great a number of books and left me besides 160 volumes of extracts written on both sides of the leaf.

He wrote a book, *On the use of the Dart by Cavalry* and others *On the Life of Pomponius Secundus, Wars with Germany, The Student* (on the education of an orator) and *On Questions of Grammar and Style.* All these books have been lost, one other, however, he wrote, which has come down to us and on this his reputation rests: This is his *Natural History* which appeared about A.D. 77 and was dedicated to the Emperor Vespasian. In the dedicatory letter addressed to the Emperor, which prefaces the first of the thirty-seven "books" into which the work is divided, Pliny tells Vespasian some interesting facts concerning the *Natural History*, and makes some quaint observations. With regard to the content of the work he says:

The truth is this, the nature of all things in this world, that is to say, matters concerning our daily and ordinarie life, are here deciphered and declared. . . . Moreover, the way that I have entered into, hath not been trodden beforetime by other writers, being indeed so strange and uncouth, as man's mind would not travel therein. No Latin author among us hath hitherto once ventured upon the same argument, no Grecian whatsoever hath gone through it and handled all. . . . Now come I and gather as it were a complete body of arts and sciences . . . a difficult enterprise it is to make old stuff new, to give authoritie and credit to novelties, to polish and smooth that which is worne and out of use, to set a glosse and lustre upon that which is dim and darke, to grace and countenance things disdained, to procure beleefe to matters doubtful and in a word to reduce nature to all and all to their own nature. As touching myself, I may be bold to say and averre that in 36 books I have comprised 20,000 things all worthy of regard and consideration which I have recollected out of 2,000 volumes or thereabout that I have diligently read and those written by 100 several and approved authors, besides a world of other matters which either were unknown to our forefathers and former writers, or else afterward invented by their posterity. . . . Moreover, considered it would be that these studies were followed at vacant times and stolen hours, that is to say, by night season only: to the end that you may know, how we to accomplish this have neglected no time that was due unto your service. The daies we wholly employ and spend in attendance about your person; we sleepe only to satisfye nature, even as much as our health requires and no more, contenting ourselves with this reward, that whiles we study and muse (as Varro saith) upon these things in our closet, we gaine so many houres to our life, for surely we live then only, when we watch and be awake.[33]

While Pliny in this preface says that he treats of 20,000 topics gleaned from 2,000 books by 100 authors, judging from the number of references and citations which he makes, he must have drawn his information from more than 100 writers. It is stated that he cites or quotes from 672 authors in all, 146 of these being Roman and 326 Greek.[34] Fortunately Pliny gives the names of the persons whom he quotes, while Aristotle and most other classical authors use the forms "It is said," or "We read."

The *Natural History*, as translated by Holland, is divided into 37 books of which the last five deal with the mineral kingdom. The book embraces a very wide range of subjects, but is essentially a Natural History, in that it is devoted chiefly to a description of the three kingdoms of nature, the discussion of other subjects being, as a rule, brought in incidentally, or for purposes of illustration. Pliny regards the world as existing for the use of man, and he tells us that his aim in writing the book is to "Benefit posterity and do good to the common life of man," and so in almost all the things which he mentions or describes, he makes a point of stating what medicinal properties or magical virtues they possess, for he says: "No man doubteth but that magic took root first and proceeded from physic, under the pretence of maintaining health, and both curing and preventing diseases."

The *Natural History* is a very large work and is really an encyclopedia (a word used by Pliny himself) of the knowledge concerning nature

---

[33] This and other quotations from Pliny's *Historia Naturalis* are taken from the translation of the work by Philemon Holland published in London in 1634. (An earlier edition appeared in 1631.) There is also a later translation of the work by Bostock and Riley published in Bohn's Library in 1857.

[34] Charles Singer, *From Magic to Science*, London, 1928, p. 12, and other writers.

possessed by the Romans at the dawn of the Christian era.   Gibbon
calls it "That immense register where Pliny has deposited the discoveries, the arts and the errors of Mankind."

Pliny's nephew, commonly known as the Younger Pliny, speaks of it
as "A work of great compass and learning and as full of variety as nature
herself."[35]   To use a simile drawn from geology, it is a great outcrop of
a conglomerate made up of an immense number of pebbles or fragments
great and small, often strange in character and full of interest, but with
little order or arrangement, brought together from many distant places,
the original sources from which they were derived having in many cases
been completely swept away by the tide of time, the whole being cemented together by a very small quantity of interstitial material contributed by Pliny himself.

Many of the statements made by Pliny in dealing with the 20,000
subjects which he discusses are, as might be expected, quite erroneous
and often ridiculous.   The book is a mine of misinformation as well as
a treasury of information.   There are an endless number of statements
concerning serpents, dragons, basilisks, and their poisons, and antidotes
for the same.   He tells of the wonderful magical properties of a great
variety of strange things, as, for instance, that if a man carry in his
shoe a dog's tongue dogs will not bark at him; of the mysterious antipathy
between the diamond and the lodestone, so that the lodestone loses its
power of attracting iron if a diamond be brought near to it.   Of how the
diamond, if it be placed upon an anvil and struck a mighty blow with a
hammer will not itself be broken but will break the anvil in two and at
the same time cause the hammer to fly to pieces, but that nevertheless
the diamond, the hardest of all known things, will be shattered if subjected to this trial after first being steeped in blood freshly drawn from
a goat.   Like Pliny, (Book 57, Chap. 4) we would say:

I would gladly know whose discovery this might be to soak the Diamond in goat's
blood; who first thought of this; or rather by what chances was it found out and
known.   What conjectures should lead a man to make an experiment of such a singular and admirable secret, especially in a goat, one of the most filthy beasts in the whole
world.   I must ascribe this discovery and all others like it to the might and benefi-
cience of the divine powers, neither are we to argue and reason how and why nature
hath done this and that: sufficient it is that her will was so and thus she would have it.

[35] *The Letters of Caius Plinius Caecilius Secundus.*   Bohn's Classical Library, London,
1908, p. 80.

For many of these weird statements Pliny is not responsible: he derives some of the wildest of them from authors whom he quotes. Concerning one of these, "Juba, the father of Ptolomaus, King of the Mauritanes, renowned for his studies and love of good letters," Thorndike[36] says

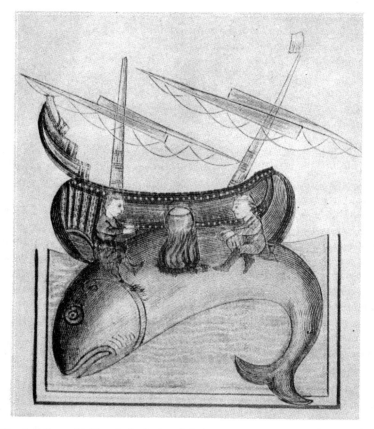

FIG. 1. Sailors mistaking the back of a whale for an Island have "landed" on it and are cooking food. (From the *Bestiàire d'Amour of Richard Fournival*, Tenth Century.)

"There hardly seems to have been a greater liar in antiquity." This is indeed giving him a high place as an exponent of mendacity among so many able competitors! Thorndike awards King Juba this distinction

[36] *A History of Magic and Experimental Science*, p. 49.

on account of a statement made by him to the effect that whales, 600 feet long and 360 feet broad have "Shot themselves out of the sea into the great rivers of Arabia."[37] But, after all, several other ancient writers have spoken of whales so large that when at rest floating on the surface of the sea, they have been mistaken for islands, a statement which forms the basis of Milton's well-known simile in *Paradise Lost* (Book I, 200–208):

> That sea-beast
> Leviathan, which God of all his works
> Created hugest, that swim the ocean stream,
> Him haply slumb'ring on the Norway foam,
> The pilot of some small night-foundered skiff
> Deeming some island, oft, as seamen tell,
> With fixed anchor in his scaly rind
> Moors by his side under the lee, while night
> Invests the sea, and wished morn delays.

Which of these, the Arabian or Norwegian whales, were the larger may be left to the conjecture of the reader.

It is interesting to note, before leaving this subject, that Pliny was of the opinion that in his time practically all the different kinds of creatures inhabiting the sea were known to naturalists, these being, he says, 176 in number, while it was impossible to discover how many species of birds or land animals there were, since there were probably many in India, Ethiopia, Scythia, and the desert regions of the world which were as yet unknown to naturalists; and he goes on to say that "It is a wonderful matter that we should be better acquainted with the former, considering how Nature hath plunged and hidden them in the deep gulfes of the maine sea." (Book 32, Chap. 11.) For our present purpose, however, the concluding chapters of the *Natural History*, numbers 33 to 37, are those of especial interest, since they deal with substances which belong to the mineral kingdom. Pliny, who regards the earth with a peculiar veneration as "The mother of all," on whose broad bosom man should find all that his needs require, in the preface to Book 33 protests at the indignities which men offer to her when they "Sticke not to search unto the very bowels of the earth," sinking their shafts even "As far as the seat and habitation of the infernal spirits," and all for the purpose, not of seeking medicine to cure the ills of men,

---

[37] *Natural History*, Book 32, Chap. 1.

but in the mad search for gold, silver and the baser metals, which when men secure serve only to corrupt their natural virtues.

> Oh! that the use of gold were clean gone: would God it could possibly be quite abolished among men, setting them as it doth into such a cursed and excessive thirst after it. Even Iron which we may say is at once the best and worst of metals, for while by its help we plough and reap, by means of it wars are waged, murders and robberies are perpetrated and the power of man for evil is multiplied.

And when the earth trembles and quakes we are too blind to see "How these are signs of the wrath of this our blessed mother" on account of such wrong and misuse.

Having duly protested against the use of metals, he mentions the various localities from which each is obtained, the methods of mining and the manner in which the metals are extracted from their ores. Gold, he says, is won in three ways. First by alluvial washing; secondly by sinking shafts, and thirdly in a way "That is so painful and toilsome, that it surpasseth the wonderful works of the Geants in old time." This consists in undermining great cliffs of gold-bearing ground or rock by means of a series of long drifts, the roofs of which are supported by timbering and then when the workings have been driven as far as they desire to carry them, they cut the props beginning at the far ends of the drifts and allow the whole to collapse. The mass of ground thus broken and disintegrated is subjected to a process of hydraulicing, washing it by means of streams of water under a great head carried in sluices and pipes, often for a distance of one hundred miles, from streams high up on the mountain sides, the water after being passed through and over the broken ground being carried off in sluice boxes in which shrubs and bushes have been placed to entangle the gold. The gold in the sluices is gathered up and the bushes are burned and the gold entangled in them is recovered from their ashes. By this process, he says, 20,000 pounds of gold is obtained every year in Spain and Portugal, adding that the greater part of it comes from Asturia and that there is no part of the world where there are gold mines so rich or which have been worked continuously for such long periods of time as in these countries. He then passes on to the consideration of silver which, he says, next to gold is the metal men seek for most eagerly. This metal generally occurs in association with lead in a mineral called galena. It also abounds in Spain. He refers to mirrors, which must be made of the best and purest silver, and wonders at the perfection of the image which they present, "Which must needs be (as all men confess) due to the reverberation of the air from the solid body of the mirror, which being beaten back again from it, bringeth therewith the said image expressed therein." He then refers to mercury, and its uses for separating gold by amalgamation, a process concerning which his statements are not correct in all respects, and then passes on to antimony, of which certain preparations are good for the eyes, to arsenic and then to litharge, cinnabar, ochres and azurite, valuable as colors.

In Book 34, he treats of copper and its various ores and alloys, and their respective uses; and of iron, referring to its wide distribution, mentioning especially the island of Elba as a locality where iron ore is being reproduced continually, a fact which is reiterated again and again by later writers, and to which reference will be made in Chapter IX, in considering the origin of metals and their ores. He says that the greatest known deposit of iron ore, the one where "The vein of this metall is largest and spreadeth itself into most lengths every way, we may see in that part of Biscay that coasteth along the sea and upon which the ocean beateth, where there is a craggy mountaine very steepe and high which standeth upon a mine or vein of yron." Here he is referring to the Cantabrian coast, and the great iron mines of the Bilbao district which have been worked so extensively down to the present day. He also mentions the medicinal and magical properties of iron and of several of its compounds. He describes at some length the lodestone, and its properties, to which further reference will be made in Chapter XIII. Tin ("plumbum candidum") and lead ("plumbum nigrum") are then considered, and Pliny says that he regards as a mere fable the statement that the former is derived almost entirely from "The islands of the Atlantic Sea" since it occurs in Spain and Portugal in alluvial deposits, and is worked in these countires. In that he is undoubtedly wrong, since it is known from other sources that while cassiterite or tin stone may have been obtained in small quantities from these latter countries, the principal supplies of tin, even in Roman times, came from Cornwall. He then passes to the consideration of stones and earths. Chief among these are the various marbles which were in such demand among the Romans; alabaster; puzzulano, the volcanic ash from which cement was made; limestone which was burned for lime to make plaster; the various raw materials used in the manufacture of glass; also millstones, whetstones and building stones of various kinds.

In the last book of the *Natural History* Pliny deals with gems and

precious stones, with their many different species and their use for
adornment, and when engraved for seals, and their medicinal and
magical properties. In connection with them he relates a host of stories
and a multitude of quaint conceits. After describing the wonderfully
beautiful Murrhinian cups and vases made of a material which was apparently fluorospar, he enters upon a lengthy disquisiton concerning
rock crystal which, he says, we may make bold to state is nothing but a
frozen water, a very hard variety of ice, found in northern countries or
on high mountains, although our friend, King Juba, states that "It
grows on an island called Neron lying beyond the Red Sea over against
Arabia." The best quality is that which comes from India and Cyprus,
but excellent crystals are also found in the high Alps. He is at a loss to
account for the fact that its crystals always have six angles, and that the
points do not always have the same form. He even mentions the
minute cavities, hair-like inclusions and rusty stains found in some specimens. Some physicians, he informs us, are of the opinion that there is
no more wholesome cautery than that obtained by concentrating the
rays of the sun on the place which requires cauterization, by means of a
sphere of rock crystal. Drinking cups were sometimes made of it, and
commanded enormous prices. He then speaks of amber, concerning
whose origin all manner of ridiculous stories have been told, making
contemptuous reference to "Such stuff which the poets have sent abroad
into the world." As a matter of fact, he says, amber is a hardened gum
from a tree which grows on the northern coast of Germany. This
substance has its value in medicine, for a collar of amber beads around
the neck of a young infant is a singular preservation against poison, as
well as against "Illusions and frights that drive folks out of their wits."
Emeralds when gazed at, refresh the eyes when weary and the longer
they are looked upon, the larger they appear to be, owing to the fact
that they cause the air about them to become green in color also. Beryl,
he says, is very closely allied to emerald if not identical with it. This,
of course, is true, but Pliny on the other hand was wrong in stating as
he does that the six-sided prisms in which beryl crystallizes are not its
natural form, but are given to it by the lapidaries who cut the mineral
into this shape in order to increase its lustre, although he goes on to say
that some are of the opinion "That Beryl groweth naturally cornered by
many faces." Chrysoberyl, chrysoprase and hyacinth are also closely
allied to, if not actually varieties of beryl. He then describes opal and
carbuncle, including under the latter name, the bright red varieties of

ruby, spinel and garnet, and quotes Archelaus as saying that some varieties have such a fiery quality that if seals are cut from them, these will melt the very sealing wax on which they are impressed. With the description of topaz and jasper he brings to a conclusion his consideration of the more important mineral species. All the other and less important minerals known in classical times are then arranged under the initial letter of their names, and referred to very briefly, mention being made of their salient characteristics, their respective virtues, and in the case of many of them, their most astonishing magical powers.

While Pliny's *Natural History* is, as we have seen, a great and rather disorderly assemblage of fact and fiction, it probably presents a true epitome of the conception of the world of nature in classical times. It is a store house which was drawn upon by writers on the subjects of which it treats for the next sixteen centuries, and supplies a large part of the material used by the authors of the lapidaries of the Middle Ages.

The first printed edition was issued, shortly after the invention of printing, by Johannes Spira from his press in Venice, appearing in 1469. E. W. Gudger[38] lists two hundred and twenty-two editions of the work, and states that there may have been a few others which he did not find. Even as late as the time of the Renaissance and down to the close of the seventeenth century it was the most widely read and most authoritative work on Natural History in existence. In Gudger's opinion, it is "The most popular Natural History ever published."

The Elder Pliny has been called "The First Martyr of Science," whether this be true it is difficult to say, but it is certain that he died in the prosecution of scientific research. This was in A.D. 79 when Mount Vesuvius broke out in its first recorded eruption and overwhelmed Herculaneum, Pompeii and other lesser cities of the plain. Strabo, who died in A.D. 25 visited Vesuvius about the beginning of the Christian era, and describes the mountain as being clothed with corn fields and vineyards to its very summit where, however, there was a flat barren stretch on which the rocks bore the marks of fire, having a fused and slaggy appearance which led him to suppose that the mountain had at one time been a volcano. There was, however, no record or tradition of any eruption having taken place. The account of the eruption which is given by the Younger Pliny, in a letter to his uncle, the historian Tacitus, is one of the most interesting literary documents in the history

[38] *Pliny's Historia Naturalis*, Isis, vol. VI, No. 18, 1924.

of geology, and will form a fitting conclusion to this brief account of the work of this distinguished man:

> You request that I send you an account of my uncle's death, and it is with the greatest willingness that I do so.
>
> He was with the fleet under his command at Misenum.[39] On the 24th of August, about one in the afternoon, my mother desired him to observe a cloud which appeared of a very unusual size and shape. He had just taken a turn in the sun, and, after bathing himself in cold water, and making a light luncheon, gone back to his books: he immediately arose and went out upon a rising ground from whence he might get a better sight of this very uncommon appearance. A cloud, from which mountain was uncertain, at this distance (but it was found afterwards to come from Mount Vesuvius), was ascending, the appearance of which I cannot give you a more exact description of than by likening it to that of a pine tree, for it shot up to a great height in the form of a very tall trunk, which spread itself out at the top into a sort of branches; occasioned, I imagine, either by a sudden gust of air that impelled it, the force of which decreased as it advanced upwards, or the cloud itself being pressed back again by its own weight, expanded in the manner I have mentioned; it appeared sometimes bright and sometimes dark and spotted, according as it was either more or less impregnated with earth and cinders. This phenomenon seemed to a man of such learning and research as my uncle extraordinary and worth further looking into. He ordered a light vessel to be got ready, and gave me leave, if I liked, to accompany him. I said I had rather go on with my work, and it so happened he had himself given me something to write out. As he was coming out of the house, he received a note from Rectina, the wife of Bassus, who was in the utmost alarm at the imminent danger which threatened her, for her villa lying at the foot of Mount Vesuvius, there was no way of escape but by sea; she earnestly entreated him therefore to come to her assistance. . . . He ordered the galleys to put to sea, and went himself on board with an intention of assisting not only Rectina, but the inhabitants of towns which lay thickly strewn along that beautiful coast. Hastening then to the place from whence others fled with the utmost terror, he steered his course direct to the point of danger, and with so much calmness and presence of mind as to be able to make and dictate his observations upon the motion and all the phenomena of that dreadful scene. He was now so close to the mountain that the cinders, which grew more abundant and hotter the nearer he approached, fell into the ships, together with pumice stones, and black pieces of burning rock: they were in danger too not only of being aground by the sudden retreat of the sea, but also from the vast fragments which rolled down from the mountain, and obstructed all the shore. Here he stopped to consider whether he should turn back again as the pilot advised him, 'Fortune,' said he, 'favours the brave; steer to where Pomponianus is.' Pomponianus was then at Stabiae, separated by a bay, which the sea, after several insensible windings, forms with the shore. He had already sent his baggage on board; for though he was not at that time in actual danger, yet being within sight of it, and indeed extremely near, he was determined to put to sea as soon as the wind, which was blowing dead in-shore, should go down. It was favourable, however, for carrying my uncle to Pomponianus, whom he found in

---

[39] In the Bay of Naples.

the greatest consternation: He embraced him tenderly, encouraging and urging him to keep up his spirits, and the more effectually to soothe his fears by seeming unconcerned himself, ordered a bath to be got ready, and then, after having bathed, sat down to supper with great cheerfulness, or at least (what is just as heroic) with every apperance of it.   Meanwhile the broad flames shone out in several places from Mount Vesuvius, which the darkness of the night contributed to render still brighter and clearer.   But my uncle, in order to soothe the apprehensions of his friend, assured him it was only the burning of the villages, which the country people had abandoned to the flames: After this he retired to rest, and it is most certain that he was so little disquieted as to fall into a sound sleep: for his breathing, which, on account of his corpulence, was rather heavy and sonorous, was heard by his attendants outside.   The court which led to his apartment being now almost filled with stones and ashes, if he had continued there any time longer, it would have been impossible for him to have made his way out.   So he was awoke and got up, and went to Pomponianus and the rest of his company, who were feeling too anxious to think of going to bed.   They consulted together whether it would be most prudent to trust to the houses, which now rocked from side to side with frequent and violent concussions as though shaken from their very foundations; or fly to the open fields, where the calcined stones and cinders, though light indeed, yet fell in large showers, and threatened destruction.   In this choice of dangers they resolved for the fields: a resolution which, while the rest of the company were hurried into by their fears, my uncle embraced upon cool and deliberate consideration.   They went out then, having pillows tied upon their heads with napkins; and this was their sole defence against the storm of stones that fell around them.   It was now day everywhere else, but there a deeper darkness prevailed than in the thickest night; which however was in some degree alleviated by torches and other lights of various kinds.   They thought proper to go farther down upon the shore to see if they might safely put out to sea, but found the waves still running extremely high, and boisterous.   There my uncle, laying himself down upon a sail cloth, which was spread for him, called twice for some cold water, which he drank, when immediately the flames, preceded by a strong whiff of sulphur, dispersed the rest of the party, and obliged him to rise.   He raised himself up with the assistance of two of his servants, and instantly fell down dead; suffocated, as I conjecture, by some gross and noxious vapour, having always had a weak throat, which was often inflamed.   As soon as it was light again, which was not until the third day after this melancholy accident, his body was found entire, and without any marks of violence upon it, in the dress in which he fell, and looking more like a man asleep than dead.

Lucius Annaeus Seneca, born about A.D. 3 and who died A.D. 65, was a contemporary of Pliny and tutor to the Emperor Nero.   He also was a voluminous writer.   Most of his works are of a philosophical character, but one of them stands in a category by itself.   This is his *Quaestiones Naturales*.[40]   It was his latest work and is a treatise on

---

[40] *Physical Science in the Time of Nero*, being a translation of the *Quaestiones Naturales of Seneca* by John Clarke, M.A., with notes on the treatise by Sir Archibald Geikie, London, 1910.

various subjects connected chiefly with meteorology, astronomy and seismology. The only portions of the interesting work which have a bearing on geology are the chapters dealing with surface and subterranean waters and with earthquakes. Being a Stoic and a moralist as well as a philosopher, he intersperses his observations on the processes of nature with moral reflections and ethical advice which gives the work a peculiar flavor. Thus the chapter entitled *On Forms of Water* is introduced by the following words:

> Having begun a mighty task in my old age, I must make up for lost time by hurrying on. The magnitude of it is actually an incentive to effort. Such studies are far superior to the historian's task of recording the deeds of robbers and butchers of mankind. The former raises us above the vicissitudes of fortune. The principal thing is to have a pure heart and clean hands, to escape slavery to self. The study of the universe exalts us to this.

Seneca's views on the two subjects above referred to will be taken up in the chapters of the present work dealing with the *Origin of Springs and Rivers* and with *Earthquakes*.

Cato, called "The Censor" (234–149 B.C.), who was a well-known Roman statesman and like his plebeian forefathers was brought up as a farmer, wrote an interesting treatise on agriculture entitled *De Re Rustica*, which contains a number of references to various building materials, tiles, bricks and cements. He also describes in detail the best method of constructing and operating lime kilns and for the purification of common salt by solution and recrystallization.

Flavius Josephus (A.D. 37 to ca. 100) refers to the "Asphalt springs" in the district about the Dead Sea, and also speaks of black lumps of asphalt which rise up through its highly saline waters and float upon its surface. In former times, according to his account, this region was very fertile and supported many towns but at the time in which he wrote the whole was a fire-swept waste on which could be traced only the blackened outlines of five of these towns. The fire, he tells us, originated from lightning.

Caius Suetonius Tranquillus (A.D. 72–123), who was a friend of Pliny the Younger, wrote a work entitled *The Meadows* dealing with Roman antiquities, laws and natural history which, however, has been lost. In his *Lives of the Twelve Emperors*, he narrates how the Emperor Augustus at his Villa at Capri gathered together a collection of immense fossil bones which were, at that time, supposed to be those of an extinct race of giants.

Galen, the renowned Grecian scholar and physicain (A.D. 129–200), visited the island of Lemnos on his travels in order to examine the site from whence came the celebrated "Lemnian terra sigillata," renowned far and wide for its supposed curative properties. It was a deposit of red earth—probably an impure hydrated oxide of iron—which was gathered with solemn ceremony by the priestesses from the neighboring city and stamped into little discs bearing the seal of the goddess Diana. Galen states that he took away with him 20,000 of these little cakes. He also mentions having visited the island of Cyprus,[41] where he saw a deposit consisting of "Sory" overlain by two others consisting of "Chalcite" and "Misty," and collected from it a great supply of these minerals for medicinal purposes. These three minerals are also mentioned by Pliny, Dioscorides and Agricola, but it is impossible to determine from the descriptions which they have left to which mineral species they should be referred. Various opinions have been expressed by different persons who have examined the several texts.[42] It seems probable that the Chalcite was an impure copper pyrite and that the others were decomposition products of this. Some, however, think that it was an ore containing zinc. Galen says that the materials which he collected from Cyprus, thirty years later had undergone such a remarkable change that he believed all these minerals were of the same nature. This change probably consisted in the decomposition, oxidation and deliquescence of the original materials with the production of hydrated sulphates.

Arrianus, the Greek historian who was born about A.D. 90, mentions the occurrence of obsidian deeply buried in the sand on the shores of the Red Sea south of Adule, to which port ships from India brought Indian steel and iron, as well as tin and gems. He says that ships came to the Persian port Omana bringing copper, which was exchanged for gold, and refers to the importation of tin, lead, gold, silver, antimony ore, coral and topaz at the ports of Munnagara and Barygaza in north-west India and to the extensive trade carried on at the port of Nelecynda in southwest India in gold, copper, tin, lead, vermilion and antimony ore, as well as in topaz, pearls, diamonds and other precious stones. He mentions also the occurrence in Taprobana (Ceylon) of pearls and various gems. (See Lenz. p. 177)

---

[41] *De Simplicium Medicamentarum Temperamentis et Facultatibus*, 9, 21.
[42] Thorndike, *A History of Magic and Experimental Science*, vol. 1, pp. 125–132.
Lenz, *Mineralogie der Alten Griechen und Römer*, pp. 116, 179.
Hoover, in his translation of Agricola's *De Re Metallica*, p. 573.
Agricola, *De Natura Fossilium*, p. 212.

Caius Julius Solinus, who flourished about A.D. 250, wrote a short work entitled *Collectanea rerum memorabilium*, which in the Middle Ages was generally known as *Polyhistor, seu De mirabilibus mundi*. It is a geographical work, presenting a sketch of the world as known to the ancients, with remarks on the origin, habits, religious rites and social conditions of the various nations. He is especially inclined to recount marvellous events connected with certain places or localities, and for this information he draws largely on Pliny and Mela, generally passing over the facts mentioned by Pliny, but reproducing his fabulous stories with further embroidery. Among the products of the various countries he speaks of the minerals which they yield, and recounts the marvellous powers possessed by agates, galactites, catochites, crystal, gagates, adamant, heliotrope, hyacinth and paeanites.

5. CONCLUSION

There has been given above what is believed to be a complete conspectus of the observations and opinions of the writers in Classical times on the subjects which are embraced by what are now known as the geological sciences. This forms the first chapter in the history of geological science and embraces a long period of some eight centuries, a period which in other fields of intellectual effort was marked by some of the greatest achievements of the human mind. When, however, the contributions of this period to geological science are considered it will be seen that these are indeed scanty and of little worth. This was not due to any lack of mental power or ability on the part of the scholars of these early times—far from it—but they had bent their minds and devoted their attention to other subjects and had not as yet discovered the true methods of scientific research. As Zittel remarks, not a single writer on these subjects in the ancient world had examined the rocky crust of the earth with a view to ascertaining its composition and the succession of the strata, still less had he any conception of the fact that the fossils which these rocks contained afforded materials from which the history of the earth might be written. The aim and object of modern geological and palaeontological study was absolutely unknown to the ancients, for baseless hypothesis and haphazard observations cannot be considered as a foundation for scientific achievement.[43]

[43] *Geschichte der Geologie und Palaeontology*, p. 11.

CHAPTER III

THE CONCEPTION OF THE UNIVERSE IN THE MIDDLE AGES

As has been already stated, it is necessary that we should have a
mental picture of the universe as it presented itself to the minds of men
in medieval times, in order that we may understand the development of
science during this period as well as the weird and fantastic explanations
often put forward to account for natural phenomena, whose causes now
seem to us simple and obvious. "No medieval writer whether on science
or magic can be understood by himself, but must be measured in respect
of his surroundings and antecedents."[1] For the world of ideas in which
we are now living is quite different from that in which they lived, one
which has been created by the rapid and untrammeled development of
the natural and physical sciences through a long period of years and
especially during the last two centuries. In the present chapter an
attempt is made to present in bare outline a picture of the surroundings
under which a student of science in the Middle Ages lived and carried
on his work—the mental climate, as it has been called, of that period,
together with a brief historical account of the causes which led to its
development.

The Roman Empire—mistress of the world—was so powerful and so
perfect in the organization of all its functions and activities that it gave
promise of enduring forever. Yet before the conclusion of the fifth
century this mighty fabric of empire broke to pieces under the impact of
the successive blows delivered by the invading hosts of those nations
whom the Romans classed together under the general designation of
Barbarians, nations which pressed down upon it from the north, and
whose numbers and martial energy were irresistible:

From the dawn of history they shew as a dim background to the warmth and light
of the Mediterranean Coast, changing little while kingdoms rise and fall in the south;
only thought on when some hungry swarm comes down to pillage or to settle.

The severed limbs of the Empire forgot by degrees their original unity. As in the
breaking up of the old society, which we trace from the sixth to the eighth centuries,
rudeness and ignorance grew apace, as language and manners were changed by the
infiltration of Teutonic settlers, as men's thoughts and hopes and interests were
narrowed by isolation from their fellows. As the organization of the Roman province

[1] Thorndike, *History of Magic and Experimental Science*, New York, 1929, vol. I, p. 4.

and the Germanic tribe alike dissolved into a chaos whence the new order began to shape itself, dimly and doubtfully as yet, the memory of the old Empire died away.[2]

This period of chaos constitutes what is generally known as the Dark Ages. The dire condition into which society had sunk and the fears which haunted men's minds even at the very heart of the Empire, in the city of Rome itself, at the close of the sixth century, was borne in upon the writer, when in Rome in the spring of 1895 he received an invitation to attend a Mass to be celebrated in commemoration of St. Nereus and St. Achilleus, two Roman soldiers, who, it is believed, belonged to the Pretorian Guard ("They of Caesar's household"), and who suffered martyrdom publicly in the time of Nero. This was to be celebrated in their chapel in the Catacomb of St. Domitilla, under the auspices of the society known as the Cultores Martyrum.

Descending with the other worshippers to the second level of the Catacomb we reached the ancient chapel. The service was most impressive. The pale gleam of the candles about the altar, the rising incense, and the low chanting of the service served to intensify the mystery of the strange surroundings. Kneeling in the semi-darkness, the movements of the officiating priests and acolytes, the genuflexions, and the solemn music, aroused a feeling of dreamy wonder whether the two apostles whose mortal remains—according to tradition—repose not far away would have been able to recognize in this elaborate ritual a development of that service once held in an upper chamber. But a highly dramatic note was struck when the celebrant, taking his seat in a marble chair in the tribune of the church read the identical homily which had been delivered by no less a person than Gregory the Great ("Apostolus Anglorum") when seated on that very spot some thirteen hundred years before, toward the close of the sixth century. The Roman Empire was then tottering to its fall, all the foundations of the earth were out of course and the striking words, calling to us across the gulf of centuries, illuminated as by a lightning flash the conditions of that far distant age when all earthly hope had passed away:

Behold the world in itself is burnt up already and yet in our hearts is still ever green.

Death is everywhere and mourning, desolation is everywhere and the strokes that smite us and the bitterness that is our daily bread:

And yet, in the blindness of our fleshly lust we are in love with the very bitterness of the world.

[2] James Bryce, *The Holy Roman Empire*, London, 1880, pp. 14, 31.

We follow though it flies us, and cleave to its tumbling ruins.

And since we cannot keep it back, we fall with that which as it falls we grapple to us.

Time was when the world held us fast to it by its delight, now 'tis full of such monstrous blows for us, that of itself it sends us home to God at last.

Bethink yourselves, therefore, that the things which run their course in time are naught.

The end of that which is temporal showeth how mere a nothing is a thing which could pass away.

The fall of the show points out to us that it was but a *passing* show and close neighbour to nothingness, even when to the eye it seems to stand fast.

Let your heart's affections wing their way to eternity, that so despising the attainments of this earth's high places, you may come unto the goal of glory which ye shall hold by faith through Jesus Christ, our Lord.

At this time Rome had already been sacked five times, and the sword of the enemy was still suspended over the city, but it was averted by the mild eloquence and seasonable gifts of the Pontiff, who commanded the respect of both the heretics and the barbarians.[3]

The Dark Ages, this long and dreary period of chaos in which, however, as in the primitive chaos, the creative spirit was moving upon the face of the waters, came to an end by the reorganization of society under the Feudal System, which is the characteristic institution of the Middle Ages. These may be said to begin about the time of the first crusade (1095–1099). While the Empire disappeared in the welter of the Dark Ages, the Church, by commending itself to some at least of the northern invaders, although besmirched by the dusk and battle, remained, in the often quoted because very apposite sentence of Hobbs "Like the ghost of the Roman Empire sitting crowned on the grave thereof." Indeed from the seventh to the eleventh century such intellectual light as existed was confined to the cloister, and that light was rather the twinkling of a few stars in a waste of darkness than the flash of coming day. In the twelfth century, however, scholars commenced to look above and beyond the confines of their ecclesiastical enclosure, owing in large part at least to an increase of mental activity which had developed in a richer and safer Europe.

The texts of Aristotle and the Grecian writers had disappeared in the chaos of the Dark Ages. Some of them, however, commenced to make their appearance in Western Europe after the fall and sack of Con-

[3] For an excellent picture of the condition of Rome and the surrounding Campagna at this time and the outstanding services of Gregory the Great in the cause of peace, see Gibbon's *Decline and Fall of the Roman Empire.*

stantinople in 1453. But another and more important influence in the
revival of learning in Europe came from quite another quarter. The
Arabs in the year 632 breaking forth from the confines of their own
country, invaded and conquered in rapid succession Persia, Syria, Egypt,
the eastern and then the western coast of Northern Africa, and then,
crossing into Europe by the Straits of Gibraltar, took possession of
Spain and advanced to the center of France, where they were fought to a
standstill at a point between Tours and Poitiers by Charles Martel in
732 and soon driven back to the south of the Pyrenees. By the tenth
century dissensions had arisen among the Arabs, and the chair of Mahomet was disputed by three deputies who reigned at Bagdad, Cairo and
Cordova respectively. When peace succeeded war each of the capitals
became a center of great wealth, luxury and display and at the same time
of learning. "Their emulation diffused the taste and the rewards of
Science from Samara and Bochara to Fez and Cordova." Universities
arose and great libraries were assembled. "The age of Arabian learning
continued for about five hundred years and was coeval with the darkest
and most slothful period of European annals."

Who does not know (says Fontani[4]) how in those gloomy and lamentable ages, in
which Europe was enveloped in the darkest shadow of ignorance, the Arabs alone and
with the greatest industry applied themselves to the promotion of the sciences.
And from the year 814 of our era, in which the renowned Al-Mamun, Caliph of Bagdad,
granted peace to the Emperor Michael III on the condition that there should be given
to him all the writings of the learned Greeks which were preserved in the Imperial
Library of Constantinople, there began the translation into Arabic of works of all
the most celebrated philosophers, nor were there found wanting men of genius who
soon became masters in every branch of science.

The studies cultivated by the Arabs were chiefly philosophy, mathematics, astronomy and medicine. The works of Aristotle and other
Grecian sages had been preserved by the Arabs, both in the form of
translations into Hebrew made by the Jews in Syria and also by translations into Arabic for Arabian scholars in Cairo, Cordova and elsewhere
in the Mohammedan world. And thus the Aristotelian philosophy,
"Alike intelligible or alike obscure for the readers in every age," as
Gibbon says, was restored to the schools in Europe through Spain,
although often in a debased and corrupted form, as complained of by

[4] Introductory chapter to Narducci's translation of Ristoro d'Arezzo's *La Composizione
del Mondo*, page lx.

Roger Bacon. "The Aristotle of the Mohammedan world, the Aristotle of Avicenna and Averrhoes, contained a whole world of new knowledge about man, about natural history, about the constitution of the universe, of which the student of Aristotle's Logic (and nothing but his Logic had hitherto been known) had never dreamed."[5] The contributions made by the Arabians to the science of medicine in its varied aspects are especially worthy of mention. It was the Arabian doctors of medicine who gave new life to the study of the natural sciences in Europe and made numerous contributions to them. For not only did the Arabs restore many of the long lost works of the Grecian teachers but they brought also many classics by their own writers, among whom in medicine and the natural sciences Avicenna may be especially mentioned.[6]

The date at which the introduction of Arabic learning into Europe began may be set down as somewhere about A.D. 900, and it continued to diffuse itself throughout the continent during the next two or three centuries. "Schools of this Arabic learning arose at Cordova, Seville, Toledo, Granada, Valencia, Murcia and Almeria, and Averrhoes had an endless race of successors.[7] There was a trend in the sciences of the Arabians as in those of most eastern peoples toward the mysterious, fabulous and occult. Astronomy was often debased into astrology. Chemistry, which owed its origin to the Arabs, turned towards the search for the Elixir of Life or for the Philosopher's Stone, and the fortunes of thousands were evaporated in the crucibles of alchemy.

The rediscovery of Aristotle had a sudden and profound effect on medieval thought, so much so that in the Middle Ages he came to be referred to as "The Philosopher," his statements being regarded as having an authority little less than that of Holy Writ. Albertus Magnus and his pupil Thomas Aquinas attempted to weld together the philosophy of Aristotle and the theological teaching of the medieval church, and out of their attempt grew the scholastic philosophy of the Middle Ages. The Arabian learning however was actuated by a liberal spirit and presented a body of teaching which it was often difficult to harmonize with the teachings of the Church, and so in the great fresco by Taddeo Gaddi on the wall of the Capella Verdi in the Church of Santa Maria Novella in Florence, representing the apotheosis of St. Thomas Aquinas as the champion of the teachings of the medieval church, Avicenna

---

[5] J. H. Bridges, *Life and Work of Roger Bacon*, London, 1914, p. 149.
[6] See Gibbon.
[7] H. H. Milman, *History of Latin Christianity*, London, Murray, 1903, vol. IX, p. 108.

and Averrhoes, as leaders of the Arabian learning, are placed among his vanquished enemies which sit in a row beneath his feet.

The first revival of learning, which took place about the middle of the thirteenth century was marked by the appearance of Dante, Francis of Assisi, Roger Bacon, Albertus Magnus, Vincent of Beauvais and Marco Polo. As remarked by Trevelyan however, the true vision of the long lost world of Hellas came only with the second or great Renaissance in the fifteenth century. When it came it gave a staggering blow to the whole medieval system, for men saw or thought they saw far back in time something more wise, more noble and more free than the world of their own experience. The early doctors and students had no such disturbing vision.

The question of the exact date which should be selected as marking the close of medieval times is one on which historians are not agreed. In his recent *Life of Sir Thomas More*, Chambers[8] puts the date of the close of the Middle Ages in England as 1529, the year when the Reformation Parliament assembled, which he says marks the close of the old and the beginning of a new era. In the history of the development of the natural sciences, the forces which brought about the Renaissance are by some writers thought to have come into action a little later than in the world of literature and the humanities. Singer[9] selects the year 1543 as the closing year of the Middle Ages in the history of medieval science, since in that year there appeared two great works distinctively modern in character and both based on critical observation and study, namely, the *De Humani Corporis Fabrica* of Vesalius and the *De Revolutionibus Orbium Caelestium* of Copernicus. It would seem that for this reason the appearance of these works should be regarded rather as marking the beginning of the Renaissance. And for the same reason the dawn of the Renaissance in the geological sciences is marked by the appearance of the writings of Agricola and especially of his *De Natura Fossilium* which was published in 1546.

With the Renaissance, a new spirit manifested itself, one which, turning from the study of tradition and the mind of man, questioned nature, "The living garment of Deity" as Goethe calls her, by close observation and experiment. The natural and physical sciences which thus arose became in course of time provided with instruments of such extraordinary power, that men could search the realm of nature from end

[8] R. W. Chambers, *Thomas More*, London, 1935, p. 368.
[9] *From Magic to Science*, London, 1928, p. 63.

to end, the microscope revealed the structure of the forms of life, the telescope swept the depths of the heavens and laid bare their secrets, the spectroscope determined the composition of the remotest star, while by radiant energy the very nature of matter itself is being studied.

The question as to the precise dates which set the bounds to these several ages in European history is one on which all historians are not, as has been said, in perfect agreement.   The reason is that the several periods fade into one another and there is no sharp line of division between them.   A tabulation of these successive periods as here adopted may, however, be set forth as follows:—

The Roman Empire $\begin{cases} \text{B.C.} \\ \text{A.D. 500} \end{cases}$

The Dark Ages $\begin{cases} \text{A.D. } \ \ 500 \\ \text{A.D. } \ \ 900\text{—Infusion of Arabian learning begins} \\ \text{A.D. 1100} \end{cases}$

The Middle Ages $\begin{cases} \text{A.D. 1100} \\ \text{A.D. 1250—First revival of learning} \\ \text{A.D. 1450—Invention of printing} \end{cases}$

The Renaissance $\begin{cases} \text{A.D. 1546—Agricola's De Natura Fossilium} \\ \text{Passing into modern times} \end{cases}$

It is not necessary for our purpose to attempt to draw a picture of social life in the Middle Ages, but rather to present in outline a picture of the material universe as reflected in the medieval mind.

In order to do so four outstanding conceptions or beliefs of medieval times will be considered:

1. The Ptolemaic conception of the universe.
2. The powers, influences and "virtues" attributed to the heavens.
3. The doctrine of the Macrocosmos and the Microcosmos,
4. The doctrine of "Signatures" and "Correspondences."

1. THE PTOLEMAIC CONCEPTION OF THE UNIVERSE

This conception of the universe was set forth by Ptolemy of Alexandria in the second century A.D., and was elaborated by Alphonso of Castile in medieval times as required by certain advances in astronomical knowledge.   In its final form as universally accepted at the close of the Middle Ages, it is expounded by one of the most interesting writers of

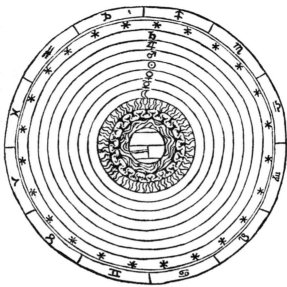

FIG. 2. The Ptolemaic system of the universe.   The crystalline sphere is very thin and the signs of the zodiac are placed in the tenth zone.   Note the channel from the ocean to the center of the earth.   (From the *Sphæra Mundi* of Sacro Busto, Venice, 1490.)

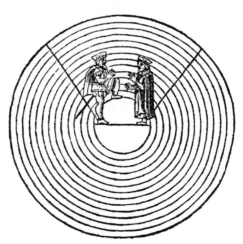

FIG. 3. The spheres with the Count of Altavilla beginning to climb them.   (From the *Scala Naturale* of Maffei.   Edition of 1607.)

the period, Giovanni Camillo Maffei of Solofra in his book entitled *Scala naturale, overo Fantasia dolcissima intorno alle cose occulte nella Filosofia*, a small octavo volume of one hundred and forty leaves, published in Venice in 1564, written in the quaint old Italian of the time and addressed to Don Giovanni di Capua, Count of Altavilla.

The world, Maffei informs the Count of Altavilla, consists of fourteen parts, namely, four elements and ten heavens. As the heavens were created merely as ancillary to the earth he includes them with the four elements as constituting "Il Mondo." These are arranged in consecutive order from the centre outwards in concentric spheres, the outer in each case enclosing that within, it as follows:— Earth, Water, Air, Fire. Then follow in their order the spheres of the Moon, Mercury, Venus, the Sun, Mars, Jupiter, Saturn, the other stars, a sphere originating certain movements in the heavens, and the Primum Mobile, which is the ultimate source of the motion of all the spheres.

These, he says, may be considered to form as it were a great stairway, up which he invites the Count to ascend.

The successive chapters of the book describe what is to be seen and learned on each successive step, and, like so many other books of the time, it embraces an enormous range of knowledge, providing succinct and definite information on an astounding number of subjects as remote from one another as the immortality of the soul, the necessity for the creation of woman, why stones and trees have the form of animals, and why men have two feet while many animals have four.

At the centre was the world, fixed and immovable.[10] This consisted of the spheres of the four Aristotelian elements, earth, water, air and fire. The innermost was the earth, the abode of man. Mantling the greater part of this was the watery sphere of the ocean and the great lakes with the rivers tributary to them. Enveloping them was the element of air. These three spheres are still recognized by geologists at the present time and are known as the Lithosphere, the Hydrosphere and the Atmosphere. But in recent times another has been added, namely the Barysphere or Centrosphere, consisting of an inner and central portion of the earth enclosed by the Lithosphere and which is now known to differ from the latter in character.

Surrounding the sphere of the air was the zone of fire, the "Flaming ramparts of the world" as Lucretius called them. This was "elemental

[10] "He hath made the round world so sure that it cannot be moved." Psalm 93: 2.

fire," not necessarily burning with visible flame but ready to burst into flame when subjected to any unusual motion. When for instance it was moved by stormy winds it blazed forth as lightning.

Out beyond these four zones were the heavenly spheres. These were transpicuous spheres revolving around the fixed and central earth, and each carrying affixed to it or embedded in it one of the planets, "These wandering fires," as Milton calls them. The innermost of these was the "Sphere of the Moon," whose motion caused the moon to pass around the earth once in each lunar month.

Next to this came the Sphere of Mercury named, on account of its rapid movement, after the messenger of the gods. Then the Sphere of Venus, usually shrouding itself in a cloudy mantle. This was succeeded by the Sphere of the Sun, "The glorious planet Sol" as Shakespeare calls it. Then came the Sphere of Mars, this planet, on account of its ruddy color, taking its name from the god of war. Next was the Sphere of Jupiter, the largest and greatest of all the planets. And lastly the Sphere of Saturn, the outermost of the planets known to the Ptolemaic world. Beyond the planetary spheres was the great Heaven of the Fixed Stars. This was so called because these stars always maintained the same position with reference to one another, and had no movement except that of their own sphere which revolved around the earth once in every twenty-four hours.

These eight comprised all the spheres which were recognized in the time of Aristotle, but later two others were added. The innermost of these two was called the Crystalline Sphere. It was brought in to account for certain apparent irregularities of movement in the starry heavens. Gorus says that this heaven is called the Crystalline Sphere because it is clear and transparent after the manner of crystal.

The tenth, last and outermost sphere was known as the Primum Mobile. Its function was to supply the driving force which, being transmitted from it to the inner heavenly spheres in succession, caused each of them to revolve in its due order and in its appointed time.

Above and beyond all these, outside of the universe, was the heavenly Empyrean, where God had his throne and dwelling place, reigning there in timeless majesty, the primal source of all motion, power, and of every "virtue" in the universe below.

This Ptolemaic system was that which Shakespeare had in mind when he speaks of the "Stars shooting from their spheres," or when he says "Two stars keep not their motions in one sphere,"[11] referring here to the

---

[11] *King Henry IV*, Part I, Act 5.

planets. It was also that on which Milton bases his conception of the universe in Paradise Lost, although he was doubtful whether the newer

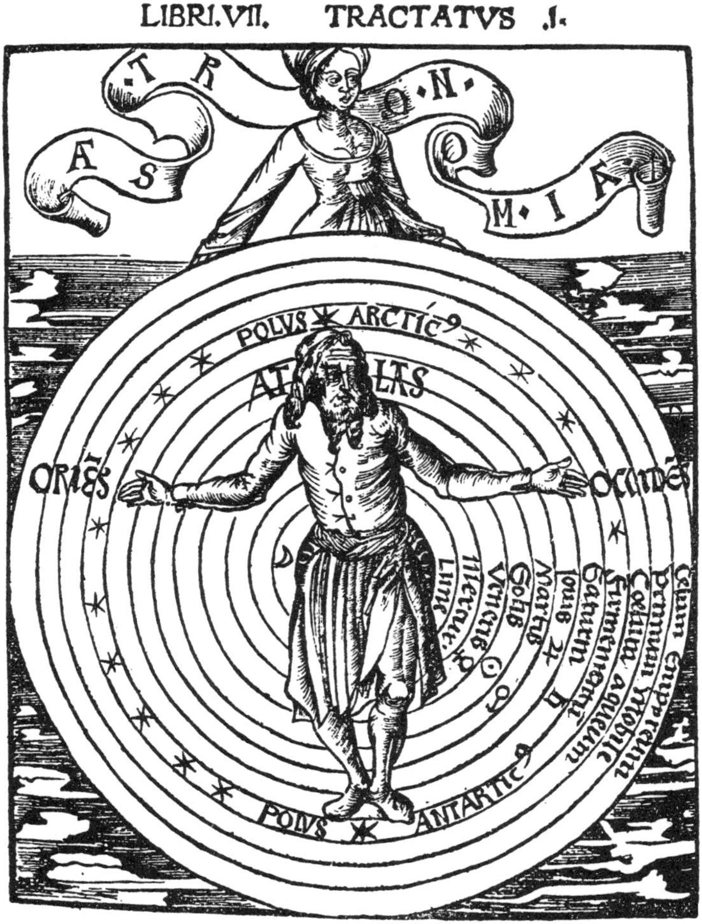

FIG. 4. The world looking down upon it from above with Atlas holding it up. There are eleven spheres because it shows the Empyrean as the eleventh. The tenth is called the Primum Mobile and the ninth is the Crystalline Sphere. (From the *Margarita Philosophica* of Reisch, 1504.)

Copernican theory might not be about to supersede it, as shown by the conversation between Adam and the Angel Raphael which is found in Book VIII.

As compared with our present Copernican conception, the Ptolemaic universe was thus small and compact, with set bounds, easily understood and comprehended by all, a little home-like universe. Any man standing upon the fixed earth at the center of the universe might, on any clear night, look upward through one planetary sphere after another, through the Sphere of the Fixed Stars, and then in imagination let his thoughts soar through the Crystalline Sphere and beyond the Primum Mobile to the very throne of God himself. Astronomical investigation however has shown that this conception of the universe was not a true one. The earth is not the center of the universe. It has been deposed from its proud place and relegated to the position of a small and relatively obscure planet revolving with the others about the central sun of a relatively unimportant solar system. "The Primum Mobile has been forever burst and into the chaos supposed to be beyond, the imagination has voyaged out, still finding no sign of shore or boundary, but only the same ocean of transpicuous space with firmaments for its scattered islands rising to view on every furthest horizon." Thus while much has yet to be learned concerning the universe, it is now known to be immensely vaster, more wonderful and more splendid than was ever dreamed of in the time of the Ptolemaic system.

2. THE POWERS, INFLUENCES AND VIRTUES ATTRIBUTED TO THE HEAVENS

The Ptolemaic theory was based on facts derived from a careful observation of the heavens carried on through many centuries and it embodied a certain measure of truth. The theologians, however, proceeded to weave into it a web of speculative fancy which had no basis in fact and which had a deep and pernicious influence on human thought. This took its origin in a book entitled *The Celestial Hierarchy* by an unknown writer, which appeared in the fifth century, and which in the Middle Ages was believed to have been written by no less a personage than Dionysius the Areopagite, who is mentioned in the Acts of the Apostles. In this it is asserted that each of the planets is the seat of one of the great spiritual powers of the heavenly hierarchy. Thus the moon is the seat or abode of the angels, Mercury of the archangels, Venus of the principalities, the sun of the powers, Mars of the virtues, Jupiter of the dominations, Saturn of the thrones, the Fixed Stars of the cherubim, and the Primum Mobile of the seraphim. The conception in its purest and most elevated form is presented in Dante's *Paradiso*, where the influence and power emanating from God, the divine source of all power whose seat is in the empyrean, passes down into the universe and is distributed among the various members of the celestial hierarchy in their respective positions in the heavens, each according to its own spiritual office or functions, and is by them, as it were, reflected and focused upon the earth which stands immovable at the center of the universe.

Thus the four inner spheres composed of the four elements and the ten outer spheres were continually exerting their influences on the earth and on mankind. Giovanni Camillo Maffei in his work, to which reference has already been made, says that the four elements are the source of certain forces resident in them and hence known as the *Elemental Forces* (a term which still survives in the modern world), but the six planets and the fixed stars, and probably also that great and mysterious outer sphere from which all motion comes, emanate or exert other and different forces or influences which diffuse themselves throughout the universe, and which are as mentioned above especially focused upon the earthly globe which forms its center. These are the *Celestial Forces*. These forces have power and act, Maffei says, chiefly upon the four elements, on bodies compounded of them, and upon the brute creation, hence the Roman augurs made use of animals in reading the will of heaven. But they have little or no influence on abstract intelligence.

The human mind, however, standing between the grosser elements on one hand and abstract intelligence on the other and being compounded of the two, may on its grosser side be subject to these celestial influences, but the mind in choosing good and setting aside evil, and in its many other activities may resist these celestial influences if malign, and rise above the influence of the stars and dominate them. As Shakespeare says, "The fault, dear Brutus, lies not in the stars but in ourselves that we are underlings." The celestial forces pervade the universe but their action, Maffei tells us, is less clearly understood than that of the elemental forces, and, authorities are not altogether agreed upon their actual scope and influence, they are therefore sometimes said to be *Occult*. Apart from or in addition to any spiritual influences, all these forces, both the elemental and the celestial, influenced the fortunes of mankind.

It was believed that each planet sent forth some special power or virtue which influenced persons born when the planet in question occupied a certain position in the sky. And while this belief has passed away, it has left its impression in the common speech of the present day, as when certain persons are said to have a "mercurial," a "martial," a "jovial" or a "saturnine" temperament, or to be "moon struck" or "lunatic."

Each of the seven metals which were known at that time was also believed to have properties resembling those of one of the seven planets, thus marking a mysterious coincidence in nature. Mercury resembled the quick-moving planet of that name, which in its turn derived its name from the swift messenger of the gods. Copper, the only reddish metal, perhaps suggested the blush of Venus, the blazing sun was like gold, the moon was the color of silver, Mars called after the god of war was associated with iron, Jupiter with tin and the distant Saturn with the dull metal lead. The signs used by the ancient astronomers and astrologers to designate the several planets are even to the present day employed on geological mining maps to designate the corresponding metals and their ores.

The occult influences of the fixed stars were believed to generate the "precious stones" or gems found within the bosom of the earth or upon its surface, probably because their brilliancy and glittering radiance reminded men of the twinkling brightness of those celestial points of light, especially as seen in the tropical skies of eastern lands. Furthermore these gems themselves were believed to be endowed with the virtues of the stars which brought them into being, and the gems found in tropical countries, where the stars shone with especial brilliance, were believed to possess these virtues in a greater measure than those found in other portions of the globe. The stellar rays indeed were believed to penetrate to the very center of the world, which was the focal point of the whole universe, producing there "A treasury of precious things."

These occult influences of the fixed stars were also believed by many to have developed those strange forms which are so often found sealed up in the rocky strata of the earth and which we now know to be the remains of animals or plants, the denizens of the world in times long past. To the stars was also attributed, by some writers, the growth of hills and mountains. They were in fact called in to explain any phenomena in nature which could not be otherwise accounted for. Ristoro d'Arezzo,[12] writing in 1282, says that no bodies, animal, vegetable or

---

[12] *La Composizione del Mondo*, testo Italiano del 1282, publicata da Enrico Narducci, Rome, 1859, p. 96.

mineral can be generated on or within the earth by a mere association of the elements, but there must be in addition an infusion of the influences of the heavens, which impress themselves upon the earthy materials, as a seal does upon wax. Thus, for instance, each sign of the zodiac as it passes across the sky leads to the generation of animals resembling it. So under the sign of Leo, lions, panthers, leopards and allied animals are born, and under the sign of Pisces, fishes are brought forth. The fixed stars also exerted a powerful influence upon the course of mundane affairs—a view supported by certain passages in Holy Writ. Did not "The stars in their courses fight against Sisera?" Did not Job refer to the "Sweet influences of the Pleiades" on human life, and did not the star lead the wise men, who could read the heavens, to the Babe at Bethlehem?

While many of the heavenly influences were benign and brought blessings to mankind, eclipses of the sun, and comets, those mysterious bodies which from time to time suddenly appeared in the sky, were viewed with great apprehension, as being the harbingers of evil, of plagues, famines or other national disasters.

In the Bayeux tapestry a comet is shown, which appeared shortly before the invasion of William the Conqueror, and which presaged the death of Harold. In the Nuremberg Chronicle pictures of comets which appeared in the sky in certain years are introduced at many places in the text as portents of the disasters which followed them.[13]

> When beggars die there are no Comets seen,
> The Heavens themselves blaze forth the death of princes.[14]

> ... like a comet burned
> That fires the length of Ophiuchus huge,
> In the Arctic sky, and from its horrid hair
> Shakes pestilence and war.[15]

There are still other signs of disaster which made a deep impression on the medieval mind, such as the reported birth of monsters. "They are called monsters, from *monstro* to show," says St. Augustine, "because

[13] In the prooemium of his *De aeris transmutationibus* (Rome, 1614), Porta speaks of the remarkable phenomena which are seen in the air, among others "Stellas libero aere vagari" and further on in the same book asks—"Quis est, qui non horrescat cruentos cometae crines, quasi futurae adversitatis praeludia?".

[14] Shakespeare: *Julius Caesar*, II, 2.

[15] Milton: *Paradise Lost*, II, 708–711.

they betoken somewhat."[16] A child with two heads is represented in
several places in the Nuremberg Chronicle as having been born in certain
years and whose appearance was followed by some dire calamity. In
fact the thought of the Middle Ages never seemed to be able to free
itself from the idea of the presence of these monsters on the earth, and
references to them crop up everywhere.

3. The Doctrine of the Macrocosm and the Microcosm

This is an ancient belief found in the *Timaeus* of Plato, and adopted
by many of the Arabian writers, which met with a wide-spread acceptance
in medieval times. It was an attempt to bring a complex universe into
some intelligible or simple relation to man himself and thus give a unity
to the whole.

According to the Ptolemaic system, as has been seen, the earth was
situated at the center of the universe. This universe was known as the
Macrocosm. Around the fixed and central earth enveloped by the
elements, the outer spheres of the universe revolved bearing the heavenly bodies on their celestial paths. Such being the case it came to
be believed, as has already been stated, that all the manifold and diverse
powers, virtues and influences, which were supposed to emanate from
these celestial bodies, or to be transmitted through them, were directed
toward and focused upon the central earth. It was also believed that
the earth had been created as the dwelling place for man. The heavenly
bodies were for its equipment or "adornament" as Thomas Aquinas
says. In the account of the creation in the *Book of Genesis* the sun and
moon "ruling" respectively the day and night, were made to "Give
light upon the earth" and to be for the measurement of time, "For
signs and for seasons and for days and years." The animals and plants
were for man's use and for his subsistance. The universe had, in fact,
been *created for man*. Man was the measure of all things and all things
existed to minister to his needs and for his pleasure. Fancy went one
step further and man came to be considered as an epitome of the universe
—as the *Microcosm*, or smaller universe.

And so Conrad von Megenberg[17] (1309–1374) begins his *Naturbuch*
thus:

God created man on the sixth day after the other creatures and made him in such
a manner that his soul and body are fashioned after the same pattern as the universe

[16] *The City of God* (*De Civitate Dei*) by Saint Augustine. A translation into English by
John Healey, first published in 1610, Edinburgh, 1909, vol. II, p. 303.
[17] *Naturbuch*, Frankfort, 1540, p. 1.

as a whole. Therefore man is often called the *Minor Mundus* or smaller world, because in him all things which exist in the greater world are found and he is a composite picture of the universe containing within himself something of all its parts. In him is a body composed of the mingled elements and a heavenly spirit created in the image of God. Growth he has in common with plants, perception and feeling in common with the animals, and reason he shares with the angels. Therefore no one should wonder that man is beloved by all things since in him all things see and recognize themselves.

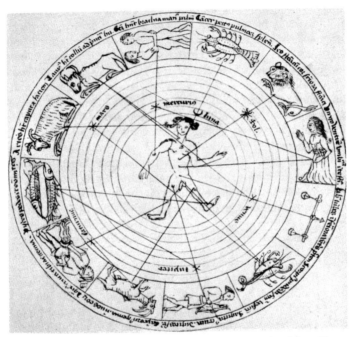

FIG. 5. The relation of the macrocosm to the microcosm reproduced from Singer: *From Magic to Science.*

At a later date the great mystic Jacob Boehme[18] wrote:

Man is the great mystery of God, the Microcosm, or the complete abridgement of the whole universe: he is the *Mirandum dei opus*, God's masterpiece, a living emblem and *hierogliphick* of eternity and time.

[18] *Signatura Rerum* or *the Signature of all Things.* Written in High Dutch 1622. (Translated into English by J. Ellistone), London, 1651. There is also a translation to be found in Everyman's Library, London, 1912.

Abraham von Franckenberg[19] gives a further exposition of the same theory. How in the Microcosm (man) is to be found something of all beings, things and powers that exist and constitute the Macrocosm. The animals of the Macrocosm are all represented in the Microcosm by the virtues and vices which characterize them. The vegetables of the Macrocosm are reflected in the Microcosm as seen in the resemblances which the organs of man bear in shape to certain flowers, fruits and vegetables. Even minerals are also represented by the powers of the several individual metals, each of which exerts and concentrates its influence on some one of the organs of the human body, while in cases of disease, stones actually grow in certain of these organs. The atmosphere also displays its influence on the Microcosm in "clouds" of melancholy or in the "falling sickness" which has its cause in lightning, and even more important are the manifold and powerful influence of the planets and the fixed stars. It was Paracelsus who changed and modified this theory and paved the way for its eventual rejection.

It is unnecessary to dwell further upon this intimate relation which was supposed to exist between the great world or Macrocosm and the lesser world or Microcosm as seen in man himself. Figure 5 sets it forth pictorially in some of its aspects, and is one of hundreds of similar representations to be found in medieval treatises on this subject.

4. THE DOCTRINE OF "SIGNATURES" AND "CORRESPONDENCES"

According to the doctrine of "Signatures" most if not all things in nature have some significant mark or "signature" which, if it can but be recognized, indicates some hidden power or property resident in the thing in question. This belief comes down from very early times. Zoroaster, Hippocrates, Theophrastus, Dioscorides and Galen make mention of it. Pliny cites several examples as for instance the stone Icterias,[20] "Resembling in color the human skin when of a sickly yellow color, and *for that reason* esteemed efficaceous against jaundice."

The doctrine was firmly held all through the Middle Ages and on into the time of the Renaissance, in the opening years of which it found an enthusiastic advocate and exponent in Giovanni Battista Porta, one of the most remarkable men of his time. He belonged to an ancient Neopolitan family, and was a voluminous writer who covered rather a wide

[19] *Gemma Magica*, Amsterdam, 1688, p. 84.
[20] *Natural History*, book 37, chapter 10.

field. His best known works and those which especially concern us here are the *Magia Naturalis*, the *De Humana Physiognomonia* and the *Phytognomonica*.

He was a man whose mind turned naturally to the study of the strange and mysterious in nature, and who endeavored to ascertain the causes of such phenomena. Consequently, in a world which was commonly believed to be haunted by demons and all manner of evil spirits, he came to be regarded as a sorcerer, an imputation which he indignantly denied. In the course of Porta's extended studies he made many important discoveries concerning optics, the sturucture of the eye, the properties of various kinds of lenses, the burning glass, the camera obscura and the magic lantern, but while he thus discovered that many strange phenomena were due to natural causes he still held that the stars and the heavens influenced the course of human affairs, which he says "Everyone believes." He was also a firm believer in the doctrine of signatures and resemblances. These signatures[21] and resemblances were to be found in all three kingdoms of nature. Thus in the animal kingdom, as taught by the physiognomists, there were to be found men in whose features, habit of body or bearing, there was a certain suggestion of the lion, the tiger, the bull, the ape, the cat or the fox, and whose mental or moral natures reproduced something of the characteristics of these respective animals. Even at the present day it is believed by many that there is a certain element of truth in this contention.

These "signatures" were also to be found in the vegetable kingdom. Porta in his *Phytognomonica*, the edition of 1591 being an octavo volume of 552 pages, devotes the greater portion of the work to the consideration and description of resemblances which a great variety of plants bear to other things in nature, such as the different heavenly bodies, the sun and moon, or various animals or parts of animals or even to certain minerals, pointing out the virtues which those plants possess corresponding to the appearances or the qualities of the objects which they resemble, "As from the appearance of men," he says, "their character may be judged, so from the appearance of plants their innate virtues may be divined." "The likeness of things showeth their secret virtues."

The work is profusely illustrated by woodcuts which show the appearance of plants indicating their secret properties. Two of these are here reproduced, one a plant (Palma Christi) whose roots resemble the human

---

[21] *Magia Naturalis*, book 1. This work was first published in Naples in 1558, and this edition was followed by many others in different languages.

hand, and which has the properties of healing diseases of the joints and
of curing gout, and the other, aconite plants whose roots resemble scorpions and lobsters, and which are a specific for the sting of scorpions.
In fact it came to be believed and taught by all authorities in medicine
that when any plant presented a resemblance to a certain organ in the
human body, a mysterious suggestion was thereby conveyed, that from
it might be extracted some medicine which would heal the diseases of
the organ in question.   The mandrake was a plant which attracted

FIG. 6. The Palma Christi plant, whose roots resemble the human hand.   (From the
Phytognomonica of Porta.)

special attention owing to the fact that its roots frequently presented
a rather striking resemblance to the human form. (See Plate 1.) It was
therefore considered to be a specific for certain diseases and to possess
magical powers.   This belief still survives in certain parts of Greece.
    The writer when visiting the island of Delos in April 1929 met with a
curious instance of the persistence of the idea that some mysterious power
was associated with this plant.   Delos was regarded as a holy island far
back in prehistoric times before the coming of the Greeks.   Near the

THE ANCIENT SHRINE ON THE ISLAND OF DELOS

PLATE I
FIGURE OF THE MANDRAKE SHOWING THE HUMAN FORM
(From the *Hortus Sanitatis*)

summit of the hill known as Kynthos on this island is a prehistoric shrine of megalithic construction, which is probably the oldest shrine in the world to be preserved in its original condition. (See Plate 1.) The rock of which the island is composed is a micaceous schist, torn to pieces and partially digested by an intrusion of coarsely porphyritic granite. At the site of the shrine the rock is traversed by a series of vertical joint planes. Between two of these, the rock crumbling away leaves a wide vertical cleft on the side of the hill. This the primitive inhabitants

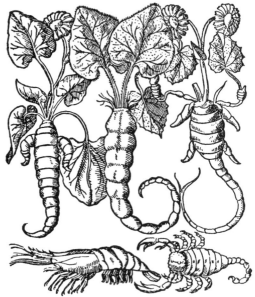

Fig. 7. The Aconite plant, whose roots resemble scorpions and lobsters. (From the *Phytognomonica* of Porta.)

roofed over with gigantic slabs of granite, accurately hewn to fit on to the granite walls on either side, forming a rather steep pentroof. On this are piled huge irregular shaped blocks of granite. At the back of the room thus formed is a rude altar composed of a block of stone about 5 feet by 3 feet in size, on the upper surface of which is a shallow depression to receive a sacrifice. This chamber is entered by a narrow door with walls of Cyclopean masonry on either side. At this shrine Ulysses and Aeneas came to worship when it was already old. The island is now bare, desolate and uninhabited except for a few government employees.

On it there is not a tree, the surface is bare rock with a sparse growth of grass and many flowers springing up through it, among them the poppy and the mandrake are especially noticeable. The mandrake was rather abundant among the ruins of ancient buildings, appearing as isolated low growing plants, in full bloom displaying groups of five-petaled flowers, pale blue to deep blue in color. Being desirous of securing some specimens of the roots of the plant showing the resemblance to the human form mentioned above, the writer attempted to dig up some of these, but found this to be impossible without special tools, owing to the extreme hardness and compactness of the stony soil. While so engaged one of the caretakers passed and informed him that it was dangerous to attempt to uproot these plants because anyone who tried to do so received an electric shock from the plant, thus indicating that there still survived a belief that some mystical or occult power was resident in the plant, although the form of this belief had been modified through the influence of modern scientific discovery.

But the mineral kingdom also had its "signatures." In chapter V treating of medieval mineralogy references are made to the occult "virtues" of many stones and especially of the precious stones or gems, whose virtues are indicated by their colors, forms and physical properties. A host of curious beliefs and stories were current with reference not only to minerals which guarded their wearers from accident and disease, but also to others which ministered to their higher needs or testified to spiritual realities—minerals which presented veritable "Sermons in Stones."

Men's minds in the Middle Ages, furthermore, were impressed with a certain sense of wonder and bewilderment at the many resemblances or similarities which existed between various objects in nature that seemed to have no connection with one another, which resemblances suggested that there might be and probably was some hidden and mysterious relation between the objects in question. Thus that there were seven metals and also seven planets, seemed to be a fact of peculiar significance, especially as has been already stated, there seemed to be a certain similarity in character between each of these metals and its corresponding planet. There were also seven notes in the octave, and partly in consequence of this coincidence we have the "Music of the spheres" mentioned as early as the book of Job, when "The morning stars sang together." There were also seven days in the week, seven virtues, seven sages, seven wonders of the world, seven gates of Thebes, seven Sleepers of Ephesus and "If Cain be avenged seven times surely Lameach seventy and seven."[22]

The significance of the coincidence in the number of the metals and the planets of course disappeared in the course of time when additional metals were discovered.

The resemblance which concretions from the clays, sands or even hard rocks of some localities often bear to various other objects, such as balls, rings, plates, fingers, feet and other members of the human body, or even to the human form itself ("Lösskindel") and to various plants and animals, and again the remarkable imitations of landscapes and sea scapes observed in the rocks of other localities also attracted much attention in medieval times. The study of these "Figured Stones" ushered in the science of Palaeontology, whose rise and development will be considered in chapter VIII.

The most voluminous and certainly one of the most remarkable contributions to the literature of "Resemblances" which appeared in the Middle Ages was the volume from the pen of Johannes Gorinus de Sancto Geminiano.[23] In the prologue he is stated to have belonged to the Order of Preaching Brothers (Dominicans) and to have been a distinguished Doctor of Sacred Theology. The work is believed to have been written about 1350 but was printed in 1477. It is divided into 10 "books" and these are subdivided into chapters. The number ten was probably selected because there were ten divisions in the Ptolemaic heavens. These "books" treat of the following subjects in succession. Resemblance, to be found in the "Heavens and the Elements," in "Metals and Stones," in "Plants and Vegetables," in "Birds and Fishes," in "Terrestrial Animals," in "Man and his Members," in "Dreams and Visions," in "Canons and Laws," in the "Arts and Artificial Things," and in the "Actions and Customs of Men."

The 1499 edition embraces 387 leaves, that is to say 774 quarto pages, printed in small Gothic type, in double columns, the text being still more condensed by the employment of contractions, and is devoted to setting forth resemblances, chiefly those presented by material things, to things of the mind and spirit. The resemblances are very often far-fetched and of little or no significance to the modern mind, but are of interest as throwing light on the mental attitude of men in the medieval times.

Certain of the resemblances mentioned by Gorus have a certain grace

[22] E. E. Kellett, *The Story of Myths*, London, 1927, p. 77.
[23] *Summa de Exemplis ac Similitudinibus Rerum*, Venice, 1499.

or beauty, as for instance that of sapphire to the act of divine contemplation. On the other hand, when he says that asbestos resembles hell fire and that bitumen bears a resemblance to charity, the statements are less attractive, while when he institutes a comparison between Christ and the rhinoceros (which he says is the same animal as the unicorn) some of the points of resemblance which he puts forward are startling to say the least. The analogy which he finds between asbestos and hell fire is based on a misunderstanding of the nature of this altogether respectable mineral. Its name is, of course, derived from the fact that although, when reduced to a fibrous condition, it looks like silk, cotton or wool, it cannot be destroyed by fire. This remarkable fact gave rise to the fable that it was a substance which when heated took fire and burned forever. This was further developed into the story that when asbestos was once ignited it was impossible to extinguish it. Gorinus speaks of a certain temple of Diana in which there was a lamp whose light came from burning asbestos, the flame of which could not be extinguished by any wind storm or by the heaviest rain. Therefore, he says, if this mineral occurring in nature, can burn with such an intense and continuous flame, how much more will the fires of Gehenna, lighted by the divine power, blaze unextinguished to all eternity.

To understand the somewhat uninviting comparison of charity to bitumen, it must be remembered that the word *charity* in its true significance does not mean *almsgiving* but rather love or well wishing for another, and so Gorinus says that it resembles bitumen because the latter burns brightly, being often used to make torches. Bitumen also is tenacious and not easily separated from anything with which it is brought in contact, and thirdly because it binds separated objects firmly together, as Noah's ark, pitched within and without, was able to withstand all the buffetings of the waters of the deluge, so Christian charity withstands all the trials and afflictions of this mortal life. Being very loyal to his own Order, Gorinus compares it to Lebanon and gives nine reasons why it resembles this mountain, so renowned in sacred history, rising pre-eminent above all the heights of the surrounding landscape, and furnishing streams of lifegiving waters to all the thirsty land around.

And so the worthy brother in his time went through northern Italy, preaching of the resemblances of earthly things to those of the hidden world which was to survive when that which now is had passed away, spreading abroad this doctrine of "resemblances," in sermons which while perhaps not of deep instructional value, were probably none the less interesting to the congregations that heard them.

Again reference may be made to the widespread belief that the earth in structure and organization bore a close resemblance to the human body. This came down from classical times. Seneca[24] held it as a "Firm conviction," but in medieval times and even later, it was widely advocated by many men of great renown, a fact referred to by Frascatus[25] thus:

Among both the ancients and the philosophers there are those who assert that the earth resembles the body of a huge animal, in which the bones are represented by the rocks, the flesh by the earth, the arteries by the rivers which are dispersed everywhere throughout the earth, while the heart is represented by the centre of the earth where there is collected the breath or vapours which carry the vital heat throughout the whole body, for the development of the juices and metals and all other things which are generated in subterranean places.

Still another manifestation of the proneness of the medieval mind to look for hidden resemblances and meanings everywhere and in all things, is seen in its fondness for and belief in allegory:

So accustomed was medieval man to allegory that he lost the ability to under-stand the plain statement of some of the books he studied, in the attempt to deduce from it some subtle moral.   Biblical exposition, of course, offered wide opportunities. When an Anglo-Saxon preacher discoursed on the visit of the Queen of Sheba to King Solomon, he saw in the two monarchs representatives of the Church and of Christ.   The Queen was the Church; she brought gold, gems and spices; by gold true belief was signified, by spices the breath of prayer and by gems the shining of good actions and holy virtues.   These treasures were borne by camels; the camels must be the heathen who, humpbacked by avarice and deformed by crime, may neverthe-less be converted by the church and led to Jerusalem.   Allegorizing earnestness made men blind to the grotesqueness of parallels between man and beast, and the preacher who interpreted "The oxen were plowing and the asses feeding beside them," as referring to the labouring clergy and the laity pasturing by their side, had no intention of writing his congregation down as asses.   Strange physical facts were invented as bases for argument.   The manner in which God hides his mysteries from man's gaze might be compared to the habit of the lion, who, knowing that the hunted will follow his foot tracks on the sand, obliterates them by waving his tail backwards and forwards.[26]

CONCLUSION

From what has been said it will be seen that in the Middle Ages all men, not only those who were unlettered but the scholars also, believed

[24] *Physical Science in the Time of Nero*, being a translation from the *Quaestiones Naturales* of Seneca by Clarke & Geikie, London, 1910, p. 126.

[25] *De Aquis Returbii Ticinensibus Comentarii*, Pavia, 1575, Leaf 20.

[26] R. W. Chambers: *Thomas More*, London, 1935, p. 79.

that they were moving in a physical world not governed by invariable laws whose action and results could be known and predicted in advance, but in one in which a multitude of mysterious and unknown forces, influences and virtues were everywhere at work. Anything might happen as the result of their intervention and anything which could not be otherwise explained was always attributed to them.

In the desperate hope of controlling these inscrutable powers men turned to magic and witchcraft. With the sun of the new learning in the early Renaissance these mists and shadows began to fade away, but they were not wholly dissipated in Europe until it had risen high in the heavens and had shone for many long decades. Indeed, even now remnants of old superstitions linger in the minds of many people from whom better things might be expected.

The mental attitude of men in Europe during medieval times in respect to these beliefs and superstitions is essentially the same as that which still survives in many, if not in most, countries in the far East. Throughout India, for instance, there is an abiding belief in the ever present existence of evil spirits, who exert their influences on the affairs of men. Even in the great religions—Hinduism, Buddhism and Mohammedanism, there still exists a substratum of primitive Animism. Sacred rocks daubed with red paint in strange designs, sacred trees from whose branches flutter myriads of shreds and strips of colored cotton or other votive offerings, are the dwelling places of malign spirits or influences which must be propitiated. Charms against the "evil eye" are sold in every bazaar, and the custom of calling in astrologers to determine propitious times and seasons in the conduct of all ordinary affairs of social and domestic life is widespread even among members of the educated classes.

In medieval times, therefore, when similar conditions prevailed it is not to be wondered at that all manner of strange and to us even ridiculous explanations were put forward and accepted to account for phenomena which now, in the light of modern knowledge, can be seen to be the results of the natural and unvarying laws of nature.

CHAPTER IV

THE "GENERATION OF STONES"

A. Theories Concerning the Ways in Which Minerals Are Brought into Existence:
  1. The Aristotelian Theory of Celestial Influences
  2. The Theory of the Petrific Seed
  3. The Theory of the Lapidifying Juice
B. The Digestive Organs of Minerals
C. Male and Female Minerals
D. Places in Which Minerals Originate and the Conditions under Which They Develop
  1. Stones Growing in the Bodies of Animals
  2. Stones Growing in Plants.....
  3. Stones Which Fall from the Air
  4. Stones Which Grow in Water
  5. Stones Which Grow in the Open Country on the Earth's Surface

It may first be noted that the ancient writers usually made no distinction between gems, common minerals and rocks. The diamond and the millstone appear together in their lists of "lapides" or stones. When any distinction is made between gems and common stones, those minerals which are clear, transparent and beautiful in color and small in size are classed as gems, while those that are dull and opaque, occur in large masses and are of relatively unattractive appearance are considered as common stones. Nothing was known in these ancient times concerning either the chemical composition or crystallographic form of minerals, although these are now considered to be the most important factors in the distinction of mineral species. Very often what we now know to be varieties of the same minerals were then considered to be different species because they differed in color or transparency, while frequently a number of entirely different minerals were classed as belonging to the same species because they had a general resemblance to one another.

The origin of stones, whether they were the gems which were such an unfailing source of delight, wonder, and even reverence, in past times or whether they were the common stones, many of which, as the millstone, the whetstone, and various kinds of clay, played so important a rôle in daily life, presented a problem for which these ancient writers found it to be extraordinarily difficult to find a solution. One author after another reiterates the same complaint and Girolamo Savonarola prefaces the ninth chapter of his *Compendium totius philosophiae* (1542), in which he treats of this subject, with the words "Est difficile aliquid certi tradere de virtute et loco generationis mineralium."

The geologist who even now, after some four hundred years of additional study, finds that so much still remains to be learned, can indeed sympathize with these his predecessors, more especially when he remembers that they sought the solution of these difficult problems, not in the open field of nature as he now does aided by all the appliances of modern science, but in the Aleian Fields of tradition and formal logic.

A. THEORIES CONCERNING THE WAY IN WHICH MINERALS ARE BROUGHT INTO EXISTENCE

During the long period of close upon two thousand years, from the time of Aristotle in the third century B.C. to the rise of the new learning at the opening of the sixteenth century A.D., three chief theories were put forward to account for the origin of stones, minerals and rocks, under the banner of one or other of which almost every writer on the subject enrolled himself.

These may be set forth as follows:

1. The Aristotelian theory of celestial influences.
2. The theory of the Petrific Seed.
3. The theory of the Lapidifying Juice.

While all three of these theories may perhaps be said to have been in existence throughout the whole period above mentioned, they in succession met with general acceptance and displaced one another in the order given above.

1. The Aristotelian Theory of Celestial Influences

While Aristotle treated with remarkable insight and feeling the realms of the animal and vegetable kindgoms, he left no treatise on the third or mineral kingdom of nature.

There is indeed a work on Stones bearing his name but internal evidence at once shows that it is a later work by another and very different hand, probably that of some Arabian author written some time prior to the year A.D. 850.[1] In his book which bears the title *Meteorologica* which he says "Is concerned with the events which take place in the region nearest to the motion of the stars," that is in the atmosphere, he

[1] Julius Ruska, *Das Steinbuch des Aristoteles*, Heidelberg, 1912.

sets forth in a general way his views as to the manner in which stones originate. His treatment of the primary nature of matter is confused and contradictory,[2] possibly owing to corruptions in the text. In the *Meteorologica*, book 1, chap. 2, it is stated that there is a primary "physical element." This is in some passages referred to as a material "substrat." In others it is stated to be a synthesis of the "four elements" referred to below. Elsewhere in his writings other views as to its nature are presented. His opinion, however, is of no particular importance in the present connection. Whatever this "substrat" may be, if it exists at all, it is clothed with or presents certain properties or qualities.

Since things are recognized by and are distinguished from one another by their properties, Aristotle considered that these properties constitute the real bases or elements of natural bodies. Four of these properties are heat, cold, dampness and dryness. Of these the first two were mutually exclusive; the same is true of the last two, that is to say, a body is not hot and cold, neither is it wet and dry at the same time. With these exceptions, Aristotle believed that these properties were present in pairs, as follows:—

Heat combined with dryness—as seen in fire.
Heat combined with dampness—as seen in air.
Cold combined with dryness—as seen in earth.
Cold combined with dampness—as seen in water.

Figure 8 taken from the *Pretiosa Marguerita*, edited by Lacinius (1546), illustrates the realtion of these four elements.

These are the four *"Aristotelian elements"*, fire, air, earth and water, which played such an important rôle in the philosophy of nature during the Middle Ages. As will be noted these are not elements at all in the sense of being different kinds of matter by the combination of which all material things are made. On the contrary they represent certain properties in their simplest combinations. The conception of the elements as concrete matter we owe to Robert Boyle (1627–1691) who set this forth in his *Sceptical Chymist*.

The composition of the Aristotelian elements furthermore was not a quantitave one. Aristotle states that in each of these elements one of of the properties preponderated or was dominant, thus fire was more hot than dry, air more moist than hot, water was more cold than moist and earth more dry than cold. The two principles were present in different

---

[2] For a detailed statement concerning the various opinions expressed by Aristotle on this subject, see Lippmann, *Entstehung und Ausbreitung der Alchemie*, Berlin, 1919, p. 139.

proportions in different cases, so that one kind of earth differed from another kind.   Aristotle goes even further than this and says: "Fire, air, earth and water, we assert, originate from one another and each of them exists potentially in each.[3]   And therefore if the relative proportion of these properties in an element, or of these elements in association with one another be changed, one body may be changed or transformed into another body.   Hence the transmutation of the metals was quite possible according to the Aristotelian theory.   An example of the Aristotelian

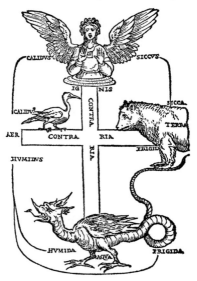

FIG. 8. The relation of the four Aristotelian elements. (From Lacinius: *Pretiosa Marguerita*.)

transformation of one element into another would be presented by heating ordinary water, when air (steam) would be produced, while if the heating be continued until it was evaporated to dryness some earth would remain.

Aristotle in the *Meteorologica* then passes on to consider the source and origin of the various phenomena of the atmosphere.   The earth's surface consists in part of land and in part of water.   When the sun

[3] The Works of Aristotle translated into English, Vol. 3, *Meteorologica*, p. 339a. Oxford, Clarendon Press, 1923.

warms the earth, the latter under its influence gives off *"exhalations"* of two kinds. The first is of the nature of vapor and this he terms the *"moist exhalation."* It rises from the bodies of water on the earth's surface and is also produced by the evaporation of water from the moist earth. This condenses into clouds.

The second kind of exhalation he terms the *"dry exhalation"* which he says is like smoke. Sometimes on looking across an arid plain, when the sun is beating down upon it, this "dry exhalation" can be seen rising as a thin smoke, but in other cases when ascending from the earth's surface it is not so distinctly visible. The dry exhalation when it has passed into the atmosphere, under certain circumstances causes thunder, lightning and other atmospheric phenomena, at other times it gathers as a thick dark cloud which condenses and forms stones, which are cast down upon the earth with great violence and hence are called "thunder bolts." Under the influence of the sun's rays these exhalations are also at times developed in the earth's interior, and at the conclusion of book 3 of the *Meteorologica*, Aristotle refers to the action of these *exhalations* while still enclosed in the terrestrial globe. He says, "Just as their twofold nature gives rise to various effects in the upper region, so here it develops two varieties of bodies that originate in the earth, namely, *fossils* and *metals."* The term "fossil" here is used, as it was in geological literature till well on into the nineteenth century, for anything dug out of the earth, although as employed by Aristotle its meaning is somewhat more restricted. He makes use of the term to include "Such kinds of stones as cannot be melted, together with realgar, ochre, ruddle, sulphur and other things of that kind." These he distinguishes from metals which include "Those bodies which are either fusible or malleable, such as iron, copper and gold." The stones contain more of the principle of dryness, while the metals, on account of the fact that when heated they become fluid, or can at least be flattened out when hammered, were believed to contain relatively more of the principle of dampness or moisture.

As stated above then, Aristotle taught that stones grew on the surface of the earth or in the earth's interior through the influence of the heat or the rays given off by the sun or other heavenly bodies, penetrating into the earth's crust and there in certain places where the conditions were favorable, causing exhalations, which, bringing about new combinations of elements, give rise to stones of various kinds.

Milton in *Paradise Lost*, book 3, sets forth the Aristotelian teaching in poetic form, as follows:—

> Where the great luminary . . .
> By his magnetic beam that gently warms
> The Universe, and to each inward part
> With gentle penetration, though unseen
> Shoots invisible virtue even to the deep
>   . . . when with one virtuous touch
> The Arch-chemic sun so far from us remote
> Produces with terrestrial humour, mix'd
> Here in the dark so many precious things
> Of colour glorious and effect so rare.

The few European writers of classical times who followed Aristotle and wrote on the subject of minerals or stones mentioned only their names, uses, and medicinal or mystical properties, but said little or nothing concerning their origin.

There was, however, an important and very interesting contribution to this and some allied subjects made by Avicenna in Persia, between the years A.D. 1021 and 1023.[4] This renowned physician had been impressed by observing that deposits of soft clay on the banks of the Oxus had become converted into a soft stone in the course of approximately twenty-three years and this he terms *conglutination*, and also by the fact that water dripping from the roofs of caverns formed stone, a fact which was interpreted by Avicenna as a solidification of the water itself, this he terms *congelation*. In such cases, he says, "We know that in that ground there must be a congealing petrifying virtue which converts the liquid to the solid, or earthiness must have become predominant in the same way as in water from which salt is coagulated, or it may be that the virtue is yet another unknown to us." A further reference to this work will be found in chapter II.

It was not until some two centuries after the death of Avicenna that the next treatise dealing with the origin of stones appeared in Europe. This was *De Mineralibus et Rebus Metallicis* by Albertus Magnus, which was written about 1260 and first printed in 1476. It is a slender volume divided into five "books." Of these the first two treat of minerals and stones, the following two deal with metals and the last with salts. In

---

[4] *Avicennae de Congelatione et Conglutinatione Lapidum*, being sections of the *Kitab al-Shifa*, E. J. Holmyard and D. C. Mandeville. (Librairie Orientaliste, Paul Geuthner, Paris, 1927.)

the first book the origin of minerals and stones is considered. These, he says, are generated by the action of a certain formative mineral virtue (mineralis virtus) acting in a favorable environment and on suitable materials in the earth's crust. This same *mineral or mineralizing virtue* he took from Avicenna, for he remarks that nothing more can be ascertained from the writings of Avicenna concerning the origin of stones except that they are generated out of earth and water by the mineral virtue.

Having made this statement Albertus, recognizing that the mere bestowal of a name upon this remarkable virtue does not advance matters to any considerable degree, goes on to say that since the nature of this force cannot be determined by direct observation, we must try to ascertain its character by analogy. Selecting for this purpose one of the other realms of nature, the animal kingdom, he states that in the generation of animals the semen issuing from the male contains and brings with it the seminary power which gives rise to a new individual; so in the mineral kingdom the celestial influences of the sun, the planets and the stars in their several courses, streaming down from the heavens, by their action upon the Aristotelian elements constituting the earth, under favorable conditions, induce changes in these which give rise to minerals, stones and metals, as well as to those intermediate bodies known to the alchemists as salts. There are also two other agencies, *instrumenta* as Albertus, using scholastic terms, calls them, which served to compact and harden these new bodies growing in the womb of nature, these are the heat and cold present in the earth itself. The former is the agent which compacts stony and earthy matter, these becoming harder when heated or baked, and the latter is active in the case of metals which become soft when heated but are made hard by cold. This seminary *mineral virtue* is thus seen to have its origin in the heavens and the heavenly bodies. The theory of Albertus Magnus is essentially the same as that put forward by Aristotle. The idea is crystallized by Johannes Velcurio in his Commentarias in Universam Physicam Aristotelis, published in 1547, in the following words:

Causa efficiens est calor celestis. Quid coelum est pater, terra vero mater metallorum intra sua viscera gestans ea tanquam foetum.

Albertus Magnus being the teacher of Thomas Aquinas and one of the most outstanding scholars of the thirteenth century, the Aristotelian theory of the origin of stones and metals through the radiation of celestial

influences upon the earth, became the commonly accepted doctrine of
the scholastic philosophers, and was accepted by almost all the writers
who treated of this subject down to the end of the sixteenth century,
and who one after the other repeated with an almost wearisome reiteration the statement that in the influences of the stars (of which the
sun was the greatest) was to be found the *vis formativa* which generated
minerals within the earth in those places where suitable conditions were
found.[5]

Andreas Baccius[6] writes:

Gems are the most precious products of nature and must be generated from the
highest source.   They come to birth like gold itself through the action of the heavens
and the stars.   They also draw their virtues from the stars.   The earth being situated
immovable at the centre of the universe has all the heavenly powers concentrated
upon it as at a focus.

Leonardus Camillus,[7] probably the greatest authority of his age on
these matters, sets forth his views as follows: Three influences must
cooperate to bring any stone into existence: first, there is that power
which proceeds from Him who is the source of all motion in the universe;
secondly, those influences which are radiated to the earth from the
planets and the celestial constellations; and thirdly, there are the qualities of the elements themselves, heat, cold, dryness and dampness.
Where these three influences are brought to a focus at any point in the
earth, there stones are generated.   In tropical countries the rays proceeding from the heavenly bodies fall more directly (that is less obliquely)
upon the earth's surface, and therefore act more powerfully.   Henec
gems are found more abundantly in tropical regions than in other portions of the earth's surface.

2. The Theory of the "Petrific Seed"

But there was another theory, or rather a number of related theories,
concerning the origin of "stones," both precious and common, which
set aside altogether the Aristotelian conceptions of the influence of the

---

[5] See Ristoro d'Arezzo, *La Composizione del Mondo* (1282), Rome, 1859, p. 108.
Konrad von Megenberg, *Das Buch der Natur.*
Gregorius Reisch, *Margarita philosophica*, Strasburg, 1504.
Hieronymus Savonarola, *Compendium totius philosophiae tam naturalis quam moralis,*
Venice, 1542.
Gio. Camillo Maffei, *Scala naturale*, Venice, 1563; and many others.
[6] *De gemmis et lapidibus pretiosis*, Frankfort, 1603.
[7] *Speculum lapidum*, Venice, 1502 (book 1).

heavenly bodies, and which were very widely taught and accepted, especially in the sixteenth and seventeenth centuries.

These theories, which for convenience have been grouped together under the present heading, differ from one another considerably in several respects but have the same central idea that minerals (or ' stones") have their origin in a "seed," which is called by some writers the *petrific seed* and by others the *saxeous* or *rocky seed*, by still others *gorgonic seed*, because like the Gorgon's head it turned all things into stone.

In the natural world, it was pointed out by the supporters of this theory, there are three distinct realms, the animal, vegetable and mineral. These are separate but present close parallels to one another. In order that their respective relations may be understood it is necessary to bear in mind the Aristotelian (or scholastic) view of the *"anima"* or principle of life which permeates the whole universe and is inherent in all things. A living thing is composed of an *anima* or soul and a *corpus organicum* which is of matter. There are three kinds of *anima*, the *anima vegetativa*, the *anima sensitiva* and the *anima rationalis*, or the vegative, the sensitive and the rational soul. The first of these is the lowest or most imperfect, being without knowledge or freedom. The second has the knowledge derived from the senses, as well as reason and liberty. This is the *anima* possessed by the brute creation. The third has the most highly developed reason as well as the lower powers and qualities possessed by the others. This is the soul of man.

The three kingdoms of nature then consist of living things, but they display, so to speak, three different intensities of life. The animal kingdom shows life in its highest development consisting of beings possessing an *anima rationalis* or an *anima sensitiva*. The vegetable kingdom has life also but on a lower plane, its subjects possess the *anima vegetativa*. A member of the animal or vegetable kingdom, however, not only possessed an anima but its body was composed of organized matter, that is consisted of, or at least contained, a complicated series of vessels and organs by which it secured and assimilated nourishment, through which it grew, and having arrived at maturity passed onto old age, and eventually died.

The members of the mineral kingdom, although by many regarded as being composed of dead and inert matter, also have in a lesser measure the *anima vegetativa*, since they show the character of living things in that they have a definite life history. They come into existence or are born, they grow to maturity and have frequently been observed to succumb to various forms of decomposition or alteration, the equivalent of diseases in the higher kingdoms of nature, which eventually destroy them.

Hartnaccus[8] sums up his discussion of the subject in the words, "Mineralia nec viva nec mortua sunt" for, he says, they have no *anima* nor can they, properly speaking, be said either to nourish themselves or to grow. But, he goes on to say, certain chemists on account of the extraordinary powers and actions ("operationes") which minerals display do not hesitate to consider them as possessed of life.

Alstedius,[9] referring to this question says:

Mineralia non habent animam manifesto viventem qualem habent vegetabilia et animalia sed latentem aliquo modo et proximam animae vegetivae. Id colligere est ex virtutibus mineralium, quae tantae sunt ac tales ut non nisi ab anima quadam proficisci queant. Quando igitur mineralia dicuntur corpora inanimata hoc accipiendum est non absolute sed comparate.

Langhansius in his dissertation entitled Vita Mineralium (Wittenberg, 1676,) which is a formal treatise in true scholastic style on the question whether stones and metals have a sort of life, reaches by the process of logical deduction the following conclusion: "Vita nulla est mineralium ac elementorum, nisi simulate." Wallerius[10] notes that minerals often assume forms and shapes that simulate or approach certain of those seen in plants, fruits, leaves, twigs, stalks, etc., which suggest a like origin. Some modern thinkers have not hesitated to endorse the view that inorganic nature has its own force or life, so-called inorganic nature being "A vast realm of finite consciousness of which our own is at once a part and example."

In order to present as clear a picture as possible of this theory of the *petrific seed*, three of its chief exponents will be selected and their views presented as nearly as possible in their own words. The first of these is a French writer, the second is English and the third a German, all writing within a period of thirty-seven years, namely between 1635 and 1672.

The first is Etienne de Clave,[11] Doctor of Medicine. De Clave re-

---

[8] *Admiranda physica*, Frankfort and Leipzig, 1684, p. 394.

[9] *Scientium omnium encyclopediae*, Lyons, 1649, vol. II, p. 144.

[10] *Systema mineralogicum*, Vienna, 1778, vol. 1, p. 4.

[11] *Paradoxes, ou Traittez philosophiques des pierres et pierreries, contre l'opinion vulgaire. Ausquels sont demonstrez la matière, la cause efficiente externe, la semence, la génération, la définition & la nutrition d'icelles*, Paris, 1635.

views the theories concerning the origin of rocks which had been put forward by the more important writers on this subject before his time, and after discussing each of them in turn, points out where in his opinion they are wrong and then[12] sets forth the theory of the petrific seed as follows: There is a fire at the center of the earth which heats all the interior of the earth by vapors and exhalations which continually pass off from it. This internal fire burns eternally within the earth as the sun does above the earth, without fuel. But it differs in its action from the sun in that while the sun produces its effects upon the earth's surface by the direct impact of its rays, the inner heat must act through exhalations and vapors by which its heat is conveyed from the centre to all parts of the earth. These vapors are derived from the rain which falls upon the surface of the earth, a portion of which sinks into the earth through pores and minute crevices, and passing downward as through a filter, it dissolves and takes up *nitre balsamique*, sulphur, bitumen and other substances and qualities. Eventually reaching the central fire, this water with its dissolved content is volatilized and returns again upward through the earth toward the surface. As it passes upwards, often now more viscous on account of its dissolved content, it possesses a power which enables it to play the part of the male seed in the generation of animals or plants, and meeting with favorable conditions at certain points in the earth's crust it sets up some action allied to fermentation which gives rise to a female element with which it unites, to form a "seed" from which a mineral or "fossil" will develop. This seed so produced expands and grows by the influence of heat, making a place for itself by means of its "esprit extensif."

De Clave goes on to say that this method of generation has a close analogy to that which takes place in the case of certain animals, as for instance some insects which are engendered by heat acting on some suitable material (for "spontaneous generation" was held to be a fact in de Clave's time). The same, he says, is also true in some members of the vegetable kingdom.

The second treatise on the theory of the *petrific seed*, is that by Dr. Thomas Shirley,[13] Physician in Ordinary to His Majesty (Charles II). The theory is set forth by him in the following words (pp. 23-128):

[12] op. cit., p. 257.

[13] *A Philosophical Essay, declaring the probable Causes, whence Stones are produced in the Greater World. From which occasion is taken to search into the Origin of all Bodies, discovering them to proceed from water and seeds. Being a Prodromus to a Medicinal Tract concerning the causes, and cure of the Stone in the Kidneys and Bladders of Men*, London, 1672.

The Hypothesis is this, viz.   That stones, and all other sublunary bodies, are made of water, condensed by the power of seeds, which with the assistance of their termentive Odours, perform these Transmutations upon Matter.

That is, that the matter of all Bodies is originally meer water; which by the power of proper seeds is coagulated, condensed and brought into various forms, and that these seeds do work upon the particles of water, and alter both their texture, and figure; as also that this action ceaseth not, till the seed hath formed it self a Body, exactly corresponding with the proper Idea, or Picture contained in it.   And that the true seeds of all things are invisible Beings, (though not incorporial); this I affirm and shall endeavour to prove. . .

The body of the Seed, or Grain (which is the Casket that contains this invisible Workman) being committed to the Earth (its proper Womb) is softened by the Nitro-sulphurous juice of the soyl; that the Vis Plastica (which is the Efficient of the Plant) may, being loosened from its Body, be at Liberty to act.   Which being done, the body of the seed, or Grain, is destroyed; according to the sacred Writ: (Except Seed, committed to the ground, dye, it produceth no fruit:) But the Architectonick Spirit being now at Liberty, ferments, by its Odor, the Liquors it finds in the Earth, converting them into a juice, fit to work the Plant out of it, which it by degrees performs. . . .

As Vegetables, and Animals have their Original from an invisible Seminal Spirit, or breath: so also have Minerals, Metals and Stones. . . .

The Seeds of Minerals, and Metals are invisible Beings; (as we have shewed, above, the true Seed of all other things are;) but to make themselves visible Bodies they do thus: Having gotten themselves sutable Matrices in the Earth, and Rocks (according to the appointment of God, and Nature) they begin to work upon, and Ferment the Water; which it first Transmutes into a Mineral-juice, call'd Bur, or Gur; from whence by degrees it formeth Metals. . . .

That Moses held Water to be the First and universal Matter, will appear from what he tells us in the First chapter of his Book of the Creation, called Genesis verse the Second, where he acquaints us, that the first material substance out of which God made this Beautiful and Orderly frame of the World, which from its Beauty the Greeks call Kosmos, was water.   His words are these; And the Spirit of God moved upon the Face of the Waters. . . . In all probability, the sense of the Expression is, that at that time, (viz. in the beginning) God infused into the bosome of the Waters, the seeds of all those things which were afterwards to be made out of the waters. . . .

The Saxeous, or Rocky Seed, contained in these Waters, (which is so fine, and subtile a Vapour, that it is Invisible; as I have before shewed all true Seedes are,) doth penetrate those Bodies which come within the Sphere of its Activity; and by reason of its Subtility, passes through the pores of the Wood, or other Body, to be changed. . . . So this Stonifying Seed, by its operating Ferment, doth transchange every particle of the matter it is joyned unto, into perfect Stone; according to its Idea or Image, Connatural with it self.

The third exposition of this theory, is that given by M. J. C. Schweiger in a Dissertation which he presented to the University of Wittenberg in the year 1665 and which is entitled *De Ortu Lapidum, Concerning the Origin of Rocks.*

It is clear (he says) that neither the stars, heat, cold, or the drying out of moisture are causes to which the generation of stones can be attributed but we have no hesitation in asserting that stones come into existence from their own proper (seeds or) seminal principles. . . Stones are generated as plants are—like produces like—gold gives birth to gold, gems to gems, stones to stones. By virtue of their seminary power, each species reproduces and multiples itself and it preserves its own species intact and perfect.

The renowned Senertus has repeatedly asserted that in stones the seed or seminary principle cannot be isolated, nor does it appear as a separate body as the seed does in most plants, but on this account its existence cannot be denied, for just as the willow, although it resembles other plants, has its seminal principle dispersed through the entire body of the tree so that a new willow tree can be propagated by planting any slip cut off from it, so the reproductive principle exists in all parts of the stone, although we have not been able up to the present time through the weakness of the human mind to discover its actual character, we believe that it can be definitely and indisputably established that stones are generated from other stones, each after its own kind.

The theory of the *petrific seed* as set forth by these three writers, is presented in the form which was generally accepted.   Many other writers in these times however put forward certain modifications of it.   The attempt to accurately define, separate and set in their proper relations to one another, the various kinds of "spirits" "humors" and "seeds," be they seminal, formative, architectonic, expansive, lapidific or gorgonic, and to bring all these into a proper relation to the harmony of nature and the teachings of Holy Writ taxed even the medieval mind which delighted in such exercises.

DeClave, in his book mentioned above, speaking of the great variety of conflicting opinions among those who treated of the subject, says:

Il y a tant de cotrastes entr'eux, qu'il est impossible des les concilier tout à fait. C'est pourquoy, dautant que pour en traitter profondement comme il seroit nécessaire, il faudroit sortir des termes & limites de la science naturelle, & avoir recours à la metaphysique voire a la Théologie, nous laisserons ce debat, & soubscrirons à la censure de l'Eglise tout ce qu'elle en voudra décider, sans en vouloir parler plus simplement que ce qui en a esté dit cy-dessus aux Chapitres précédents.

DeClave's book, while written with admirable clearness, presenting as it does the diverse opinions of many authors with arguments for and against their many and conflicting views, none of which have any but the most meager basis in study or observation of the facts or phenomena discussed, calls forth in the modern reader a sense of mild though genuine amusement; but must have led to a feeling of bewilderment on the part of not a few contemporary readers, as indicated by a few lines pencilled by some unknown writer of former times on the fly leaf of a copy of de Clave's book now in the possession of the writer, which runs as follows:—

> Sunt bona, mala quaedam
> Alia mediocria
> Et nunc erudimini
> Qui donc est infaillible?

3. The Theory of the Lapidifying Juice

This theory gradually replaced the theories of the "*petrific seed*" as these had superseded the earlier theory of Aristotle. The odd term *Lapidifying Juice* does not commend itself to the modern reader, but the theory itself represented a distinct advance toward a true explanation of the problem of the origin of "stones" concerning which, although very much has now been definitely and finally established, much still remains to be learned by further study and research. Lapidifying juice is, of course, a literal translation of the Latin term *Succus Lapidificus* which in this, or in some synonymous form such as *Succus Lapidescens, Humor Lapidescens, Sugo Petroso, Spirito Petrifico,* or *Semence pétrifiante,* occurs everywhere in the literature of the time which deals with the genesis of "stones." If this petrifying principle occurs in a vaporous rather than a liquid form, it is termed *Aura Lapidifica, Aura Bituminosa, Aura Petrifica* or *Vento Petroso.* As stated by Gimma[14] it "Like the Gorgon turns all things into stone."

The theory of the lapidifying juice may be briefly set forth as follows:— There is circulating through the earth's crust a *succus* or fluid body which under certain conditions has the power of turning various substances into stone. It passes not only through open cracks and fissures in the earth's crust but also traverses the minute crevices and pores of the rocks. Its action is seen in very striking manner in caverns where it gives rise to great icicle-like pendants of marble which hang from the roof, or to corresponding columnar masses which rise from the floor, presenting a great variety of wonderful forms. Not only is it found in the rocky crust of the earth but it is present in the waters of the sea and, in a gaseous form, in the atmosphere as well. Palissy[15] says: "En la mer il y a trois espèces d'eaux, la commune, la salée & la vegetative ou congelative." It deposits stone in the sea, and in its vaporous form is present

---

[14] *Storia Naturale della Gemme, delle Pietre è di Tutti i Minerali,* Naples, 1730, vol. I, p. 78.

[15] *Discours admirables de la Nature des Eaux et Fontaines,* Paris, 1580, p. 366.

in the air where at times it condenses into stones which fall upon the earth   It is found even in the human body where in the bladder, kidneys or other organs, it gives rise to masses of stone.   In fact the passage of this *succus* through the earth and its enclosing envelopes of water and air, which constitute the macrocosm, may be compared to the circulation of the blood through the human body or microcosm.   The primal force which gives rise to the movement of this *succus* in the earth's crust is generally considered to be the same as that which actuates the circulation of the blood in the body of man, as well as the greater movements of the heavens—the Great First Cause.

There is a difference of opinion among the various writers on the question of the nature of the *succus lapidescens*.   Some with Du Hamel[16] consider it to be merely water holding rocky matter in suspension, "Nec succus lapidificus aliud est, quam aqua saxeis ramentis praegnans," which forms stone by the precipitation of its rocky content.

Fallopius[17] says that the *succus* may be present in waters in two different conditions, namely *confusus* and *optime commixtus*.   If it be confusus, which may probably be best translated as "commingled" with the water it will act in the manner described by Du Hamel and by the redeposition of "ramenti" give rise to rocks which are generally of a porous nature. If on the other hand the succus is optime commixtus, which may be translated as "dissolved," it will produce rocks which are compact and solid.

Fallopius adds that there is another variety of very pure succus which by its own solidification forms gems and precious stones.   In this he is supported by Gimma, who says that the succus when mingled with water gives rise to alabaster.   If the succus passes through clay or other porous materials it hardens these to stone by forming a cement which fills up the interstitial spaces.   Paracelsus[18] says it turns sea foam into masses of porous rock, referring to the pumice sometimes found floating in great abundance in the sea near certain volcanic vents.

Furthermore, since there are many different sorts or varieties of stones, the succus which produces them cannot always be identical in character. Fallopius[19] is of the opinion that there must be at least three varieties— one from which clear crystalline gems and minerals are formed; a second,

---

[16] *De Meteoris et Fossilibus*, Paris, 1660, Lib. II, p. 185.

[17] *De Medicatis Aquis atque de Fossilibus*, Venice, 1564, p. 97.

[18] See Aldrovandus, *Musaeum Metallicum*, Bonn, 1648, book IV, p. 818.

[19] *De Medicatis Aquis atque de Fossilibus*, Venice, 1564, p. 98b.

less pure in character, which gives rise to the common opaque stones, and a third which is present in the waters of the sea and which produces coral and similar objects by turning certain marine plants to stone, and which when very pure gives birth to pearls. Mattioli[20] points out that there must be many kinds even of that purest succus which crystallizes into gems, in order to explain their great variety in color, lustre and brilliance.

Evidence of the existence of different kinds of succus is also seen in banded or stratified rocks whose different beds are of different colors. That these different succi in some cases will not mingle with one another but remain separate, like oil and water, is shown by the streaks of different colors often seen on mountain sides in the same cliff or rocks.[21]

Baglivi[22] who devoted much attention to the question of the origin of rocks and who studied a number of occurrences in the field, chiefly in Italy, where rocks were to be observed apparently in the act of growing, has set forth his observations and conclusions in an important paper. He is of the opinion that minerals or stones grow chiefly when they are in a soft or porous condition. While in this state fluids or solutions carrying nutriment pass through their pores and deposit the material required for their growth by a process closely analogous to that described by Cardan as evidence of the presence of an *Anima Vegetativa* in the growing stone, that is by an "intrinsic" growth "Per intus-susceptionem" as in plants and animals. He does not believe however that the presence of this porous structure is any indication of the presence of life. By the deposition of material in this manner the minerals or rocks grow in volume and if soft are hardened. When the rock becomes hard it in many cases continues to grow by the deposition of material on its external surface, "Per appositionem partis ad partem," as the Schoolmen say. The agencies by which rocks grow, and soft materials, such as clay, become hardened into rock, are three in number. The first is a sticky bituminous *Humor* which rises from the earth's interior, having its origin in the deep-seated subterranean fires. The second is represented by certain saline substances and solutions, and the third is the succus lapidificus or aura lapidifica. This succus, he says, is nothing else than water heavily charged with stony matter. The aura is a wind carrying a similar burden.

[20] *Commentaires sur Dioscorides*, Lyons, 1579.
[21] Gimma: Vol. I, p. 71.
[22] *De Vegetatione Lapidum*—in his *Opera Omnia*, London, 1714, p. 497.

The theory of the petrific seed, as already stated, gradually faded away, being replaced by that of the lapidifying juice as the generally accepted explanation of the origin of stones. But while this theory did not establish itself until a late date it was foreshadowed in the writings of Seneca and Vitruvius back in classical times.

It was Agricola however who first clearly set forth the theory of the *lapidifying juice*. This was one of the many services which he rendered to geological science, services which were due to the fact that he was the first of the writers on this science who set aside the body of quaint, interesting and far-fetched speculations and ratiocinations of the Schoolmen and based his opinion on his personal observation of the earth's crust itself, more especially during his long residence in the mining districts of Saxony. His work *De Re Metallica*, which treats of mining and metallurgy, is known to all, and another work of his, *De Natura Fossilium*, may be regarded as the first handbook of mineralogy in the modern sense of the term, so much so that he is commonly called the "Father of Mineralogy."

It is however in still another work of his, not so well known, that he treats of the "succus lapidescens" and of the part which it plays in the origin of minerals and rocks. This is entitled *De Ortu et Causis Subterraneo. um*, the first edition of which was issued at Basel in 1546 from the press of Froben. In this Agricola ridicules the view of Aristotle that common stones, gems and metals all derive their origin from the action of the stars. The theory of Albertus Magnus concerning their origin, he says, reminds one of the dreams of Hermes nor, he goes on to say, do stones contain any seed by which they can reproduce their kind, the forces to which they owe their origin lie altogether outside of the stone itself. In presenting his own views, he states that he will pass by all speculations with reference to remote first causes and come at once to the immediate causes of the origin of stones. These are, he says, first, heat and cold, and secondly, the succus lapidescens. If the term succus lapidescens be translated, as it can properly be for this was Agricola's conception of it, by the English expression "mineral-bearing solution," it will be seen that Agricola's teaching is that minerals and stones are produced either by the deposition of the mineral matter which the succus lapidescens holds in solution, owing to heat which will cause such solutions to dry up and deposit the stony matter which they hold in solution, or by cold which will cause water to become chilled and so deposit its dissolved material. Such solutions also pass through masses of uncompacted earthy material and transform them into solid rocks by supplying a cement which fills up their minute interstices.

Agricola also recognized that there is a class of rocks which have been produced from rock detritus resulting from the abrasion of rock masses by the agency of running water and which have become cemented by the deposition of interstitial material through the action of the mineralizing solutions which have percolated through them.   Agricola's feet were on the path which was to lead to the true solution of this age-old problem of the origin of the rocks of the earth's crust.   He had correctly determined the origin of rocks which are known as the clastic rocks and of those which owe their origin to chemical precipitation.   He however knew little or nothing concerning the origin of the great classes of the igneous and metamorphic rocks, these presented problems which required for their solution a much deeper knowledge of natural processes than Agricola possessed, processes which are still far from being thoroughly understood even after centuries of additional study.

B. The Digestive Organs of Minerals

A number of writers, among whom Cardan[23] was one of the chief, pointed out that usually gems and other transparent minerals, when examined by transmitted light, show within their substance many minute cavities, pores, little tubes, fibers or cloudy streaks, while the common opaque minerals and rocks often display porous, streaked or fibrous structures.   These, they claim, represent a simplified form of the organic structures seen in animals and plants, in keeping with the lower type of life which minerals display and that through the instrumentality of these structures, minerals nourish themselves and grow as do the members of the higher kingdoms of nature.

Baccius,[24] who was a supporter of the Aristotelian theory, vigorously opposes the contentions of Cardan and his followers, "You might as well say that Caesar was alive in a statue which bears his form or that men or dragons lived in the clouds because these at times take forms which resemble them," and having thus disposed of this interpretation of the nature of these imperfections in gems he proceeds to offer an, if possible, more ridiculous explanation of their origin but one consonant

[23] Hieronymus Cardanus, *De Subtilitate*, Nuremberg, 1550, Book 7.

[24] *De Gemmis et Lapidibus pretiosis*, translated into Italian with a very full commentary by A. Wolfgang Gabelchover, Frankfort, 1603, p. 26.

with the Aristotelian theory that gems and other stones owe their origin
to the influences of the heavenly bodies, namely that the imperfections
in question might be caused by the influence of certain stars which exerted
a contrary influence to those by whose action the gem was being developed.

This question as to the manner in which minerals grew was one which
was keenly discussed by those who held the theory of the petrific seed.
The followers of Cardan held that the life and structure of the mineral
being closely analogous to that of the plant, growth took place "Per
intus-susceptionem," as the Schoolmen say, that is by the absorption of
nutriment through pores in the mineral and the distribution of this
nutriment by means of a circulatory system of fibrous canals and cavities,
as in the case of a plant.   Another school of which Scaliger and Steno[25]
were two of the most distinguished members, while accepting the theory
of the petrific seed, held that, while stones grow, they do so in quite a
different manner from that taught by Cardan, namely by a deposition of
new matter upon their exterior surface, that is, "per appositionem partis
ad partem."

This latter view was considered to receive support from the fact that
stones can often be observed to present a banded appearance, as if
they had been built up of successive layers.   This mode of growth,
however, was not considered to afford such direct evidence of life although it was pointed out by some writers that many trees which undoubtedly possessed life showed this structure resulting from the deposition of successive layers of woody matter about their trunks,

C. Male and Female Minerals

But many of these ancient writers held not only that the members of
the mineral kingdom originated from seeds, but that these various
minerals have the power of reproducing their kind.

Since among animals and plants it was recognized that new individuals
usually arise through the union of the two sexes, analogy suggested the
presence of a like distinction in the mineral kingdom.   Male and female
stones are mentioned by a number of writers in widely separated ages.
This distinction is made by Theophrastus in his *History of Stones*, which
is the earliest treatise on mineralogy that is known.

---

[25] *De Solido intra Solidum naturaliter contento.*, Florence, 1669, p. 39.

In carnelians (he says), that species which is pellucid and of a brighter red is called the female and that which is pellucid and of a deeper red with some tendency to blackness, the male. The Lapis Lyncurius is distinguished in like manner, the female being more transparent and of a paler yellow; and the Lapis Cyanus is in the same manner divided into Male and Female specimens: the Male being deeper in colour.

Pliny[26] also says that "Diphris is of two kinds, the male and the female, wherein may be perceived very distinctly the members that distinguish the sex by reason of a certain line or vein of the stone."

Valentini[27] mentions "Lapides Judaici" of both sexes, the former stouter and the latter more slender in shape. These "stones" which are referred to by all the old writers, are the club-shaped spines of certain cretaceous echinoderms found abundantly in the limestones of Palestine and which were brought to Europe by the Crusaders and pilgrims on their return from the Holy Land. The Balas Ruby (Spinel) was generally considered to be the female of the carbuncle. Whether the writers who employ these terms do so merely in the sense that there is a variety of each mineral which by its intense color or stout habit displays a masculine type, while another variety by its paler color and more delicate form suggests a feminine character, or whether they actually believed that these two varieties are two sexes by whose union new individuals are produced, is not clear. Probably most of them had no definite opinion on the subject. Albertus Magnus,[28] however, after a full discussion of the subject, reaches the conclusion that minerals have no power of reproducing themselves, others however held the opposite view.

Sir John Mandeville, for instance, was of the opinion that the two sexes came together in the production of new individuals, at least in the case of diamonds, for in the edition of Marbodus which was issued under his partronage (or by him) he made certain additions and alterations in the text, one of which has reference to the diamond, on which he wrote two long chapters. In one of these, speaking of the diamonds of India, he says that the most diamonds occur:

Es Montagnes et es roches ou il y a minières d'or et croissent ensembles malles & femelles, et se nourrissent de la rousée du ciel et concivent et engendrent et font des petits delez eulx qui moulliplient et croissent tous les ans. Je l'ay mainteffois essayé, car si on les garde avec un peu de la mine de la roche où il croissent qui tienne à la

---

[26] *Natural History*, Book 37, chap. 10 (Holland's translation).
[27] *Natur und Materalien Kammer*, Frankfort, 1714, p. 54.
[28] *De Mineralibus et Rebus Metallicis*, Cologne, 1569, Book I, chap. 3.

pointe de desseure et on les moille souvent de la roussée de may, ils croissent tous les
ans et deviennent les petits bien grans et bien gros selon leur natur.[29]

It may here be of interest to note that shortly after the Royal Society
was founded the council of the Society directed that a series of questions
concerning certain matters which had long been under discussion should
be sent out to certain persons who were considered to be best qualified to
answer them from actual knowledge based on careful observation, and
the first question on the list was addressed to "Sir Philiberto Vernatti,
Resident in Batavia in Java Major," the question being "Whether diamonds and other precious stones grew again after three or four years in
the same places where they had been digged out?" The answer returned
was: "Never, or at least as the memory of man can attain to."[30]

Rueus[31] also refers to another instance:

There is a certain wonderful fact that should not be passed over in silence. It
was told me many years ago by a certain lady whose word could be trusted, that
there was a Lady "Heurensis" descended from a celebrated Luxemburg family who
had two diamonds which she had inherited and carefully preserved, which frequently by a miracle of nature produced other diamonds and that anyone who
watched them at certain times would see that they passed through birth throes and
produced offspring similar to themselves. The cause of this phenomenon, if indeed
one may be permitted to engage in philosophical speculation concerning it, I should
consider to be that the celestial power works its way into the parent stones, which
power we may call the Adamantific Virtue, and which first hardens and transforms
the surrounding air into water, then changes this into some substance allied to diamond and then finally alters this into the very diamond itself.

Commenting upon this story Worm[32] remarks:

If this is a fact it presents a strong argument for the existence of a reproductive
power in stones, whereby they may propagate themselves and reproduce their kind.
But he (Rueus) says that it was related to him by a lady whose word was worthy of
confidence, but he does not say that he saw this wonder with his own eyes.

Pliny[33] says that Theophrastus and Mutianus believed that certain
stones reproduced themselves. This statement although often quoted
by later authors is not correct so far at least as Theophrastus is concerned.
Pliny is not always strictly accurate in his statements concerning the

[29] Quoted in Pannier, *Les Lapidaires francais du Moyen Age*, Paris, 1882, p. 192.
[30] Tho. Sprat, *The History of the Royal Society of London*, 1667, p. 158.
[31] Franciscus Rueus, *De Gemmis*, Zurich, 1556, p. 4.
[32] *Museum Wormianum*, Leyden, 1655, p. 103.
[33] *Natural History*, Book 36, chap. 18.

opinions expressed by authors whom he cites.   What Theophrastus says is that certain stones *are reported* to possess this remarkable property but that he is very doubtful whether this assertion is correct.   His words are: "The greatest and most wonderful of all the qualities of stones—if indeed the statement be really true—is that they bring forth young."   Pliny[34] in another passage mentions a stone called Peantides or Peantes as reported to become pregnant and bring forth young.   Albertus Magnus[35] says of it:

> Peranites lapis est generatus de Mecheton et est feminei sexus.   Nam certo tempore dicitur concipere et parere consimilem lapidem naturalem, valere autem dicitur pregnantibus.

and a similar statement concerning it is repeated in the later lapidaries.

Leonardus, however, writing in A.D. 1505 while reproducing the above statement adds:

> But though this is written by some it does not please me, I rather think that such writers have fallen into an error in misunderstanding the words of the ancients.   For when they say that such a stone can conceive they do not mean that it can conceive but that it affords help to women in their conception and bringing forth.   Which of these opinions is true I leave to the judgment of others.

Still another mineral which was reported to possess the same remarkable property is that which in Latin bore the name Gasidenum.   Of it Pliny says, "Haec quoque gemma concipere dicitur, & intra se partum fateri concussa, concipere autem trimestri spatio."   That is to say, the mineral conceives every three months and the new stone can be heard to rattle within it when the parent stone is shaken.

There is, however, one stone which is of especial interest in this connection and which is described in almost every lapidary or treatise on mineralogy which appeared from the time of Theophrastus down to the beginning of the nineteenth century.   It is the renowned aetites, etites or eagle stone.   It owes its interest to the fact that it was believed to be a stone which could be seen in the very act of giving birth to its young.   Pliny in one of the references to this stone which are found scattered through his *Natural History* states that there are several varieties of it differing in color and other respects, that it is found in several countries, and that there are two genders, male and female.   The stone is stated to

---

[34] *Natural History*, Book 37, chap. 10.

[35] *De Mineralibus*, Book II. chap. 14.   This statement of his is quoted verbatim, from the *De Virtute Universale* of Arnoldus de Saxonia.

be hard, round in shape, and usually smooth on the exterior, resembling a small coconut. Many specimens are remarkable in that when shaken they give forth a rattling sound owing to the fact that they enclose another smaller rounded stone, to which the name Calimus was given. If the larger stone, the etites, is broken open this smaller stone within it rolls out, and the new aetites is thus born.

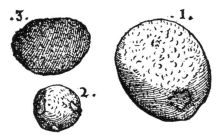

FIG. 9. Three specimens of Aetites from Conrad Gesner's collection. (From *De Rerum Fossilium . . . Figuris.*)

FIG. 10. Aquilina gravida di altra aquilina che e la doppiamente gravida. (From Bausch: *De Lapide Haematite et Aetite.*)

When in ancient times a story of this kind came to be generally believed, it is interesting to note how the tale gradually grows through the addition of interesting details and embellishments by successive writers. With regard to aetites we learn from one writer that in a certain place which he visited his slumbers were disturbed by frequent reports which he heard at intervals during the night. Next morning on investigating the cause of these he found that this district was one in which these aetites were abundant, and that during the night they burst open with a loud report, setting free the little stone or stones which they enclosed. Nierenberg[36] states that in the province of Rio del Plata in Paraguay the natives call these aetites "Coccos," but he calls them "ova solis." They grow in the ground and when they have come to maturity, are cast forth from the earth and being disrupted with a loud report, scatter about a shower of various kinds of gems such as amethysts and topazes. So soon as the natives hear this sound they hurry to the spot to collect these precious stones; frequently, however, their haste is in vain for they find nothing of any value. Another[37] writer says that in the night the stone gives forth a sound, which people say is the birth cry either of the mother stone or of the child which is born. Still another author observes that nature is a stern godmother since the birth of the child necessitates the death of the mother.

The stone derives its name "eagle stone" from the fact, so all the ancient writers say, that eagles place it in their nests when they are about to lay their eggs. In the book entitled *De Lapidibus* (or *De Mineralibus*) which is attributed to Aristotle but which was by Avicenna (see p. 41) it is stated that the female eagle lays her eggs only with difficulty, and when the male eagle finds that she cannot produce her eggs, he at once betakes himself to India where this stone is to be found, and returning thence with a specimen in his beak, places it beneath his mate, who through its influence lays her eggs with ease. Isidore of Seville says that she cannot lay her eggs unless she has one of these stones in her nest. Others say that these stones raise the temperature in the nest, so that the eggs are hatched out more readily. In order to enhance the attractiveness of this story, the male eagle is always said to resort to some country at once remote and romantic. The Hortus Sanitatis states that the stone is found in the mountains of India between Chirras and Saradi not far from the shore of the ocean. But the eagle sometimes brings it from Persia, where it is also known to occur.

Pliny (Book X, chap. 3), after enumerating and describing six kinds of eagles, goes on to say:

Of the six kinds before rehersed the three first and the fifth have in their nest a stone found called Aetites. This stone is medicinable and singular good for many diseases and if put into the fire it will never a whit consume. Now this stone, as

---

[36] Quoted by Bauchius in his *Schediasmata Bina Curiosa de Lapide Hematite et Aetite,* Leipzig, 1665, Part II, p. 23.

[37] Peterman: *Reisen im Orient, II,* p. 132. Quoted by Ruska in *Das Steinbuch des Aristoteles,* p. 4.

they say, is also with child, for if a man shake it, he shall hear another to rattle and sound therein. But that vertue medicinable above said is not in these stones, if they be not stollen out of the very nest from the airie.

Many wonderful properties were attributed to these aetites, properties which became more varied and more wonderful as time went on. It is interesting to find these stories from Pliny re-echoed down through the centuries in Europe. Dioscorides, Mattioli, Isidore, Marbodus, Rueus, Brasaulos, Mylius, Caussinus, Albertus Magnus, Leonardus and a host of others repeat them.

Bernard Palissy[38] however writing in 1580 and always in advance of his times, summarily dismisses the theory that this eagle stone reproduces

FIG. 11. The eagle carrying an aetites to its nest. (From *Hortus Sanitatis*.)

itself, and puts forward another explanation of its origin in the following words:

I believe that it is nothing but a petrified fruit and that which it contains is merely the nut, which having shrunk up somewhat rattles against the sides when the stone is shaken.

This aetites is now known to be nothing more than one of those concretions which are frequently found in rocks of aqueous origin composed of consecutive layers, often differing somewhat in character and com-

[38] *Discours Admirables de la Nature des Eaux et Fontaines*, First Edit., Paris, 1580, p. 232.

position, deposited around some nucleus, often a grain of sand or some small fragment of a fossil.   If one of these layers is more soluble than the others and percolating waters pass through the rock after complete consolidation of the whole, this more soluble layer is redissolved leaving the central portion of the concretion loose within the external shell so that it rattles when the concretion is shaken, giving to it among the Germans the name of "Klapperstein."

Another type is represented by the well known geodes which are found in certain beds of the Keokuk of the Central Mississippi Valley.   Their cavities whose origin is uncertain have been lined with calcite, quartz or chalcedony, sometimes associated with pyrite or chalcopyrite which at times are altered to limonite.   Fragments of these minerals becoming detached from this inner lining cause a rattling sound when the geode is shaken.

D. THE PLACES IN WHICH MINERALS ORIGINATE AND THE CONDITIONS UNDER WHICH THEY DEVELOP

The ancient writers who busied themselves with the problem of the origin of minerals and rocks were in all cases much impressed with the fact that "stones" are often found in the act of growing in various parts of the human body.   These "Calculi" are described and figured by many authors.   Two of the treatises which set forth most clearly and in the greatest detail the views concerning the origin of stones which were held in the seventeenth century, were written by men who were particularly interested in these calculi, on account of the light which they were considered to throw on the origin of all the other "stones."   The first of these, a small book of 143 pages, has already been mentioned.   It bears the title, A Philosophical Essay, declaring the probable Causes whence stones are produced in the Greater World . . . , being a Prodromus to a Medicinal Tract concerning the Causes and Cures of the Stone in the Kidneys and Bladders of Men, written by Dr. Thomas Shirley, London, 1672.

The other work is entitled *Traité des Pierres qui s'engendrent dans les Terres & dans les Animaux, ou l'on parle exactement des Causes qui les forment dans les Hommes*, by Nicolas Venette, Amsterdam, 1701.   Venette, who was "Docteur et Professeur du Roy et Doyen des Medecins de la Rochelle," took an especial interest in this subject on account of the fact that he himself suffered from stone in the kidney.   He states that he had observed that stones grew in caverns in the viscera of the earth, as well as in the bodies of men, and thinking that the cause was similar or identical in both cases, he determined to make a thorough study of the subject and to ascertain, if possible, the cause of the growth of such stones. Dr. Venette's studies led him to support the theory of the lapidifying juice. He with Shirley and all the older writers thought that they saw in nature some force or forces at work, which tended to produce stones not only in the human body, and in the bodies of many different animals, but also in the tissues of many trees and plants in the vegetable kingdom, as well as in at least three of the elemental zones which enfold the world. Whether stones were formed in the outer zone of fire, the "flaming ramparts of the world" as Lucretius calls it, is not known, but they believed that there was direct evidence that they were formed in the air, in the water, and in the lowest zone, that of the earth.

Hieronymus Savonarola,[39] following the view which Albertus Magnus expressed in his De Mineralibus writes:

Est deficile aliquid certi tradere de virtute & loco generationis mineralium. Non enim est dubium aliquam esse virtutem activam ad generandum & efficiendum lapides, sicut etiam sunt virtutes ad caetera generanda, sed haec virtus multum latet maxime quia ubique videntur lapides generari silicet in aere, in nubibus, in aqua, in terra, in plurimis locis & etiam in corporibus animalium, unde non habetur locus unus generationis eorum sed fere ubique generantur.

Stones then grew almost everywhere in nature and under the most diverse conditions. The places where they grew may be tabulated as follows:—

In the Animal Kingdom: (a) The human body, e.g. calculi, bones. (b) The bodies of animals, e.g. "Allectoria," "Bezoar," "Celonites," "Lyncis."

In the Vegetable Kingdom: In certain trees, e.g. Cacholong, Amber.

In the Zone of the Earth: In caverns, quarries and the open country, e.g. Stalactites, Aetites, Boulders.

In the Zone of the Water: In the sea and lakes, e.g. Corals, Pebbles and Sands on the shore, iron ore in the lakes, etc.

In the Zone of the Air: As Thunderbolts, e.g. "Ceraunius," Belemnite, etc.

1. Stones Growing in the Bodies of Animals

It is unnecessary to say anything further here concerning the calculi which develop in various parts of the human body since the character and results of these are only too well known to all, but we may pass at

---

[39] *Compendium totius philosophiae,* Venice, 1542, Book IX, p. 213.

once to consider briefly certain "stones" which were found, or which were supposed to be found in the bodies of certain animals. These stones attracted much attention from earliest times, chiefly on account of the mystical or medicinal properties which they were believed to possess. Aristotle,[40] writing of stones which grow in the bodies of animals, observed that:

> The story about the crane is untrue: to wit that the bird carries in its inside a stone by way of ballast and that the stone when vomited up is a touchstone for gold...
>
> Fishes do not thrive in cold places and those fishes suffer most in severe winters that have a stone in their head, as the Chromis, the Basse, the Sciaena, and the Braize, for owing to this stone they get frozen with the cold and are thrown up on the shore.

Many of these stones are mentioned by Pliny as being widely known among the Romans in his day, or as being mentioned in the writings of the older Greek writers whose works have long since disappeared.

With the advent of Arabian learning the Arabian physicians introduced into medieval Europe another of these stones unknown to the Greeks or Romans and which was believed to have extraordinary powers to counteract all sorts of poisons, and to be a specific against the plague. This was the Bezoar stone, which speedily acquired a reputation as an antidote that surpassed all the others.

Camillus Leonardus in his Speculum Lapidum (1505) describes some fifteen of these "stones" which were found in the bodies of animals, and others not on his list are mentioned by other writers. The names of those described by Camillus are set forth in the following list:

Alectoria—in the intestines of capons
Asinius—in the wild ass
Aquilinus—"in a certain fish"
Bezoar—in an animal resembling a goat
Chelidonius—in the stomach of a young swallow
Cimedia—in the brain of a fish of the same name
Draconites—in the head of a dragon
Doriatides—in the head of a cat
Hyena—in the eye of a hyena
Kenne—in the eye of a stag
Lyncurius⎫
Lyncius ⎬—in the urine of the lynx
Limacie—in the head of an animal of the same name

[40] *Historia Animalium*, Book VIII.

Margarita—in the pearl oyster
Radaim—in the head of a cock
Vulturis—in the head of a vulture.   This is sometimes called Quadratus.

Of these stones mentioned by Camillus, the Bezoar was the most re-nowned.   It was found in the stomach of a mountain goat, or goat-like animal, that lived in the mountains of Persia and India, this variety was known as the Oriental Bezoar.   In 1568 Petro de Osma went to Peru[41] (he calls it the "Island of Peru"), to seek this valuable stone in the Andes. He there found an animal resembling that which had its habitat in Persia and India, and in its stomach found bezoar stones which came to be known as occidental bezoar.   These bezoar stones varied in color and

FIG. 12. Bezoar Stone—cures cases of poisoning.   (From *Hortus Sanitatis.*)

were generally smooth and round in shape.   They were usually hard but sometimes more or less spongey, in character being conposed of concentric layers of some secretion from the body of the animal, usually deposited about some minute foreign body which served as a nucleus, and some of them contained a certain admixture of hair which the animal had licked off its hide, or little bits of other indigestible material which it had swallowed.

In the *Hortus Sanitatis* (1497) some interesting woodcuts accompany the brief descriptions of certain of these stones which grow in the bodies of animals.   Six of these are reproduced in figures 12 to 17.   Valentini in his Natur- und Materialien Kammer, Frankfort, 1714, also gives pic-

[41] See Johann Wittich, *Bericht an den Wunderbaren Bezoardischen Steinen*, Leipzig, 1592.

tures of certain animals each with "his stone" (see figures 18 and 19).

Bezoar was considered to be a most powerful antidote to all poisons and a specific against the plague which ravaged Europe in these times,

FIG. 13. Vulturis or Quadratus, from head of a vulture.   (From *Hortus Sanitatis*.)

FIG. 14. Bufonites, from head of a frog.   (From *Hortus Sanitatis*.)

but the name gradually came to take on a wider significance, in fact, the name bezoar came to be a general designation for an antidote to poison.

The oriental bezoar stones brought from Persia were the most highly valued of all and sometimes sold for ten times their weight in gold.   They

were considered as suitable gifts for royalty and three specimens were
presented by the Shah of Persia to the Emperor Napoleon, a little over
a century ago.   Owing to the high price of these stones, imitations of

FIG. 15. Kenne, from eyes of a stag.   (From *Hortus Sanitatis*.)

FIG. 16. Alectorius, from intestines of a capon.   (From *Hortus Sanitatis*.)

them were often made and sold as specimens of the true stone.   Many
of the old writers mention these fraudulent imitations and give directions
as to how they may be recognized.   On May 25th, 1630, at a Court of

the "Master, Wardens and Society of the Art & Mystery of the Apothecaries of the City of London" certain "pretented Bezear stones sent by the Lord Mayor to be viewed were found to be false and counterfeit and fitt to be destroyed and the whole table (court) certified the same to the Lord Mayor and adjudged the stones to be burnt."[42]

Later on, in 1715, Frederich Slare,[43] a Fellow of the College of Physicians and also a Fellow of the Royal Society of London, in a book which he dedicates to the Royal Society, sets forth the results of an extended investigation which he had carried out to determine the character and medicinal properties of bezoar stones. In this he shows that in many cases at least the claim that they have been taken from the stomach of

FIG. 17. Draconites, from head of a dragon.   (From *Hortus Sanitatis*.)

animals is untrue, that they have no medicinal properties and that those who buy them waste large sums of money in "Exchanging good silver for clay and dirt."

A number of these stones which were reputed to have been formed in the bodies of animals or men have curious legends related concerning them. One of the strangest of these stories is that connected with the lapis cervinus sometimes called Kenne or Keme (see figure 19). These

[42] C. J. S. Thompson: *The Mystery and Art of the Apothecary*, London, 1929, p. 186.

[43] *Experiments and Observations upon the Oriental and other Bezoar Stones, which prove them to be of no use in Physick*, London, 1715.

FIG. 18. The oriental bezoar and his stone. (From Valentini: *Der Volständigen Natur- und Materialien Kammer.*)

FIG. 19. The stag and his tear. (From Valentini: *Der Volständigen Natur- und Materialien Kammer.*)

were believed to be the tears of stags hardened into stones.   Pliny[44] says that there is no one who is so ignorant that he does not know that there is an intense enmity between stags and serpents.   The stags kill and eat the serpents, for they know that thereby they can greatly increase their strength and renew their youth, but having done so they are always seized with an intense thirst.   With this preliminary introduction we may let Bauhin[45] take up the tale:

When now the stag has devoured the flesh of those serpents, driven mad by thirst, it darts forward at full speed to find some pool or river in which to plunge itself, that by the cooling water the mighty heat which has seized upon it after devouring these noxious animals may be assuaged.   It remains standing in the water, drinking nothing, until that burning heat has been tempered and passed away.   Standing immersed in these waters, taught as it is by nature, like Tantalus thirsty in the midst of the billows, it does not drink, for if it should taste the least drop of water, it would fall dead upon the spot.   In the meantime tears slowly ooze forth from its eyes, which little by little grow thicker in the corner of its eyes, are congealed there into stones the size of a chestnut, or of an acorn.   When it feels itself relieved of the poison, stepping out of the water, it turns aside into its own lairs, and to remove the stones which are an obstacle to vision it rubs its head against the trees, or as others say, in the act of stepping out from the river, the stones fall from its eyes to the ground.   When they are found the merchants of Sicily and of the East sell them at a high price as an efficaceous remedy against any poison.   For this is the Belzahard, that is to say, "The Antidote Stone," held in such high esteem among those who possess it, that they have not the slightest fear of any poison whatsoever.

The story was a very favorite one in the Middle Ages, but in the seventeenth century its authenticity was commencing to be doubted. A very quaint woodcut in which the stag is shown shedding his stony tears appears in the *Hortus Sanitatis* (see figure 15).

Another stone about which there is an interesting tradition, although it is not mentioned by Camillus, is bufonius or bufonites (see figure 14). The belief was that this stone grew in a toad's head.   It is to this bufonites that Shakespeare refers in "As you like it" when Duke Frederick says:

> Sweet are the uses of adversity
> Which like the toad ugly and venomous
> Wears yet a *precious jewel* in his head.

Ermolao Barbaro the commentator on Pliny (1495) observes that it is not mentioned by any of the ancients, that one may look in vain for

---

[44] *Natural History*, Part 2, chap. viii, p. 321 (Holland's translation).
[45] *De Lapides Bezaaris*, Basle, 1625, p. 33.

a reference to it in the works of Pliny, Galen or Dioscorides, and that it was an invention of writers in the Middle Ages, among whom was Albertus Magnus and others of his time. The statement that it is not mentioned by Pliny is apparently however not quite correct for this writer speaks of a stone which he calls Batrachitae, which seems to be the same stone as bufonites, stating that it bears this name because it has the color of a frog, but he does not say that it came from a frog's head. Sir Ray Lankester in a chapter entitled *The Jewel in the Toad's Head* in his *More Science from an Easy Chair*, London, 1920, writing of this stone says, "It is commonly called Bufonius but was also known as Borax, Nosa, Crapondinus, Crapandina, Chelonitis and Batrachites. It was also called Grateriano & Garatronius after a gentleman named Gratterus, who in 1473 found a very large one reported to have marvelous power."

Like the true bezoar it was believed to be a specific against poison and also valuable in the treatment of epilepsy. Aldrovandus[46] referring to this stone and its several varieties as described by certain writers, says that in order to ascertain whether such stones really did exist in the heads of toads he had made a somewhat extended personal study of the subject and had killed many frogs and examined their skulls. These he found were soft at the time the animal was killed but after a time acquired a stony hardness, as he observes many other soft things do in nature. He concluded that the bufonites is merely a hardened portion of the frog's skull. As a matter of fact, says Lankester, there is no stone or "jewel" of any kind in the head of the common toad or of any species of toad, common or rare. In the *Hortus Sanitatis* there is a picture of a man extracting one of the "jewels" from the head of a gigantic toad (see figure 14). Sir Ray Lankester quotes a physician[47] who, writing in *Notes and Queries*, says that it was commonly believed that these stones were thrown out of the mouths of old toads (probably the tongue was mistaken for the stone) and that if toads were placed on a piece of red cloth they will eject their "toad stones" but will rapidly swallow them again before one can seize the precious gem. This physician goes on to say than when he was a boy he procured an aged toad and placed it on a red cloth in order to obtain possession of "the stone." He sat watching it all night but the toad did not eject anything. "Since that time" he says, "I have always regarded as humbug ("badineries") all that they

---

[46] Ulysses Aldrovandus, *Musaeum Metallicum*, Book 1, p. 814. Bologna, 1648.

[47] *Notes & Queries*, 4th Series, Vol. VII, 1871, p. 540.

relate of the toad stone and of its origin." He then describes the stone which passes as the toad-stone and this description, Lankester says, exactly corresponds with the toad-stones which are well known in the collections of old rings. Lankester says further that he has examined a number of these old rings preserved in the British Museum and in the Ashmolean Museum at Oxford, two are of calcedony with the figure of a toad roughly carved on the stones and the others, slightly convex "stones" of a drab color, are the palatal teeth of a common fossil fish called Lepidotus.

It is interesting to note in this connection that some animals swallow stones which, remaining in the stomach, serve to aid digestion. That this is also true of some ancient animals long since extinct is mentioned in the following note by Sir Arthur Smith Woodward, which appeared in the *London Times* of July 21st, 1931, concerning certain specimens in the British Museum. Referring to a fossil plesiosaurian presented to the Museum by the late Mr. Thomas Codrington, he writes:

It was the first specimen of a fossil reptile in which stomach stones had been noticed. It is a fragment of an extinct sea reptile, a Plesiosaurian, not a crocodile, and near it may be seen in the museum the complete stomach contents with stones of another Plesiosaurian discovered by the late Mr. Alfred N. Leeds in the Oxford clay near Peterborough. In the same museum there is a very fine specimen of one of the oldest known crocodiles from the Lias of Wurtemberg showing stones mixed with the contents of the stomach. The habit of swallowing stones to aid digestion among reptiles thus dates back to the earliest times.

2. Stones Growing in Plants

The best known examples of these is Tabasheer, an opaline material found in the joints of the bamboo and in the East held in high esteem for medicinal purposes.

According to the *Hortus Samitatis* the mineral onyx is of vegetable origin and is derived from the tears of the tree called onica, which when they harden turn into onyx and on this account an onyx gives a fragrant odour when placed in a fire. The *Hortus* goes on to say that "the stone is frequently seen to be adorned with wonderful pictures which it explains had been painted while the tears were still soft and plastic."

3. Stones Which Fall from the Air

Another class of stones in these ancient times were believed to grow in the air and to fall from the skies, especially during storms, or on dark nights. Among these *Glossopetra* or *Glossopetrus* may first be mentioned. The earliest reference to this stone so far as is known is in a passage in Pliny's *Natural History*, which runs as follows:—

> Glossi-petra resembleth a man's tongue and groweth not upon the ground but in the eclipse of the Moon falleth from heaven and is thought by the magitians to be very necessary for pandors and those that court faire women: but we have no reason to believe it, considering what vain promises they have made otherwise of it; for they bear us in hand that it doth appease winds.[48]

From this passage it is evident that glossopetra was known long before Pliny's time and that it was believed to fall from the sky on moonless nights.

FIG. 20. The tree Onica whose tears harden into the mineral onyx. (From the *Hortus Sanitatis*.)

Isidore[49] and Camillus Leonardus[50] respect Pliny's statement, adding that the magicians, in their art, attribute magical powers to these stones and hold that by means of them the occult influences of the moon may be evoked. Boccone says that in Ottranto it is called the Thunder Stone ("Pietro del tuono") because of the belief that it fell from the sky during thunder. Cardan says "Id est fulmen ipsum." Benedictus Mazotta[51] goes still further and states that this stone which falls with

[48] Book XXXVII cap. 10 (Holland's translation, London, 1634).
[49] *Etymologiae*, Book XVI, cap. 15.
[50] *Speculum Lapidum*, Book II.
[51] *De Triplica Philosophia*, Bologna, 1653, Book 1, p. 23.

lightning has the shape of a tongue because this is the form of the flash of fire which encloses it.

These fossils present rather a striking shape being triangular in form, flat, and with a hard, smooth and polished surface, usually showing minute crenulations along the edges on two sides, while on the third side the stone is rough, unpolished and thicker than elsewhere. On account of their shape they were usually called, and frequently believed to be, serpents' tongues, birds' tongues, or ducks' tongues, which had been turned to stone. Aldrovandus,[52] De Boote[53] and others, however, believed that these objects neither fell from the sky nor were parts of animals which had been petrified but that they were *"sui generis,"* that is to say, they were minerals which grew in the earth where they were found. An elaborate academic discussion of the whole question of the nature and origin of glossopetra may be found in a special work on this subject by Reiskius,[54] who reaches the same conclusion as Aldrovandus and De Boote.

The earliest figures of glossopetra which appear in geological literature are those given by Conrad Gesner in his *De Rerum Fossilium, Lapidum et Gemmarum maxime figuris et similitudinibus Liber*, published in 1565. The drawings are poor, as are most of those in this quaint old book, but they are interesting as coming from the first work on geology in which illustrations were employed to elucidate the text. Several varieties are figured. Gesner mentions that in his specimen marked A the edge shows distinct serrations and that this and specimen B resemble the teeth of fishes, while C1 is more like a bird's tooth. In C2 he says the root is wanting, while C3 is called by some people "Cornu Serpentis" or a serpent's horn and is taken from a picture sent to him by his friend the very learned doctor of Nuremberg, Georgius Sittardus.

One of the best known localities for glossopetra was the island of Malta, where St. Paul was shipwrecked on his way to Rome and where, as he and his companions were making a fire to warm themselves, a viper fastened itself to his hand and was by St. Paul shaken off into the fire. Reiskius tells us that in his day it was believed by the inhabitants that after the event the Apostle by a curse deprived all the snakes on this island of their venom, thus making them quite harmless, and that the glossopetrae are the teeth of these snakes which were turned into

[52] *Musaeum Metallicum*, Bologna, 1648, p. 601.
[53] *Gemmarum et Lapidum Historia* (edited by Adrianus Tollius), Leyden, 1647, p. 340.
[54] *De Glossopetris Luneburgensibus*, Leipzig, 1684.

stone by a further act of the Apostle, as other snakes were in Britain at a later time by St. Hilda. Hence these glossopetrae also came to be known as "Linguae Melitenses" or "Linguae Sancti Pauli."

Another version of this tradition is given by Marie Pompée Colonne,[55] namely that when St. Paul cursed the serpents of this island, after he had been bitten by one of them, the country was as a result no longer able to produce these creatures, and that nature "pour se divertir" and in memory of the great miracle which had been worked in Malta by the Apostle, gave birth within the earth to great numbers of tongues and eyes of these animals modelled in stone. Reiskius reproduces two statements issued at Malta in 1654 "permissu superiorum" certifying these statements of Colonne to be true and recommending the "Lapillus S. Pauli" to the faithful and others owing to the miraculous powers of healing with which they were invested.

The resemblance of glossopetra to the teeth of certain sharks had, as already mentioned, been noted by Conrad Gesner as far back as 1565 and many who refer to this fossil during the next hundred years note the same fact.

Paul Boccone, "Gentilhomme Sicilien" in his *Recherches et observations naturelles*, Amsterdam, 1674, a work dedicated to Cosmo de Medici, refers to the view held by many, that glossopetra is a stone which grows in the earth like other minerals and that the specimens of various sizes represent their several stages of growth, but goes on to say that he inclines to the opinion that they are for the most part the teeth of the Carcharias or dog fish, or of allied species of fish, on account of their close anatomical resemblance to these, although he could not be absolutely certain of this since while the jaw bones of these animals are found in the deposits with glossopetra, he had not been able to find a specimen in which these were actually in place in the jaw. A picture from his book showing the appearance of a typical glossopetra and a shark's jaw with its teeth in place is reproduced in figure 21.

Steno[56] in his celebrated work entitled *De Solido intra Solidum naturaliter contento dissertationis Prodromus*, refers to an objection to this explanation of their origin. He says:

One great difficulty is presented by the vast number of teeth which every year are brought out of Malta, seeing that almost no ship goes thither but it brings back some

[55] *Histoire naturelle de l'Univers*, Paris, 1734.
[56] Florence, 1669, p. 61.

of them.   For my part I can find no other answer to that difficulty but this: 1. That
such dog fish have each six hundred teeth and more: and that as long as they live
they grow new teeth.   2. That the sea driven by winds is wont to thrust those bodies
it meets with, toward some one place and there heap them together.   3. That dog
fish come in shoals and consequently the teeth of many of them may have been left
in one place.   4. That in the lumps of earth brought from Malta, together with the
teeth of numerous dog fish, there are found shells of various molluscs, so that should
the number of teeth incline one to attribute their production to the earth, yet the
structure of these teeth and the large number of them in each animal, as well as the
resemblance of the earth itself to that which is found on the sea bottom and the
other marine bodies found in the same place, all alike favour the contrary opinion.

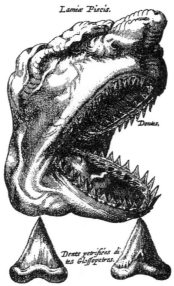

FIG. 21. Glossopetrae.   (From Boccone: *Observations Naturelles*.)

Further evidence by investigators served to establish the fact that
glossopetrae were in reality the teeth of fossil sharks.

The great demand for these objects was due to the wide-spread belief
that they possessed great medicinal value and magical powers, especially
in rendering those who used them immune to poison, or in acting as an
antidote in the case of persons who had already been infected with
poison of any sort.   For this purpose fragments of them might be worn
in rings, provided they were inserted in the ring in such a way that they
came in contact with the body of the wearer.   Or they might be hung

about the neck or bound upon the arm of the person. Or again they might be reduced to powder and taken internally mixed with wine, water or some other liquid; all this is set forth "permissu superiorum" in the documents which were issued in Malta in 1654, and to which reference has been made, as well as by many writers on this subject both before and after this date.

But the times were changing and old opinions were fading away, among men of education at least, and so Leibnitz after a careful summing up of the views which had been held concerning the virtues of glosso-petra and the evidence that could be induced in support of these views, concludes that the most useful purpose which this much prized object can subserve in connection with the human body, is as a dentifrice, seeing that when reduced to a fine powder the hardness and sharp-ness of the constituent grains gives to it a special value for this purpose. So that this mysterious "stone" which was supposed to have dropped from the heavens on dark moonless nights imbued with strange myste-rious powers, has indeed fallen from its high estate.

Another fossil which was generally believed to have fallen from the sky was Belemnites, which on that account was often described as a thunderbolt. As it bore some resemblance to the human finger and was found on Mount Ida, many of the older writers referred to it as Dactylus Idaeus. It occurred very abundantly in many localities in the shaly beds of the Lias, which disintegrated readily when exposed to the action of the atmosphere, leaving the belemnites which it contained scattered about in great numbers. The unusual shape of these fossils, resembling the sharply pointed head of a spear or arrow, together with the prevalent idea that they were found in especial abundance in places which had been struck by lightning, gave rise to the belief that they had fallen from heaven. So far as can be gathered from the perusal of the accounts of them given by many writers, the idea that they were connected in some way with lightning seems to have been suggested by the translucent character and pale yellow color of certain belemnites, "resembling Falernian wine." With the approach to modern times this fantastic idea was gradually abandoned. Some writers came to consider them as probably the teeth of great fishes, Langius[57] writing in 1708 expressed his view that they were a kind of stalagmite on account of their long, narrow and rounded shape and their concentric structure. It was demonstrated, however, in relatively recent times that they were the

[57] *Historia Lapidum figuratorum Helvetiae*, Venice, 1708, p. 133.

hard inner support of a certain extinct type of cuttle fish analogous in a general way to the "bone" in the cuttle fish of modern times.

Brontia and Ombria are really fossil echinidea, or sea urchins, but were also believed by many of the ancients to have fallen from the skies. Pliny[58] refers to them in the following terms:—

Brontia is shaped in manner of a Tortoise head: it falleth with a crack of thunder (as it is thought) from heaven; and if we will believe it, quencheth the fire of lightning. . . . Ombria, which some call Notia, is said to fall from heaven in storms, showers of raine, and lightning, after the manner of other stones, called thereupon Ceraunia and Brontia: and the like effects are attributed to it.

Pliny here is evidently setting forth an opinion held by many about his time concerning the truth of which however he himself has grave doubts.

Agricola[59] classes brontia and ombria together and says that the common people consider them to have been cast down upon the earth's surface by lightning when they are called brontia or in rainstroms when they are called ombria.   Later in Germany brontias were called "Donnerkeile" and the ombrias "Regensteine."   His description of the common ombrias show that they are echinidea of the type of Cidaris.   Gesner[60] gives figures of two specimens of ombria sent to him by Kentmann which are the first that were published.

Plot[61] referring to these says:

After the stones some way related to the *Celestial Bodies* (e.g. Asteriae or Star Stones, meaning six-sided elements of the stalks of crinoids) I descend next to such as (by the vulgar at least) are thought to be sent to us from the *inferior Heaven*, to be generated in the clouds and discharged thence in time of *Thunder* and *violent showers:* for which very reason and no other that we know of, the ancient *Naturalists* coined them suitable names, and called such as they were pleased to think fell in Thunder, Brontiae, and those which fell in Showers, by the name of Ombriae.

Ceraunius (or Ceraunia).   This name, derived from the Greek word meaning thunder or lightning, is employed by Pliny to designate four different kinds of stones, of which the last is one which he says is highly prized by the Parthian magicians and which is found only in places which have been struck by lightning.   To this class belong those stones to which this name is given by later writers who believed that the stones

[58] *Natural History*, Part 2, chap. XXXVII, p. 10 (Holland's translation).

[59] *De Natura Fossilium*, Book V.

[60] *De Rerum Fossilium . . . Figuris*, p. 62.

[61] *The Natural History of Oxford-shire*, Oxford, 1705, p. 91.

in question had fallen with the lightning from the heavens. Some of these writers divide the carauniae into two classes—the first comprising the stony, and the second the metallic varieties. To the cerauniae undoubtedly belong the true *meteorites*, as they are now called, many of which have been actually seen to fall from the heavens.

In this class, it has been conjectured, may be included certain shapeless stones worshipped in ancient shrines and reputed to have fallen from heaven, as for instance, that in the temple of Cybele at Pessinus and "the image of the great Goddess Diana which fell down from Jupiter" in the temple of Ephesus. It may be noted in connection with the latter that

FIG. 22. Two stones falling from the heavens during a storm. The smaller pointed one is a glossopetra and has killed a man, the larger is a ceraunia and is splitting open a tree. (From Reisch: *Margarita Philosophica*.)

the expression "which fell down from Jupiter" as used in the Acts of the Apostles (Chap. 19, verse 35) gives a wrong impression of the meaning of the original Greek text, the words used merely indicating that the "image" was believed to have fallen down out of a clear sky. In the eastern end of Kaaba in Mecca on the inside of the wall about 5 feet from the ground, the "Black Stone" is embedded. It is probably also an aerolite which before the time of Mahomet flashed through the atmosphere and fell in the vicinity of an encampment of Arabs, who took it up carefully and made it, if not their god, at least an object of extreme reverence.[62]

[62] Eldon Rutter, *The Holy Cities of Arabia*, New York, 1928.

Another and even more interesting class of stones which were believed
to have fallen from the heavens were those known in Germany as the
Donnerkeil, Donneraxt (thunder axe), Stralhammer, Stralkeil, Stralpfil.

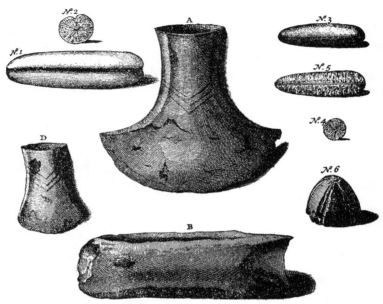

FIG. 23. Cerauniae.   (From Rumphius: *D'Ambonische Rariteitkamer*.)

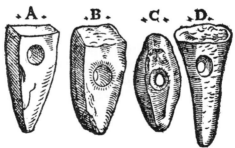

FIG. 24. Cerauniae or thunder stones.   (From Gesner: *De Rerum Fossilium ... Figuris*.)

These were peculiar in having a more regular shape than the others to
which reference has just been made.   Some have an elongated oval form,
many others are pyramidal in shape, while others again resemble small

hammers or axes. Still more remarkable and puzzling was the fact that many of these latter were perforated with a hole "wo sein aequilibrium ist"[63] as Valentini says, and which looked as if it had been intended for the insertion of a handle. These objects were in some cases made of a hard and tough stone and in other cases were of iron or some other metallic substance. Kentmann[64] describes three Donnerkeils and a Donneraxt which form part of his collection of "fossils." One of the former he says is stated to have fallen from the heavens at Vienna in 1544 and was by the force of the lightning flash which accompanied it, driven through a house and buried to a depth of twelve ells in the earth beneath. Albinus[65] mentions certain others which fell in Saxony. Among these was one

which fell when in 1561, on May 17th a bluish black Donneraxt, shaped like a hammer, five fingers long and almost three wide and pierced with a round hole, was driven through a windmill at Torgaw. It was harder than the Stolpische Stein or basalt. Again another blackish gray Donnerkeil was found some years ago at Eilenburg which in its fall split to pieces a great strong oak tree. And once again a pointed light gray Donnerkeil fell at Siptitz a village near Torgaw, and also split an oak, which stone the farmers dug out and brought to the tax collector at Torgaw.

It has been suggested that the belief which obtained in Germany that these "Donneraxten" had fallen from the heavens received some support from the fact that the god Thor was generally represented as armed with an axe. It is to be noted, however, that Pliny mentions Ceraunia having this form. Gesner[66] also describes these objects at some length, giving what are probably the first figures of them which were ever published, including among them the Donneraxt which fell at Torgaw from the collection of his friend Kentmann.

No mention is made of these objects falling from heaven before the time of Avicenna. It was he who first introduced the idea of their celestial origin into Europe. It came, with so many other fables and superstitions from Arabian sources, and with that great stream of Arabian learning which flowed into Europe in the latter half of the Middle Ages and which did so much to vivify Europe and to usher in the Renaissance.

These cerauniae came to be considered to be possessed of certain supernatural powers. They were carried about on the person or kept

[63] *Natur- und Materialien Kammer*, Frankfort, 1714, p. 54.
[64] *Nomenclaturae Rerum Fossilium*, Zurich, 1565, p. 30.
[65] *Meissnische Bergchronika*, 1590, p. 153.
[66] *De Rerum Fossilium . . . Figuris*, Zurich, 1565, p. 62.

in houses to ward off lightning or danger from storms. They were also believed to effect cures in the case of certain diseases, when taken internally admixed with the proper menstrua. Since these stones possessed such remarkable powers it was of great importance to be able to distinguish those which were genuine from those which were spurious. The following method of determining this, is set forth in a curious passage in Gesner. If the stone to be put to the test, he says, is taken and wound round about with a fine thread so that no part of the stone is visible and the thread is nowhere doubled, and the stone thus covered is laid in hoar frost, the thread will grow damp if the stone is genuine.

As time passed on the wonderful stories concerning these various stones which were supposed to have fallen from the heavens underwent a progressive change, and the truth concerning their origin gradually revealed itself. It is interesting to trace the ebb and flow of opinion in the works of the writers of the sixteenth, seventeenth and eighteenth centuries. The controversy centered itself on the question of the origin of those metallic bodies which bore such a marked resemblance to objects of human workmanship and which are now known to be the tools and weapons of primitive man. Cardan, Aldrovandus and De Boot doubt whether these things fall from the heavens, while other contemporary writers state without hesitation that they do.

Agricola who was a contemporary and a fellow countryman of Kentmann and Gesner, but who stands out from his contemporaries as one in whom the spirit and methods of modern science were already beginning to manifest themselves to a marked degree, scouts Avicenna's statement that the Arabs were of the opinion that the best German sword-blades were forged out of these iron "Donneraxten." Agricola in referring to this assertion says:

This fable concerning the German swordblades is one which has been imposed upon the credulous Arabians by self-seeking traders. In Germany it is well known that iron does not fall from heaven but that we dig it out of the earth.[67]

Valentini[68] thinks that it is much more probable that these so-called Donnerkeils, like other stones, have their origin in the earth and that they owe their peculiar forms to the fact that they are "sports of nature." König[69] in discussing the subject expresses a certain indignation at the

[67] *De Ortu et Causis Subterraneorum*, Basel, 1546, Book V, p. 78.
[68] Op. cit. 1714.
[69] Emanuel König, *Regnum Minerale*, Basel, 1703, p. 244.

manner in which the "New Philosophy" deals with this and allied questions.   He says:

> It is difficult to decide how it is that the Ceraunii have forms identical with the various objects made by man . . . What do these forms signify unless "Archaeus sive spiritus fulminans" fashions and makes them for his own pleasure out of the metallic or stony material drawn up from the earth into the clouds?   But because the new philosophy hisses off the stage both these forms and the Archaeus, it teaches that they have their origin in the fortuitous concourse of their constituent particles.   If nature in the lightning clouds makes stones resembling various instruments which fall to the earth, does it not in the same way make for its pleasure pictures of our various domestic things in agates and other gems, and these same figures in different places and at different times, a fact which forsooth dispels the explanation that they are formed by the fortuitous concourse of matter.
>
> Many people deny that Ceraunii are generated in the clouds but say that they come into existence on the earth where lightning strikes and are made of the aggregation of sand grains, as stones are.   Among these is Agricola (De Ortu Subterraneorum, book V) where he ridicules the credulity of Avicenna in writing that in Persia there fall from the heavens with bright flashes of light, brazen bodies resembling barbed arrows which will not melt in a furnace.

And finally John Woodward[70] writing of these objects says:

> I come next Sir! in pursuit of your commands to give some account of the Stone Weapons and Instruments—Now tho' these carry in them so plain tokens of art and their shapes be such as apparently point forth to any man that rightly considers the use each was destined to, yet some writers of fossils, and of great name too, have been so sanguine and hasty, so much blinded by the strength of their own fancy and prepossessed in favour of their schemes and notions, that they have set forth these bodies as natural productions of the Earth under the names of Cerauniae.

By the middle of the eighteenth century, however, the new philosophy had so far prevailed that these mysterious "Donnerkeile" and "Donneraxten" came to be universally recognized to be what they are, implements of human workmanship, the tools and weapons of primitive man.

At the opening of the nineteenth century the question of the origin of meteorites was a subject which was being very actively discussed.   Four views were held each of which was supported by a group of men of learning and distinction.

These were as follows:

1. That meteorites were of terrestrial origin and had been thrown up into the heavens by volcanoes or by hurricanes.   Laplace suggested

[70] *Fossils of all Kinds Digested into a Method suitable to their Mutual Relation and Affinity,* London, 1728.

that they might have been ejected by the Lunar volcanoes, while Biot and Poisson submitted mathematical demonstrations proving that fragments of rock might be projected from the moon on to the earth's surface.

2. That they were produced by the fusion of terrestrial material through the action of lightning in the places where they were found.

3. That they were concretions formed in the atmosphere.

4. That they were masses of matter which were strangers to our planet.

To sum up then, these stones which were believed to have fallen from the heavens were recognized as being very varied in character. Some of them, such as the sea urchins, belemnites and glossopetra are now known to be the remains of fossil animals and are of purely organic origin. Others were stone or metal implements or weapons fashioned by men of the stone, bronze or iron ages, while still others were true meteorites which really did fall from the sky, and are of extra-terrestrial origin. Concerning the last mentioned class, while modern research has immensely increased our knowledge of their nature and origin: the words of an inscription which is said to have been engraved on a meteorite which fell at Ensishem on November 7th, 1492, and which was placed in the church of this town, are still true: "De hoc multi multa, omnes aliquid, nemo satis"[71] (Concerning this stone many have said much, everyone something, no one enough.").

4. *Stones Which Grow in Water*

Thales of Miletus, born 640 B.C., and who is said by Lactantius to be "The first of those who enquired after natural causes," taught that water was the first principle of all things and that it could be, and in nature was, converted into earth. Van Helmont (1577–1644) was one of the ablest supporters of this dogma and Boyle believed it. It was not until 1770 that Lavoisier in an elaborate paper communicated to the Academie des Sciences showed that the various proofs which had been adduced as evidence of this transmutation were fallacious and that there was no shred of evidence for the opinion that water can be converted into rock.

---

[71] Joseph Izarn, *Des Pierres tombées du ciel ou Lithologie atmosphérique*, Paris, 1803, pp. 328, 421.

Conrad Gesner, *De Rerum Fossilium . . . Figuris*, Zurich, 1565, p. 66.

That certain stones grew in the water was known and widely commented upon from very early times and especially in those wonderful "petrifying waters" which issued from certain springs and seemed to turn to stone all objects with which they came in contact. There is an endless repetition in book after book of stories about these springs—of the bird which Strabo tells about, which flying over one of them lighted on the surface of the pool and splashing the water over its wings was turned to stone, and that other story of the Emperor Frederick who visited one of these springs and who, in order to test its petrifying qualities, drew off his glove, and having made a mark upon it, ordered that it be immersed in the water up to this mark, and after two days it was found to be turned into stone where the water was in contact with it while elsewhere the original leather remained unchanged.

The fact that in many places masses of stone could be seen to be growing up from the bottom or to be in the course of deposition on the banks of lakes or streams, led to the belief that the water itself was turning into stone.

Water itself can turn into stone, either congealed by cold in falling drop by drop, or else solidified by the mineralizing virtue.[72]

Or again:

Around the island of Elandia in the Getic sea huge masses of marble are produced in the sea itself so that the island seems to be surrounded by columns and pyramids built by hand.   Whence it appears that even in[73] sea itself marble grows "sui generis."

Ovid also says, "The Cicones have a river whose waters, if drunk, turn the vitals into stone, make marble of everything they touch.

But it was in great caverns that this growth of limestone, deposited from waters dripping from the roof, was seen in its most striking form, as stalactites pendant like icicles from the roof, or as stalagmites growing upwards from the floor, "dripstones" as they came to be called in later German and English geological literature.   The curious imitative shapes often assumed by these great masses, taking the form of churches, or temples, animals or the human form, as seen by the uncertain flare of the torches which were used to explore the dark recesses of the caves, suggested the existence of some mysterious formative power at work in the dark recesses, such as was thought at the time to be active in other realms of the world of nature.

[72] Gregorius Reisch, *Margarita Philosophica*, Strasburg, 1504, Book IX.
[73] Hieronymus Cardan, *De Rerum Varietate*, Basel, 1581, p. 162.

Thus the Abbe Passeri[74] in a letter to a friend describing his visit to caverns in Monte Cucco near Gubbio, writes:

A short distance further on we found a little mound made of the whitest alabaster and a number of other bodies having the form of cylinders or sharp cones rising from the floor or growing out from the walls in the form of branches or knobs, white and transparent. Some of these bodies were of immense size and had the form of rude columns so that without much labour they might be chiseled into truly artistic shape. I broke off some of the smaller ones and found that while all of them were formed of the same species of alabaster, some were pierced with a little hole from top to bottom . . . . Another had the form of a fine torso of a statue. The walls were covered with a more opaque variety of this stone moulded into what looked like bunches of grapes, of which I send you a specimen so that you may see that it is composed of an impure and opaque alabaster made up of a succession of thin layers which could not have been produced by water but which drew their nourishment from the rock itself, as was also the case with the marble cylinders. I saw in other places immense masses of a purer variety of the alabaster in forms resembling the most beautiful fountains in our gardens. Others were like ernomous statues which presented little short of a perfect shape and which when a light was placed behind them were seen to be translucent. The cave contained so many of these wonderful forms that, Illustrious Friend, you cannot imagine how many castles in the air they conjured up. . . . What thoughts are called forth in one's mind by such a strange visit when one tries to reach a conclusion concerning the way in which nature had produced this immense cavern within a mountain. I looked about carefully to see if there was any ore in the cave but came to the conclusion that there was none. Neither could I see anywhere evidence of human handiwork . . . I cannot imagine how large caves of this kind come into existence when as you believe, such mountains are uplifted from the earth's surface, leaving these immense caverns beneath them.

He believed then that this stalagmitic deposit, displaying these curious and suggestive forms, was not a deposit from water at all but drew its nourishment from the rock itself and grew as plants do, from some hidden seed.[75] For, he says, at the creation, the command "Increase and multiply," was not given to mankind alone but to all things, which were originally created with their respective seeds and continue reproducing their respective species and shapes.

Pliny in his *Natural History* refers to *marble*, including under this name not only the varieties of the rock to which this term is now restricted, but also other rocks which are now known by other names. Among these was the rock now known as basalt, but which Pliny classed as

[74] *Della Storia de'Fossili dell'Agro Pesarese*, Bologna, 1775, p. 165.
[75] See also Tournefort, *Description du Labyrinthe de Candié avec quelques observations sur l'accroissement et sur la Génération des Pierres*. Mém. de l'Academie Royale des Sciences, 1702, p. 217.

black marble.   In fact, as has also been mentioned, the ancients following
Pliny gave the name marble to all rocks which were hard enough to be
used for building purposes and which could be cut and polished so as to
present a smooth, shining or glistening surface.

There is a curious passage in *Paradise Lost* (Book III) in which Milton
uses the word in its derivative sense applying it to the clear, sparkling
and lucent atmosphere, when Satan ascending into paradise:

> Winds with ease
> Through the pure marble air his oblique way
> Amongst innumerable stars.

This "black marble" or basalt was also believed by many to have been
formed from water.   For at the Giant's Causeway in Northern Ireland
and at other places it could be seen rising up out of the waters of the sea
in six-sided forms identical in shape, they say, with those displayed by
rock crystal which was at that time believed to be pure water frozen so
hard by the intense cold of the high mountains where it was found, that
it was impossible to thaw it out again.   Collini speaks of them as "Ces
colonnes de Basalte, qui doivent être regardées comme des Cristaux d'une
grandeur extraordinaire."[76]

In his *De Rerum Fossilium. . .Figuris*, Gesner gives a full page plate
representing a group of basalt columns, and states that the drawing
from which it was executed was given to him by Kentmann and so
strongly was Kentmann impressed by what he considered to be the
striking resemblance between quartz crystals and basalt columns, that
he has allowed his imagination or enthusiasm to carry him away and, to
make the resemblance more perfect than it is in nature, he has represented the individual basalt columns with pyramidal terminations (See
figure 25).   This idea that basalt was deposited from water gained at a
later time additional force and impetus under the teaching of the Wernerian school and was not finally dissipated until well on in the nineteenth
century.

That the sands on the sea shores were crystallized out of the waters of
the ocean was also at one time generally believed.   Marie Pompée
Colonne in his very interesting book entitled *Histoire Naturelle de*

---

[76] M. Collini, *Journal d'un Voyage qui contient différentes observations minéralogiques
particulièrement sur les Agates et le Basalte*, Mannheim, 1776, p. 340.   Among other writers
who described basalt columns as crystals may be mentioned Tremblay in 1656, Henckel in
his *Pyritologia* (1725) and Welch in his *Systematic Description of the Mineral Kingdom* (1764).

l'Univers which was published in Paris in 1734, has a chapter entitled
De la Génération de Sable. Accepting the old idea of the four elements,
fire, air, earth and water, he says that sand is merely a kind of salt in
which the element of earth is *dominant.* The sea, he says, produces
both salt and sand along its shores. Sand under the microscope looks
like glass, and glass is made by fusing sand and salt together. Sand in
fact is merely a salt which is "Un peu plus terrestre que le sel ordinaire."

This opinion that sea sand was chemically precipitated or crystallized
out of the waters of the sea, was also set forth by no less an authority

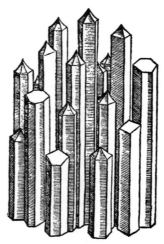

FIG. 25. Columns of Basalt. (From Gesner: *De Rerum Fossilium . . . Figuris.*)

than the great botanist Linnaeus, who in addition to his epoch-making
works on botany wrote an extended treatise on the mineral kingdom.[77]
He begins his treatment of the *Arenatae* or siliceous group with the
words: "They are formed from the most minute drops of water which
fall as rain from the upper air and grow together into bodies of a sand-
like material," and elsewhere "All still-standing fluids in nature gradually
crystallize into a fine sand-like material. Such crystalline material
when it is deposited from sea water is called *sand* when it is thrown up
on the sea shore and becomes dry."

[77] *Vollständiges Natursystem des Mineralreiches* (translated from the 12th Latin edition)
Nuremberg, 4 volumes, 1777–1779.

Curiously enough, the clastic origin of even pebbles and gravel was not recognized. Henckel[78] waxes quite eloquent in his despair at his inability to ascertain its origin:

O Caillou, caillou! Quelle est la matereil qui t'a produit? C'est la pierre la plus commune & par conséquant une de celles qui méritent le plus d'être examinées . . . mais plus il nous importeroit de le connoître, moins nous avons de phénomènes qui puissent nous faire remonter à son origine.

Je sçais seulement que nous n'avons aucun vestige, aucun exemple, aucune expérience qui puissent nous faire connoître qu'il s'en former de nouveaux, par conséquent son origine doit remonter aux tems les plus reculés & il la doit sans doute à des circonstances qui n'existent plus et que nous ne connoissons paz. (They must have been produced under conditions which are unknown to us and under circumstances which no longer exist).

Bourguet,[79] a writer who was a contemporary of Henckel, has much more definite views on this subject.

The primitive particles of which the earth was made must have been moving in a straight line. This movement, however, owing to the "diverses résistances qu'ils rencontrèrent les uns à l'égard des autres" would gradually develop into a circular motion forming "tourbillons." Great whirlpools of this primitive matter formed the sun, the earth and other planets, in the case of the earth it produced the concentric layers, seen in the earth's crust and which we now call beds or strata, the primitive particles settling down according to their specific gravity. Smaller whirlpools acting on soft materials gave rise to pebbles and gravel seen in the primitive rocks; there was still material which took the form of sand or finely crystalline material.

Even as late as the opening of the nineteenth century some doubts concerning the origin of pebbles and gravel still survived. De Saussures[80] writing in 1803 refers to these in a clear and excellent exposition of the whole question, and gives the true explanation of their origin. Pebbles, he says, occur in the valleys of rivers which run through the mountains where the rivers issue from them. But they are also found in places far from rivers where it would seem no waters had ever run, and even on the mountain slopes high above the rivers, as is the case also with the remains of certain animals which live in the sea. These facts

---

[78] *Idée Générale de l'Origine des Pierres, fondées sur les Observations et des Expériences* (translated from the German), Paris, 1760, p. 410. It is appended to his volume entitled *Pyritologie ou Histoire Naturelle de la Pyrite.*

[79] *Traité des Petrifications*, Pt. 1, Paris, 1742, p. 25.

[80] *Voyages dans les Alpes*, Paris, Vol. 1, p. 123; Vol. II, p. 107.

have led some people to say that pebbles found in such places must have fallen from heaven, or that they had come into existence in the places where they were now seen, for after all "Nature" could as easily produce rocks which were rounded, as those which are angular in form. But naturalists, De Saussure goes on to say, without desiring to limit the power of "Nature," were practically unanimous in declaring that these pebbles owed their shape to the fact that they had been transported to their present position by water and that they had been rounded by mutual attrition during the process. He adds, however, that there are some stones, such as geodes and calcareous concretions, which display a rounded shape and concentric structure, and which have quite a different origin having been deposited by water in successive layers in cavities or about some central nucleus. De Saussure further notes that careful observation shows that the gravel in river beds is identical in petrographical character with the rocks composing the mountains on either side of the valley, and that they can be seen, starting with fragments of angular form at the source of the river to become more and more rounded the farther they are carried down stream. On the other hand those deposits of gravel which are seen high up on the mountain sides were different in composition from the country rock on which they repose, they had evidently been carried from some distant point, and he attributed their rounded shape to the intense attrition to which they had been subjected during the rapid retreat of the waters at the time of the great "débâcle," when the oceans were drained away from the continents. We now know that the gravels to which he refers were of glacial origin and were formed during Pleistocene times. And so this problem of the origin of pebble deposits which so perplexed the early geologist has now been fully and finally solved.

Among those stones which had their origin in water none attracted more attention or were the subject of more discussion and difference in opinion among early writers than coral. To go back to the earliest writer on these subjects in Europe, Theophrastus,[81] we find that after describing sapphire and haematites or blood stone, he writes:

> To these may be added Coral for its substance is like that of stones. Its colour is red and its shape cylindrical in some sort resembling a root. It grows in the sea.

Then comes Pliny who regards coral as a sea shrub which is soft, bearing berry-like fruit which however, so soon as it is withdrawn from

[81] *On Stones*, Chap. LXVII.

the water becomes hard and turns to a red color. This same idea is expressed by Ovid,[82] who was a contemporary of Pliny: "So coral hardens at the first touch of air, whereas it was a soft plant beneath the water." It is also referred to by Ovid in Book IV of the *Metamorphoses*, where in the story of the release of Andromeda by Perseus, an explanation of the origin of coral is given as follows:

The maiden also now comes forward, freed from her chains, she, the prize as well as the cause of his feat. He washes his victorious hands in the water drawn for him; and that the Gorgon's snaky head may not be bruised on the hard sand, he softens the ground with leaves, strews seaweed over these and lays on this the head of Medusa, daughter of the Phoroys. The fresh weed twigs, but now alive and porous to the core, absorb the power of the monster and harden at its touch and take on a strange stiffness in their stems and leaves. And the sea-nymphs test the wonder on more twigs and are delighted to find the same thing happening to them all; and by scattering these twigs as seeds, propagate the wonderous thing throughout their waters. And even till this day the same nature has remained in corals so that they harden when exposed to air and what was a pliant twig beneath the sea is turned to stone above.

Dioscorides repeats the same story and contrasts coral with iron which when highly heated and plunged into water becomes hard, while coral is hardened by being withdrawn from the water of the sea and exposed to the air. Bartholomaeus Anglicus and Marbodus tell the same tale, while at the opening of the sixteenth century Camillus Leonardus[83] says that coral is a stone which grows in the sea in the habit of a bush without leaves. Aldrovandus[84] after recapitulating the views of the earlier writers says with his characteristic directness, "So far we have been dealing with silly prattle now let us consider the matter seriously," and goes on to state that corals on the sea bottom have a woody character, but being exposed to the action of the "lapidifying juice" present in the sea water, become petrified by this agent as they grow. Nichol[85] says that it is a plant which when growing in the sea has a green color but when brought into the air "Turns to a very noble and beautiful red, blushing at the injurious hand that offereth violence to its secret, tender and spreading growth." Guisonny[86] has no hesitation in stating that in his opinion "coral is a mere mineral composed of much salt and a little earth" which is precipitated from sea water and grows in a manner

[82] *Metamorphoses*, Book XV, v. 416.
[83] *Speculum Lapidum*, Book II.
[84] *Musaeum Metallicum*, Book III, p. 285, 1648.
[85] *A Lapidary or History of Pretious Stones*, Cambridge, 1652, p. 161.
[86] See Boccone, *Recherches et Observations Naturelles*, Amsterdam, 1674, p. 22.

which is closely analagous to that of the *metallic tree* which is produced when a globule of mercury is placed in a solution of silver nitrate. Anselmus de Boodt writing about the same time as Aldrovandus expressed the belief that the coral became petrified only after the death of the organism, which might take place either in the ordinary course of nature or because of an undue vehemence in the action of "lapidifying juice."

It is unnecessary to refer to the opinion of any other of these early writers, for while many refer to corals, their opinions all lie within the range of those already quoted, corals were true stones of a somewhat aberrant form, or imitative shapes, or they were plants, which either while alive or at their death became petrified.

In 1753, however, the Sieur de Peyssonnel[87] in a manuscript entitled *Traité de Corail* which in that year he presented to the Royal Society of London, showed that these beautiful objects were neither stones nor vegetables but animals, and that "Like oysters and other shell fish nature had impowered them to form themselves a stoney dwelling for their support and protection each according to his kind." The work of Peyssonnel, Jussieu and Ellis established beyond a question that corals belonged to the animal kingdom all appearances to the contrary notwithstanding, and a controversy which had lasted for at least two thousand years was at length brought to a close.

5. Stones Which Grow in the Open Country on the Earth's Surface

There are also in the writings of ancient authors many cases recorded which were believed to afford indisputable evidence that stones grow, or can be found growing, in certain places in the open country on the surface of the earth.

P. Francisco Lana, for instance, in an early volume of the *Philosophical Transactions of the Royal Society of London* (Vol. II, p. 465) in a short article entitled *The Formation of Crystals* tells of a visit which he paid to a certain tract of pasture in England in which there was a strip of the land on which the growth of grass was sparse and the soil less fertile then elsewhere. On it however by a careful search there could be found many hexagonal crystals doubly terminated by hexagonal pyramids, and which, from his description, were evidently quartz. He was told that these were produced by the condensation and crystallization of the dew at night time, for if on any evening all the crystals were gathered up from any given plot, next morning their place would be found to have been taken by others, providing the intervening night was one on which there was "a serene and dewy sky." He expresses the opinion, however, that they probably owed their origin to the coagulation of the dew by "Nitrous streams, for this is the natural coagulation of water and makes 6 sided figures in snow." It was to the presence of this nitrous material, he conjectured, that the semi-barren character of the land was due, and he goes on to state that having dug up some of the earth where the crystals grew he "Drew a salt from it which had the taste and figure of nitre, though some grains were square and others pyramidal."

The opinion that quartz crystals were merely ice frozen by such an intense cold that it was impossible to thaw it out again, was generally accepted by mineralogists in early times, a belief which is embodied in the name *crystallos* which was given to this meneral and which was derived from the Greek word meaning ice.

Pompée Colonne[88] gives instances where rocks can be observed to be growing out of the ground. The first of these is at the village of Fontainebleau. Here the surface of the country is strewn with blocks of rock, the largest being about the height of a man. He relates that while looking at these, a man standing near him said: "Sir, these rocks which you see have grown here as trees do and they have roots in the earth like trees," and when asked how he knew this, the man replied that in his portion of the country near Champagne there were certain quarries where a quartzose rock was worked. When the working was carried to a certain depth, no satisfactory provision being made for drainage, water filled the workings and prevented further work. But in the course of a period of five or six years the rock grew again so that it appeared above the level of the water and quarrying could be resumed.

The belief that rocks grow survives in the minds of some uneducated persons even to the present day. The writer encountered a curious instance of this when making a geological survey of the Haliburton-Bancroft area of Northern Ontario some years ago for the Geological Survey of Canada. The district is on the margin of the settled portion of the Canadian Shield and is of course heavily glaciated. He had engaged a local farmer to row him around a large lake for the purpose of examining the outcrops exposed on its shores and was returning at the

[88] *Histoire Naturelle de l'Univers*, vol. 1, pt. 2, 1734.

close of the day from the shore of the lake to the farmer's house. At one
point there was seen lying directly across the path, a long glaciated roche
moutonné of gneiss and perched on its smooth bare summit was an
enormous almost spherical boulder of gneiss which must have weighed
twenty tons. It formed a striking spectacle sharply outlined in the
level rays of the setting sun. Wishing to see how the strange sight
impressed the farmer, the writer said to him, "That is a curious rock,
how do you suppose it got up there?" The farmer at once replied "It
grew" (or rather he said "it growed"). "How do you know that?" the
writer replied. "Well," said the farmer, "I have lived here for fifteen
years and come up this portage frequently, and every year I can see that
the rock is a little bigger than it was the year before." He had been
struck by the remarkable sight and in looking for an explanation could
find no other than that the immense boulder had grown there like the
trees in the surrounding forest, and this idea having fixed itself in his
mind, he fancied that each time he saw it the boulder was somewhat
larger. As a matter of fact it was really getting a little smaller as its
surface scaled off under the action of the weather.

Baglivi[89] presents a series of *Observations* in each of which he cites a
case in which he believes that there is distinct evidence that rocks are
actually growing at the present time. Two of these are at quarries in the
vicinity of Rome, one the travertine quarries at Hadrian's Villa and the
other those of peperino near the Villa Marii. St. Augustine in his work
on the Trinity speaks of a plastic force innate in all things, "Natura est
vis quaedam rebus insita ex similibus similia procreandi," so that all
things, not only animals and plants but rocks also are continually reproducing themselves.

These stone-generating forces indeed were believed to be so widespread
and potent in various places on the earth's surface that even animals and
men were frequently found turned into stone.

König[90] gives a whole page of instances derived from various authors.
But perhaps the most striking, not to say the most terrible case, cited by
any author is that described by Cornelis de Judaeis in his *Tabula Asiae*,
where near Samogedi in Tartary certain men when grazing their sheep and
camels, were changed into stone as they stood. Aldrovandus in his

---

[89] *De Vegetatione Lapidum*. In his collected works printed at Lyons in 1714, pp. 499
et seq.
[90] Regnum Minerale, Section II, 1703, p. 269.

Musaeum Metallicum, Book IV, p. 823, gives a picture of this strange occurrence.

A similar, if not indeed the same incident is mentioned by Giovanni Botero,[91] where he states that on the left bank of the river Sur in that part of Scythia inhabited by the Tartar Hordes, there is a plain on which is to be seen a great number of what appear to be statues of men, horses and camels which it is believed represent a Tartar host changed into stone on account of their sins, by a miracle similar to that in which Lot's wife was turned into a pillar of salt.   Giovanni Lorenzo Anania (whose name is suggestive in this connection) in his *Fabbrica del Mondo* also gives an account of this same occurrence and says that this petrifaction overtook a whole Horde of Tartars and that a Tartar Horde consisted of 10,000 men.   Valentini, however, in his *Opere Mediche* states his belief

Fig. 26. Tartars with their sheep and camels near Samogedi, turned into stone.   (From Aldrovandus: *Musaeum Metallicum*.)

that the appalling change of this great body of men into stone figures was not due to a miraculous intervention of Providence but was the result of an "Errore di Natura" and brought about by "Un vento pietroso."

Other cases of men and animals becoming petrified on the occasion of certain earthquake shocks, Gimma[92] thinks may be due to the escape of petrifying exhalations, but that there should be anywhere exhalations sufficient in volume to petrify 10,000 men with their animals is more, he thinks, than we can be expected to believe.

Helmont narrates the story of the Tartar Horde also and says that a whole army consisting of men, women, camels, horses and dogs were transmuted into stone and remain so to this day, "a horrible spectacle,"

[91] Relationi Universali, Book II, paragraph 1.
[92] *Della Fisica Sotteranea*, Tome 1, p. 67.

and this happened in the year 1320 in the district between Russia and Tartary in latitude 64°, not far from a Fen of Kataya, a village or horde of the Biscardians. He thinks it must have been due to "A strong hoary petrifying breath of ferment making an eruption through some clefts of the earth, the land being stony underneath and the winds having been silent for many days."[93]

And finally there may be mentioned that terrible disaster recorded by Seyfrid[94] which overtook the city of Biedoblo in the country of the Moors, five days' journey to the south east from the city of Tripoli, when on a certain night in the year 1634, the city with all its inhabitants, their cattle, trees and gardens was suddenly turned to stone, a fact that can be vouched for, since the city still stands there to be seen by all who wish to visit it.

[93] See Shirley, *A Philosophical Essay declaring the probable causes whence stones are produced,* London, 1672.

[94] *Medulla Miribilium Naturae,* Sulzbach, 1679, p. 543.

CHAPTER V

MEDIEVAL MINERALOGY

1. Introduction

In medieval times no distinction was made between minerals, rocks and fossils as these words are now understood—all were brought together into one class and were known as "Fossils" from the Latin word *fossilis*, that is, a thing dug up out of the earth.   With the advance of knowledge the distinction between these three classes came to be recognized and the three sciences of mineralogy, petrography and palaeontology arose devoted to the study of these three groups respectively.

In the present chapter those fossils which are now known as minerals will be considered and as the views held in the Middle Ages with reference to the origin of minerals and rocks or stones, has already been reviewed at some length in chapter IV, in the present chapter consideration will be given to the views held in the Middle Ages concerning minerals apart from the question of their "generation."

2. SOURCES

Our information concerning medieval mineralogy is derived from the several sources already mentioned in Chapter I. First, and of least importance, are the medieval encyclopedias. These works aim at setting forth the whole range of human knowledge at the time in which they were written and among other things treat, in a general way, of minerals. The information which they give is compiled from other and older writers whose works in many cases have now disappeared and are known only from the quotations given in the encyclopedias. A large part of the information, such as it is, has its origin in Pliny's *Natural History* or comes from Theophrastus. They also copy from one another as they appear in successive order.[1] The earliest of these is that extraordinary collection of miscellaneous information which was the encyclopedia of the Dark Ages by St. Isidore of Seville,[2] who was born in Cartagena in 570 and died in 636. He was for some thirty years Bishop of Seville. His descriptions of minerals are taken largely from Julius Solinus, and he devotes especial attention to the etymology or derivations of the names of minerals, as indicated by the title of his work. These are for the most part quite incorrect and in many cases ridiculous, as when he says that aurum (gold) is derived from aura (air) because it is through the air that the splendor of the metal is reflected to our eyes.

Another early encyclopedia is that of Rabanus Maurus who was born in 776 at Mainz. He was Abbot of Fulda and almost continuously from 804 to 842 was head of the celebrated cloister school in that centre of learning. In the sixteenth chapter of his book, which is entitled *De Universo*, he treats of minerals but follows Isidore and adds little or nothing of his own. He tries to bring everything into relation with Holy Writ and his treatment is frequently very recondite, fanciful and mysterious.

Another writer who deserves special mention here is Alexander Neckam (1157–1227), who was one of the most remarkable English men of science of the twelfth century. In him we have a curious link between the history of science and ordinary secular history, for he was the foster-brother of King Richard the First of England (Coeur-de-Lion). In a

---

[1] See Karl Mieleitner: *Geschichte der Mineralogie im Alterthum und im Mittelalter* (*in Fortschritte der Mineralogie, Kristalographie und Petrographie*. Herausgegeben von der Deutschen Mineralogischen Gesellschaft, Vol. 7, Jena, 1882).

[2] *Etymologiae sive Origines*, Venice, 1483.

chronicle formerly existing among the manuscripts of the Earl of Arundel it is recorded that "In the month of September 1157 there was born to the King at Windsor a sone named Richard and the same night was born Alexander Neckam at St. Albans, whose mother gave suck to Richard with her right breast and to Alexander with her left breast." He received his early education at the Abbey School of St. Albans and later studied at the University of Paris, where he became a professor in 1213. His relation to the King who was a lover of learning explains, in part at least, the brilliant position which Neckam achieved in later life. While still bound by a reverence for authority, he sought for something more satisfactory than the teaching of the schools and frequently shows a leaning toward experimental science. Neckam in his book has a long satirical discourse on the logical teaching of the schools of the University of Paris and especially on the quibbles and falsities into which, in the scholastic teaching, logic continually degenerated. Wright who edits Neckam's book and supplies an interesting and valuable preface to the volume, remarks that Neckam gives a number of examples of this:

He shows how, by the train of reasoning employed in the schools, a man—Sortes the man of straw employed in the language of the scholastic disputations—might be proved to be a stone, or a rose, or a lily, or any other object ; how what Sortes or any other individual said was at the same time true and false; how Sortes at the same time knows something and knows nothing; and a number of other similar quibbles.

Neckam's book[3] was intended to be a manual of the scientific knowledge of the time but has an additional interest from the introduction of many contemporary stories and anecdotes illustrating the conditions and manners of the age. He draws from almost every scientific fact which he mentions a moral or religious lesson or application, although these are often far fetched and their truth by no means very apparent. His book abounds in stories of animals and plants. He mentions the metals briefly and refers to a dozen minerals or stones but does not describe them. Concerning mercury he remarks:

Quicksilver is necessary in gilding; at first the substance of the gold appears to be totally absorbed by the quicksilver, but afterward, by the agency of fire, the quicksilver is consumed and the colour of the gold comes out in all its brightness. So the mind is not gilt with the gold of wisdom without the agony of tribulation and some-

[3] *De Naturis Rerum*, with a poem of the same author *De Laudibus Divinae Sapientiae.* Edited by Thomas Wright, London, 1863, p. ix (Rolls Series).

times the beauty of wisdom appears to have entirely disappeared in the tribulation until it is submitted to the solace of the Holy Ghost, represented by fire, and then the strength of wisdom returns to its brightness.

Like most people in the Middle Ages, Neckam believed that gems and precious stones possessed extraordinary virtues. Thus in referring to agate he repeats the statement made by Marbodus, to the effect that the agate (*achates*) carried on the person rendered the bearer amiable, eloquent and powerful, and he explains the story of Aeneas having a faithful companion named Achates by supposing that he carried with him an agate stone whereby he acquired the love of many people and was rescued from many dangers.

At a considerably later period in his life, Neckam wrote a poem called *De Laudibus Divinae Sapientiae*, which is a metrical paraphrase of the *De Naturis Rerum*, but in it many of the stories are omitted and in places very considerable additions of new material are made.

In the interval between the writing of the two works Neckam had evidently continued his studies and enlarged his knowledge of mineralogy for in the *De Laudibus* after naming the metals he mentions thirty seven minerals, most of them precious stones, and refers to their distinctive virtues.

Another great Englishman among the ranks of the medieval encyclopedists is Bartholomaeus Anglicus, whose book *De Proprietatibus Rerum*, written about 1240, was widely read and had a great influence in its time. He treats of a series of stones and metals arranged in alphabetical order closely following Isidore of Seville.

Another of the great encyclopedias is the *Speculum Mundi* by Vincent of Beauvais written between 1240 and 1264. In the general prologue he says that there are a multitude of books and time is short and furthermore man's mind cannot gather in and retain all that the books set forth, therefore he had determined to condense all knowledge into one great work. It is indeed a great compilation in which he quotes from 450 writers by name, some of whom are now known only by the "excerpts" which he gives from their writings. In the brief portion dealing with minerals and rocks he seems to have added nothing to the observations of former writers. The work went through many editions: that in the Osler Library, which is one of the earliest, was printed "not later than 1478." Among the writers whom he quotes most frequently is the almost unknown writer of a very early encyclopedia, Arnoldus Saxo,[4]

---

[4] See Sarton, *Introduction to the History of Science*, Vol. II, p. 592, for references to the literature concerning this author.

whose work *De Finibus Rerum Naturalium* was written between 1220 and 1230. In it the section on stones consists of an annotated list in alphabetical order of 81 names, and is in turn taken from Albertus Magnus, "Evax," Dioscorides and certain Arabic writers.

Another writer whose work is but little known is the Dominican Thomas Cantimpratensis, who was born at Brussels in 1201 and died between 1263 and 1280. He resided for many years at Cambrai, from which place he derives his name. His encyclopedic work entitled *De Natura Rerum*, was never printed, but several manuscripts of it exist, one of the most perfect being in the Staatsbiliothek in Munich. He describes about 70 minerals, the number varying somewhat in the different manuscripts, and enlarges on the occult properties of the precious stones, but adds nothing to the work of earlier authors from whom his information is derived. According to Valentine Rose[5] a number of the trade names for certain stones, as for instance corneolus, granatus and rubinus, first appear in this book.

The encyclopedia attributed to Berengarius and entitled *Lumen Animae* may also be mentioned.[6]

In addition to these encyclopedias there appeared in the Middle Ages a number of popular works on natural history, many of them compilations by men who cannot be ranked as scholars. Some of these works were very widely read, and foremost among them is a very celebrated work entitled the *Book of Nature* ("Buch der Natur") by Conrad of Megenberg. The author of this work was born in 1309 and died in 1374. His father was Constable of the castle of Megenberg near Schweinfurt. He studied at Paris and Vienna and was later appointed to the important position of Canon in the Cathedral of Regensburg. This is the first Natural History which appeared in the German language. It is a compilation drawn largely from the work by Thomas Cantimpratensis or Thomas of Cambrai, already mentioned, but much material has also been taken from Aristotle and Galen, coming through Arabian channels. There are four copies of fifteenth century editions of Conrad of Megenberg's book in the library of the British Museum, the oldest of which was published in Augsburg in 1475. On the title-page of each of these it is stated that the work was translated ("transferiert") out of Latin into German by Conrad von Megenberg. The Latin text referred to was evidently that of Thomas Cantimpratensis. In the portion dealing

---

[5] See Mieleitner, *Geschichte der Mineralogie in Alterthum und im Mittelalter*, p. 476.

[6] A critical description of this work will be found in Thorndike's *History of Magic and Experimental Science*, Vol. III, pp. 547–559.

with gems, these are arranged alphabetically in order of their Latin names, and introduced with the words "In this 6th chapter of our book we will speak of gems, their color, the peculiar properties which characterize them and how we can increase their powers." 82 stones are mentioned, a number of these being quite fabulous as for instance the Terobolen, which are said to be "Stones which come from the orient, some of them presenting the form of a man and others that of a beautiful young woman. If they are brought near to one another they send forth flames and fire." Also the Dyadochos, a stone which when cast into water causes certain spirits to appear who will answer questions which are put to them. The book also contains many wonderful stories about animals and plants, which made it very attractive to the general reader, its popularity being enhanced by numerous illustrations, so that it passed through many editions. The book is worthy of study as presenting an admirable picture of the medieval conception of the world of nature. Megenberg himself, however, recognized that many statements in his book, even when drawn from what were considered to be sources of the highest authority were open to doubt, and in many places he does not hesitate to go so far as to state that he does not believe them.

The Herbals of the late Middle Ages and the Renaissance form another source from which some knowledge of the medieval mineralogy may be obtained. In 1484 Peter Schöffer of Mainz printed one of these in Latin which bore the simple title *Herbarius*. There was such a demand for this, that early in the following year he issued a similar but much larger book in German written by Johann von Cube the "Stadtarzt" of Frankfort, to which the author, in the preface, gives the name *Hortus Sanitatis* or, in German, *Garten der Gesundheit*. The book as the name indicates was intended to treat more particularly of the medicinal virtues of plants. In addition to the plants, however, a few animals and minerals, from which materials of therapeutic value were supposed to be obtained, are included. The work appeared in a great number of subsequent editions, some in German, others in Latin. It was also translated into Dutch, Italian and French. It appeared in a greater and in a smaller form known as the *Hortus Sanitatis Major* and the *Hortus Sanitatis Minor*. The smaller form is the older and was never printed in Latin but always in some other language. Most of the later editions are divided into three sections dealing with plants, animals and minerals respectively, completely illustrated and supplying information concerning the character and medicinal value of each and all. Some of these illustrations are reproduced in Chapter IV. In the French edition of 1529 mention is made of 140 metals, stones and gems.

Who the author of the work was is really not known with certainty. By some it is believed to be Johann Wonnecke, his designation "von Cube" being derived from the fact that he came from Caub a town on the Rhine between Bingen and Coblenz. Chouland[7] thinks that the book is really a compilation from the works of a number of earlier writers. In any event it was the most popular illustrated book on natural history of the time.

A number of other herbals, based upon the *Hortus Sanitatis* and embracing minerals of supposed therapeutic value, appeared shortly after it. Among these may be mentioned that of Rösslin[8] (1533) which passed through six editions in thirteen years. The second edition of this work mentions eighty metals and gems and is illustrated. Another by Lonicer, also illustrated, appeared in 1551 and treats of 85 metals, stones and gems.

The lapidaries will now be considered.

3. MEDIEVAL LAPIDARIES

The most important source from which our knowledge of medieval mineralogy is obtained is the medieval lapidaries. They are treatises, usually quite short, which deal exclusively with metals, stones and gems, and in which as in the sources already referred to, nothing is said of the composition of the stones, and very little concerning their physical characters, but their medicinal, magical or mythical characters are set forth, these being, at the time when the lapidaries were written, properties which such stones were almost universally believed to possess, and which caused them to be looked upon with interest, if not with awe.

The lapidaries were very popular not only among the educated classes of the community but also among the middle classes, as shown by the fact that there were many of them and they were in great demand, and when written in Italian were frequently translated into the vernacular languages of many of the leading European countries. Probably the goldsmiths and jewellers offered them for sale as setting forth the value of their wares, and the people bought the gems not only for purposes of

---

[7] *Graphische Incunabeln für Naturgeschichte und Medicin*, Leipzig, 1858. (Verlag der Münchner Drucke, Munich, 1924, p. 44.)

[8] *Kreutterbuch von aller Kreutter, Gethier, Gesteyne und Metal Natur, Nutz, und Gebrauch*, Frankfort.

adornment but for the magical powers which they were supposed to possess. How many of these lapidaries have survived is unknown. Some of them were never printed but are known to exist in the form of manuscripts only, in one or other of the great libraries of Europe. Among the more important of these lapidaries the following may be mentioned.[9]

1. *The Lapidary of Aristotle.*[10] It is now known that this book, although it bore the name of Aristotle down through the centuries, was not written by him but is of Persian origin from the pen of some unknown author who lived about the middle of the ninth century. It describes 72 stones and is of especial value in that it affords an insight into the mineralogy of the Arabs, which had an important influence on the development of the science of mineralogy in Europe at a later date.

2. *The Lapidary of Marbodus.* This is in Latin hexameters and was written by Marbodus, also called Marbodacus or Marboeuf, Bishop of Rennes, between 1061 and 1081. It is by far the most important of the early medieval lapidaries and is referred to again on page 149.

3. *Albertus Magnus: De Mineralibus.* One of the best and most comprehensive of the western medieval lapidaries, it was written about 1260. There are several works, dealing in whole or in part with minerals, which are attributed to Albertus. Of these, that entitled *De Mineralibus* is the most important and is at the same time undoubtedly a genuine work of his. The first edition was printed at Padua in 1476. Other editions have somewhat different titles: one printed at Oppenheim in 1518 bears the title *Liber Mineralium Domini Alberti Magni*, another in which the text is identical was printed at Cologne in 1569 with the title *De Mineralibus et Rebus Metallicis Libri Quinque.* It is a compilation from earlier writers with the addition of some facts derived from the author's own observations. While Albertus preserves the medieval attitude of mind, he shows unmistakable signs of a belief in the value

[9] Brief references to other less well known lapidaries of Persian, Spanish and other origin will be found in the *Introduction to the History of Science*, by George Sarton (Carnegie Institution of Washington, 1931), Vol. II, pp. 48, 780, 842, &c.

See also L. Pannier, *Les Lapidaires français du Moyen Age*, Paris, 1882.

Karl Mieleitner, *Geschichte der Mineralogie in Alterthum und im Mittelalter*, Jena, 1922.

Joan Evans, *Magical Jewels of the Middle Ages and the Renaissance particularly in England*, Oxford, 1922.

[10] J. Ruska, *Das Steinbuch des Aristoteles mit literargeschichtlichen Untersuchungen nach der arabishen Handschrift der Bibliothèque Nationale herausgegeben und übersetzte*, Heidelberg, 1912, p. 45.

See also Chap. II of the present work in which the authorship of this work is discussed.

of research and personal observation.   He tells us that he has wandered far to visit places rich in mines that he might test the natures of metals. Seventy stones are mentioned, arranged alphabetically, and Albertus says that gems differ from other stones in that in them the watery element preponderates over the earthy constituent and he attempts to classify them by color.   He, however, while saying little of the medicinal value of gems, enlarges at length on their mystical and wonder-working powers and virtues.   There is scarcely an ill that flesh is heir to, for which he does not indicate some stone that will act as a protector.   He also shares the general opinion of the time that when these gems or stones have certain signs or figures engraved upon them, such "sigils" greatly enhance the power of the stone.[11]   He states furthermore that in many cases at least, the sigils were not carved in the stone by any human agency, for the gems on which they are found are too hard to be worked by any tool; they are of celestial origin, having been impressed upon the growing stone by the mighty agency of the stars which is also displayed in those gigantic starry pictures which are seen in the heavens and are called constellations.

   4. The Lapidary of Alphonso X.   This is a beautifully illustrated lapidary the only recorded manuscript of which is in the Escurial Library.[12]   It is said to be a Chaldean lapidary of unknown date, which was translated into Arabic by Abolays and from Arabic into Spanish by Garci-Perez.   This latter translation was finished in 1278.   Evans, who gives an extended review of the work, states that it deals with 360 "stones" among which are included substances such as sulphur, soda and even sponges.   Meyerhoff and Foster say that it describes 280 "stones," with brief references to some of their physical properties, uses and medicinal values.[13]

   The basis of classification is color "Cast into the 12 signs of the Zodiac."

   The treatment of the virtues and powers of the stones as influenced by the stars and changing according to the position of the planets indicates the complexity of the connection made by Arabic science between

---

   [11] See Joan Evans, *Magical Jewels of the Middle Ages and the Renaissance.*   Oxford, 1923, also

   Lynn Thorndike, *A History of Magic and Experimental Science,* Vol. II, Chap. LIX,

   [12] It was published in facsimile by Blasco in Madrid in 1881.

   [13] See H. A. Meyerhoff and Mary Louise Foster, *The Power of Stones—a Thirteenth Century Manuscript.*   Preliminary list of Abstracts of Papers to be read at the meeting of the Geological Society of America held in Cincinnati, O., in December 1936.   In this abstract it is stated that the book has just been translated into English by M. L. Foster.

minerals and the celestial forces.   The court of Alfonso the Learned of
Castile (1223–1284) was as typical a court of the Arab Renaissance of
the thirteenth century as that of Lorenzo de Medici was of classical
learning in the fifteenth century.   Jews, Mohammedans and Christians
there met on equal terms; his court was the Academy where an epitome
of the science of the Mediterranean world of the thirteenth century was
to be found.[14]

5. *Volmar: Das Steinbuch.*[15]   An old German lapidary written about
1250, which mentions some 38 stones.

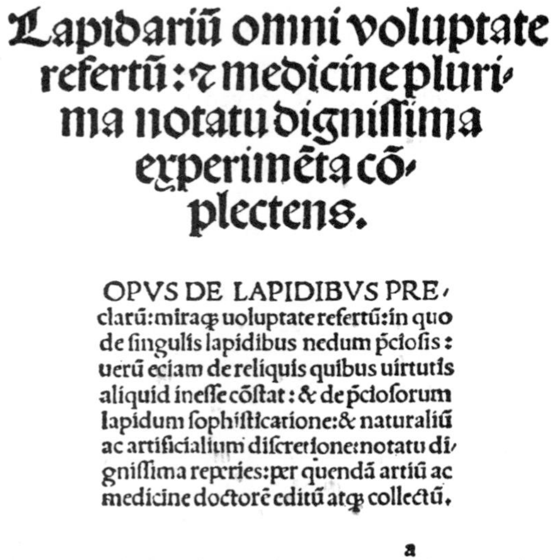

**Lapidariū omni voluptate
refertū:⁊ medicine pluri=
ma notatu digniſſima
experimēta cō=
plectens.**

OPVS DE LAPIDIBVS PRE=
clarū:miraqȝ uoluptate refertū:in quo
de ſingulis lapidibus nedum pćioſis:
uerū eciam de reliquis quibus uirtutis
aliquid ineſſe cōſtat:& de pćioſorum
lapidum ſophiſticatione:& naturaliū
ac artificialium diſcretione:notatu di=
gniſſima repȝries:per quendā artiū ac
medicine doctorē editū atqȝ collectū.

a

Fig. 27. Title page of Steinpreis: *Lapidarium omni Voluptate refertum.* (N.B. The
upper and lower (blank) parts of the page have been cut off.   It is reduced slightly in size.)

6. *Steinpreis:*[16] *Lapidarium omni Voluptate refertum.* This is an
interesting lapidary of which very few copies are known to exist.   Dolch-
Langer knows only three copies, those in the Glasgow Hunterian Museum, the British Museum and the Munich Staatsbibliothek.   There
is a fourth copy in the Adams collection at McGill University.   It is a
small quarto volume of 28 unnumbered leaves, is believed to have been

[14] See Joan Evans, *Magical Jewels*, Oxford, 1933, p. 43.

[15] *Das Steinbuch, Ein altdeutsches Gedicht von Volmar, mit Einleitung, Anmerkungen und
einem Anhange*, Herausgegeben von Hans Lambel, Heilbronn, 1877.

[16] *Lapidarium omni Voluptate refertum: medicine plurima notatu dignissima experimenta
complectens*, Vienna (printed by Io. Winterberger).

written by Steinpreis and was printed shortly after the year 1500, probably about 1510. In the opening paragraph the author states that he has two objects in writing the book, first to give such information concerning the various minerals as will enable them to be recognized, and second, to set forth their origin, virtues and properties.

The book is divided into two parts. The first consists of twelve short chapters setting forth and establishing the fact that the precious stones have peculiar virtues and properties, how these manifest themselves, how they may be strengthened and how they should be venerated. He inveighs against those ignorant and foolish persons who, although quite unfitted to express any opinion on the subject, are wont, in their bestial stupidity, to blaspheme God by stating that these gems possess no such virtues. These opinions he proceeds to confute in the true medieval manner. Let such men listen to the words of Aristotle, the head and chief of all philosophers, let them hear what the divine Bonaventura says upon this subject, and what the divine Plato asserts. The words and the authority of these great men establish and definitely prove that these stones have the mystic virtues which they are asserted to possess

The second part treats of 117 different stones in alphabetical order. This portion consists largely of quotations from Albertus, Serapion, Pliny, "Dyast," Evax and Avicenna. Indeed a large portion of it is a direct and verbatim copy from Albertus Magnus.

7. *Erasmus Stella*:[17] *Interpraetamenti Gemmarum Libellus.* This lapidary was written in 1517 by Dr. Erasmus Stueler. He was born about 1450 and studied at Leipzig and Bologna. In 1501 he was appointed town physician and later became burgomeister of Zwickau in Saxony, which was at that time an important mining centre. It is a book of quarto size and comprises 21 unnumbered leaves. It shows (figure 28) a distinct advance upon the older lapidaries in that the alphabetical arrangement of the minerals is abandoned and they are grouped into four classes according to their color: white, green, red, and blue or black. Thirty-three minerals are described and a number of others are mentioned by name. The descriptions given of the minerals are longer and more complete than in most of the older lapidaries, and more attention is devoted to the varieties of the several minerals and to the localities where they are found. Furthermore he has little belief in the magical properties attributed to gems by the older writers and in the

[17] *Erasmi Stellae Libonothani Interpraetamenti Gemmarum Libellus unicus*, Nuremberg, Fredericus Peypus, 1517.

chapter having the title *In Universum de Gemmis* he says that while no
one in his sound mind denies that gems have medicinal and curative
properties, he would find it very difficult to concede that they possessed

ERASMI STELLAE
LIBONOTHANI
INTERPRAE=
TAMENTI
GEMMA=
RVM LI=
BELLVS
VNI=
CVS.

Sola falus feruire deo,
Sunt cœtera fraudes,

FIG. 28. Title page of Erasmus Stella: *lapidary*.   Nuremberg, 1517.   (Natural size.)

any of those powers attributed to them by superstitious persons, such as
those of bringing to their wearer joy, sorrow, tranquility or security.
He merely mentions such powers in describing the several gems in order

that the reader might know that they had been attributed to the respective stones by the ancient writers, but not because he himself shared such opinions, which were ridiculous if not indeed actually irreligious.

In the chapter on rings (in which gems were mounted) he refers to the custom of wearing the ring on the third finger of the left hand, which he says was derived from the Egyptians who believed that a vein passes from this finger directly to the heart.

8. *Camillus Leonardus: Speculum Lapidum.* The last of the medieval lapidaries to be mentioned is that of Leonardus. This will be referred to again on page 155.

It is a significant fact that in all the lapidaries which appeared prior to the sixteenth century the minerals, when any definite system of classification was adopted, were arranged and listed in alphabetical order. This was owing to the fact that their authors knew practically nothing of the composition of the minerals and little or nothing of their actual physical characters—with the exception of color. They knew scarcely anything concerning the form of minerals, or of their hardness, specific gravity, cleavage, lustre or fusibility and nothing of their optical characters, on which knowledge alone a proper classification could be based.

The number of species enumerated differs largely in different lapidaries. Marbodus mentions 60, Albertus Magnus 93, the Steinpreis *Lapidary* 117, Leonardus, who gives a longer list than any other writer, catalogues 279 separate names. The list given by Leonardus includes almost all those given by the other writers. Steinpreis differs most widely from him in enumerating 38 names not mentioned by Leonardus, while omitting a much larger number of those found in the lapidary of that author.

The Lapidary of Marbodus. It may be of interest here, in order to present an adequate idea of the character and contents of these medieval lapidaries, all of which cover essentially the same ground although some of them do so more thoroughly than others, to select two, those of Marbodus and Leonardus, and describe them in some detail. The work of Marbodus is the earliest lapidary of the Middle Ages, and also the one which is quoted most widely. As has already been stated, Marbodus was Bishop of Rennes and wrote his lapidary, in 734 Latin hexameters describing sixty stones, between the years 1061 and 1081. More than 100 manuscripts of this lapidary are known, and it was translated into French, Provençal, Danish, Hebrew and Spanish. After the invention of printing, 14 editions[18] appeared between 1511 and 1740 and still others at later dates. To a student of the present time it seems strange that a treatise on minerals should be written in verse. Janus Cornarius who includes the lapidary of Marbodus in his edition of the *Macer Floridus* herbal (1540) says that this was done in order that the contents of the book might be more easily remembered.

Since there are certain differences in different texts, it may be stated that those which are used in the present notes are the first printed edition published at Vienna in 1511[19] and the third edition which was printed at Freiburg in 1531.[20]

The poem commences with the following lines:

Evax, king of the Arabs is said to have written to Nero,
Who after Augustus ruled next in the city,
How many species of stones, what name and what colours,
From what regions they came and how great the power of each one.

A reference to Evax king of the Arabs in almost the same words occurs in the Hellenistic lapidary which is attributed to Damigeron, to which reference has been made in a former chapter, and which is one of the principal sources on which Marbodus has drawn. This Evax is also referred to by Pliny in the twenty-fifth book of his *Natural History* in the following words, "Evax a king of the Arabians, wrote a book as touching the virtues and operations of Simples, which he sent unto the Emperor Nero." Nothing further, however, is known of this person so frequently mentioned in the medieval lapidaries.

The sixty stones of which Marbodus treats are, in the first edition of his work presented without any definite order, but in the third edition they are arranged alphabetically. A critical examination of the list shows that they may be classified in the following groups:

1st group:

Alabandina—staunches a flow of blood
Androdamas—dispels anger
Apistos—a heavy black rock with red lines—when heated retains the heat for seven days
Calcophonus—preserves and strengthens the voice

[18] For list and description of these with their respective dates of publication see Beckmann, *Marbodi liber Lapidum seu de Gemmis*, Göttingen, 1799, p. xi.

[19] *Enchiridion Marbodei Galli de lapidibus pretiosis*, Vienna, 1511.

[20] *Marbodei Galli Poetae Vetustissimi de lapidibus pretiosis Enchiridion, cum scholiis Pictorii Villingensis*, Freiburg, 1531.

Chalazias—cannot be heated, cold by nature

Chrysopasius—shines brightly at night and is dark during daylight

Diadochus—when thrown into water causes spirits to appear who may be interrogated

Dionysias—prevents intoxication

Galactites—cannot be heated

Gagatromeus—secures victory in conflict, a stone for soldiers

Hephestites—thrown into boiling water will freeze it, calms sedition

Hexacontalithos—small stones showing 60 different colours. Found in Libya in the land of the Troglodites

Hieracites—similar properties to Galactites

Lipares—gives success in hunting, the hunter can secure any animal

Medus—gives sight to the blind

Molochites—prevents accidents

Ophthalmius—cures blindness

Orites—cures wounds, gives protection against the pestilence

Pantherus—variegated in colour and hence has the virtues of many other stones

Peanites—a female stone which at certain seasons brings forth young, assists women in childbirth

Sagda—originates at the bottom of the Chaldean Sea and rising through the water, attaches itself to the bottom of a ship which, through its influence, is quickly and safely brought to port. It is however impossible to detach the stone from the bottom of the ship except by cutting away the wood to which it affixes itself.

These 26 stones, making up approximately one-third of the whole, are mythical—or at lest the description of them, when any is given, is so trivial that it is impossible to connect the name with any particular mineral.

2nd group (This comprises six "stones"):

Alectorius

Calcedonius

Chelonites

Dragonites

Hyaenia

Lyncurius.

These all come, or are supposed to come, from the body of some animal. They are described in Chapter IV.

3rd group:

Cerauneus—falls from the Heavens, protects from lightning

Corallus—coral; preserves from tempest, gives abundant harvests, drives away demons

Aetites—aids childbirth

Unio—pearl, possesses many virtues, cures fever and epilepsy

The four members of this group are really not minerals.   They are referred to at some length in Chapter IV.

4th group:

Achates—protects from serpents, makes men strong, handsome and eloquent
Amethystus—preserves from intoxication
Calcedonius—gives success in all undertakings
Chrysoprasus—may have virtues but if so these are unknown
Corneolus—cools anger, staunches blood
Crystallus—valuable for nurses
Enhydros—(see below)
Heliotropia—guards the body from poison and the soul from error.   This stone
   was by others stated to make its wearer invisible.   Dante in his Inferno
   sees the damned running about under a hail of fire, "No hope of hiding hole
   or Heliotrope."
   The story in Boccaccio's *Decameron*[21] told by Madame Eliza on the Eighth
   Day centres around the adventures of certain persons who "travelled to the
   Plaine of Mugnone to find the precious stone called Heliotropium, the
   vertue whereof is so admirable: as whosoever beareth it about him, so long
   as he keepeth it, it is impossible for any eye to discern him, because he
   walketh meerely invisible."
Iris—when the sun shines on it, it gives forth the colours of the rainbow .
Jasper—seventeen varieties, each of which possesses its own properties.
Onyx—produces hate, evil visions, strife and bloodshed.   The evil effects of the
   onyx are counteracted if one wears a sardonyx at the same time.
Prasius—has no special virtues.

These fourteen stones are all varieties of quartz—some coarsely crystalline and others cryptocrystalline and distinguished from one another
chiefly by difference in color.   They attracted attention probably because most of them were comparatively common minerals and because
of their distinct and striking colors.

Achates—Marbodus, interpreting Virgil's "Fidus Achates" literally,
ascribes, as already mentioned, the escape of Aeneas from his many
perils to his having always carried an agate with him.

Enhydros—consists of quartz or of calcedonic or opaline silica which
is deposited from heated waters and in which there are cavities partially
filled with water, so that the inclusions of water can be seen within the
substance of the transparent quartz or calcedony.   This remarkable
stone attracted widespread attention among the ancients.   Marbodus
says: "Enhydros perpetually sheds tears which drop as from an over-

flowing spring. It is," he observes, "extremely difficult to see how this
comes about, for if the substance (water) is derived from the stone itself
why does not the stone become smaller and turn completely into water.
But if on the other hand the water finds its way in from the outside and
flows out again, how is it that the external water is not prevented from
entering by the water which is in the act of coming out?"

5th group:

> Adamas—dispels evil dreams and saves its wearer from the influence of poison
> Asbestos—(see below)
> Beryllus—preserves and develops conjugal love and has other remarkable virtues
> Carbunculus—shines in the dark   (See below)
> Chryselectrus—golden in colour, resembling electrum (amber).   The devouring
> flame is said to be its substance.   For, held near a fire, it quickly kindles.
> Chrysolitus—drives away evil spirits
> Gagates—cures certain diseases, overcomes demons and magic arts
> Haematites—cures various diseases of the blood and has other valuable prop-
> erties
> Hyacinthus—red, yellow or blue in colour.   (Some specimens were probably
> varieties of amethyst and others corundum.)   Confers safety on the traveller
> and secures to him his just requests.
> Magnes—(see below)
> Pyrites—fire stone, burns the hand that roughly presses it
> Saphirus—(see below)
> Selenites—shows an image of the waxing and waning moon.   Cures diseases
> of the chest
> Smaragdus—includes emerald, malachite and probably other green stones.
> Enriches its wearer and makes him eloquent, wards off epilepsy and has
> many other virtues.
> Topazius—staunches a flow of blood, if thrown into boiling water causes it to
> become cold.

The fifteen minerals of this group, with the exception of the fifth, bear
the names of well-known mineral species.   The designation Adamas is
given by Marbodus to the diamond, which derives its name from its
unconquerable virtue.   Marbodus mentions four countries in which it is
found, India, Arabia, Cyprus and Philippi in Greece.   The specimens
from these four countries differ from one another in character.   The
Indian Adamas is a diamond.   The others were in all probability varieties of quartz.   Marbodus repeats the statement of Pliny that if the
stone is placed upon an anvil and struck with a hammer with all possible
violence, the stone will remain unharmed but the anvil and hammer will
fly to pieces.   He also reiterates the story that if a diamond is placed in

the warm blood of a stag, it will break to pieces. Asbestos is stated to have the color of fire and to be found in Arcadia. When once lighted, like the fires of hell it burns forever, hence its name. No magical or medicinal properties are assigned to it by Marbodus. He states that there are twelve varieties of Carbunculus. Under this name were included ruby, Balas ruby (spinel), brilliant red garnets and probably other stones resembling these in color. He says that the stone gives out light, shining in the darkness like a glowing coal, and that it comes from the country of the Troglodites. The statement that it shines by night is repeated by Lonicerus[22] and also by Rösslin,[23] who says that it turns night into day. Marbodus says that Saphirus is the "Gem of gems," possessing vast powers, medicinal, magical and spiritual. Heavenly blue in color, it is endowed with transcendant celestial powers. Not only does it preserve its wearer in health of body and mind but shields him from all error and breaks the chains of prisoners and sets them free. But Marbodus goes much farther and claims that it turns aside the anger of God and secures from Him a favorable answer to prayer. This statement, which represents, as it were, a shocking climax to the long series of talismanic powers which are assigned to the various gems in this interesting little book, aroused the disapproval and even the indignation of many chruchmen, who endeavored to throw doubts upon the authorship of the work as being from the pen of a Bishop of the Christian church.[24] But, as Thorndike[25] observes, practically everyone at that time believed that marvellous powers had been divinely implanted in gems, and such being the case, why should not God be more easily reached through the medium of gems, since he had endowed them with their marvellous virtues?

In the third edition of the Lapidary of Marbodus which was published in 1531 there are glosses appended to each chapter, also some introductory and concluding verses by Georg Pictorius. At the end of the book there is a short poem by the latter addressed to the millstone. Being evidently somewhat weary of the continued enumeration of the magical and mystical powers attributed to the precious stones in the work of Marbodus and probably harboring some secret doubts as to the reality

[22] *Kreuterbuch*, Frankfort, 1573, p. lviii.

[23] *Kreuterbuch*, Frankfort, 1535, p. lvii.

[24] Pannier: *Les Lapidaires Français du Moyen Age des XII, XIII et XIV siècles*, Paris, 1882, p. 17.

[25] *A History of Magic and Experimental Science*, New York, 1929, Vol. I, p. 782.

of the powers in question, Pictorius turns as it were for relief to the common millstone, and addresses a concluding poem to it as a stone, which is indisputably full of virtue for mankind, and which renders inestimable services to the human race.

The Lapidary of Camillus Leonardus. While the Lapidary of Marbodus was written at the very beginning of the Middle Ages, in fact a few years before the date which we have taken as marking the commencement of medieval times, the second lapidary which has been selected for more detailed consideration appeared when the Middle Ages were passing into the Renaissance. This is the *Speculum Lapidum* of Camillus Leonardus. The first edition was published at Venice in the year 1502 and this was followed by others printed at Venice in 1516, at Augsburg in 1533, at Paris in 1610 and at Hamburg in 1717. All these were in the Latin language, but an English translation (omitting Book III) was published in London as late as 1750. The book also figured in one of the most shameless cases of piracy in the whole history of letters, a well-known author Lodovico Dolce, having made a literal translation of it into Italian,[26] which he published at Venice as his own work, without making the slightest reference to the fact that Leonardus was the author of the book, in fact on page 3 he refers to it as "mia fatica."

Leonardus was physician to Caesar Borgia to whom also he dedicates his book, so that as King[27] observes, he "ought to know something about poisons."

His lapidary is divided into three books. The first treats of the nature and origin of stones, their beauties, colors and virtues. In the second there is a formal description of 279 minerals arranged in alphabetical order. As has been already mentioned, Leonardus includes in his lapidary almost every mineral mentioned by any previous writer, which gives to his book a special interest and value. Many of the minerals are, however, as is the case in all the medieval lapidaries, merely names to the modern student, and cannot be recognized as designating any mineral now known to be a true mineral species. Some are merely varieties of the same species, others are certainly altogether fabulous. In the third book the author treats of the figures engraved upon gems and other stones by the ancients, of the particular virtues of the engraved stones, how they absorbed the influence of the planets and constellations

[26] *Libri tre di M. Lodovico Dolce ne i quali si tratta delle diverse sorti delle Gemme che produce la Natura*, Venice, 1565.

[27] *The Natural History of Gems or Decorative Stones*, London, 1867.

and why a stone engraved with any of the twelve signs of the zodiac is
supposed to take its virtue from that sign, and what this peculiar virtue
is.   He mentions that the Israelites in the wilderness were the first who
distinguished themselves in this art of engraving gems and that the
Romans were the greatest masters of it.

Speculum Lapidum Clariſſimi Artium Et Medicine Doctoris Camilli Leonardi Piſaurenſis.

⟨⟨Valerii Superchii Piſaurenſis Phyſici Epigramma.

⟨⟨Quicquid in humanos gemmarum parturit uſus
Terra parens:uaſti quicquid & unda maris.
Qualibet exiguo claudis Leonarde libello
Mirandum/& feræ poſteritatis opus.
Quod poſitis Cæſar interdum perlegat armis:
Seruariq̃ ſuas imperet inter opes
Et tibi pro meritis æquos decernar honores:
Conſulat & famæ tempus in omne tuæ.

FIG. 29. Title page of the *lapidary* of Camillus Leonardus.   Venice, 1516.

In his lapidary Leonardus follows Albertus Magnus in adopting the
Aristotelian theory of the origin of stones through the influences of the
heavenly bodies.   Gimma,[28] however, states that owing to the inclusion

---

[28] *Della Storia Naturale delle Gemme, delle Pietre e di tutti Minerali ovvero della Fisica
Sotterranea,* Naples, 1730, Vol. I, p. 55.

in it of certain ideas and certain statements drawn largely from Arabian sources, the book fell under the condemnation of the Church and was entered in the *Index Expurgtorius*.

The fear that the book might exercise a dangerous influence upon some of its readers also called forth the following warning on the part of the English translator:

> But tho' what I have said, in regard to the Use and Excellence of this little Treatise, is incontestibly the Truth; yet I must give the Reader a Caution in the Perusal of it, which is this: That the Author living in an Age when Superstition universally prevail'd, and when the Study of Astrology, Palmistry, Charms, Spells, Sigils, &c. was greatly in Vogue, but which, in our Days, is entirely out of Use, at least is laid aside by the Learned: I say, the Author, falling in with the Maxims of the Age wherein he lived, has assigned such Virtues to particular Stones as will not be allowed by the Moderns; as that such or such a Stone shall give the Possessor of it, Courage, procure him Victory over his Enemies, make him successful in Love, in Litigations at Law, and other Undertakings, with other Fancies of the same Kind, which have long since been exploded. He, however, gives us this Caution, that in his Description of the Virtues and Properties of Stones, he has inserted nothing but what he has collected from the Writings of the most learned Men that have treated of the Subject; so that he exhibits nothing, or but very little, as his own Opinion, nay, sometimes he banters and ridicules the extravagant Fancies of those whose Sentiments he quotes: So that when the ENGLISH READER meets with these odd Whimsies, he is to look on them in their proper Light, and to give a due Attention to the more weighty and important Design, and Use of the Book.[29]

While Leonardus, as he himself states, garnered his material from a succession of older writers, he shows some indications of having come under the influence of the newer methods of study which were about to be advocated by Agricola and his followers, in that he treats of certain physical properties of minerals, such as "Perspicuity and Opacity, Hardness or Softness, Gravity and Lightness, Density and Porosity" and of the importance of these for the recognition of various stones. In fact the *Speculum* of Leonardus, which was one of the most widely read lapidaries of the time, in its successive editions bridged over the transitional period between the old and the new mineralogy, since the first edition appeared in 1502, or forty-four years before the publication of Agricola's *De Natura Fossilium*, while the last or English edition coming from the press in 1750, brings us nearly to the time of Werner. When Boyle wrote his *Skeptical Chymist* in 1677, the belief in the four elements of Aristotle was already seriously shaken. Boyle considered

---

[29] *The Mirror of Stones*, Camillus Leonardus, M.D., London, 1750, pp. IX, X.

them as worthy of his trenchant criticism, so that about the time that the last edition of the *Speculum Lapidum* of Leonardus issued from the press, the belief in the Aristotelian theory of the origin of minerals and stones through the action of the celestial bodies flickered out. In fact, the third book of the *Speculum* was omitted in the English translation, since, as the translator remarks, "We judged it wholly impertinent to trouble our readers with speculations not agreeable to right reason nor indeed consistent with our religion."

That Leonardus did not repose implicit reliance on all the traditional stories concerning the virtues of stones, may be gathered from what he says concerning the stone Lippares:

> Lippares, or Liparia, is a stone to which all Kinds of Animals come of their own Accord, as it were by a natural Instinct. Some say, that he who has this Stone, needs no other Invention to catch wild Beasts; it is frequently found in Lybia. Others say, that it has a wonderful Virtue in defending Animals. For when a Beast is pursued by Dogs and the Hunters, he hastens to find out this Stone, to which he flies as to his Protector and Defender. For so long as the Animal looks upon the said Stone, neither the Dogs nor the Huntsman can see him, which if it be so, is indeed very strange; yet it is affirmed by the Learned; and as to this, I believe the saying of Pliny is very true, That there is no Lie so impudent which is not vouch'd for by Authority.[30]

Little or nothing is known of the life of this interesting author beyond what he himself tells us in the letter in which he dedicates his work to the "Most Illustrious and most glorious Prince Caesar Borgia." In it he states that he was a citizen of Pesaro and that he was busily engaged in the "Practice of Physic and Speculation." He goes on to say that:

> Being governed by the desire to promote the Benefit and Utility of Mankind we have composed this little Treatise on the Nature of such Stones as contribute to the Health or Usefulness of Men, tho' at the expense of late hours, much labour and diligent enquiries, and tho' the materials of it were dispersed thro' many volumes by various authors. We have, however, with the utmost care, labour and attention, collected such things as have been handled in the writings of the most famous men into this small Tract, which we have entitled The Mirror of Stones: in which, as in a Looking-Glass we may behold their Nature, Powers and Sculptures and attain to the knowledge of many things.

The book was evidently highly prized by those interested in minerals and gems even at so late a date as 1750, and the translator's preface to the English version throws an interesting light not only on the scarcity

---

[30] English Translation, p. 117.

of the work at the time the translation was made, but also on the high esteem in which it was held, as well as the difficulties in the path of a book collector in that age:

If the value of a book is to be rated by the scarcity of it, I am apt to think that there is not a Librarian in Europe can shew one of equal bulk that has a better title to the choice of the curious than this *Mirror of Stones*. For though the number of its Pages are but 244 in a small Octavo and printed in large Letters yet thére is wrote on the cover of that which by a peculiar Favour I am possessed of, "This is a scarce book and has been valued at 100 Pistoles."

This is equivalent to eighty pounds. He goes on to say:

A certain Nobleman who is pleased to honour me with his Friendship sought for it in vain in the most noted Libraries in England; but being determined to have it if there was one in Europe, sent a gentleman to France, where he was to make the best Enquiry he was able among the Booksellers, and to search every Library where there was any probability of its being lodged: and if his Enquiries should prove unsuccessful there, he was to proceed to Italy, and so on to other countries till he should find it. After a long and expensive search he at last was so happy as to light upon two of them, which he purchased, tho' at an exhorbitant Price and brought them to his noble Master, who was so pleased with the purchase that he not only paid him generously for his time and expenses, but over and above as a Gratuity and Reward for his Diligence presented him with a Bank Note for £30.

It will be noted that these medieval lapidaries, although they are the most important source from which information concerning the mineralogy of the Middle Ages is derived, really supply relatively little knowledge concerning the actual character of the minerals of which they treat. The names used to designate certain minerals in the Middle Ages, in many cases at least, refer to other minerals than those which now bear the same names. Thus the term carbuncle in medieval times was used to designate a bright red hard transparent mineral, it might be a ruby, pleonast (or Balas ruby), an almandine, pyrope or allied variety of garnet or some other mineral resembling them. Smaragdus included not only emerald but malachite, green feldspar or other minerals which had a bright green color. The name sapphire was used also for lapis lazuli and probably also for a number of other blue minerals.

The lapidaries are essentially handbooks of magic and medicine. It was on account of their supposed magical and medicinal properties that minerals were prized, and it is thus interesting to note that the science of mineralogy arose from the study of magic and medicine, although it made no real advance toward the status of a true science until the rise

and widespread development of the mining industry in Europe.   All
the old writers followed authority blindly and it was observation alone
that could break this pernicious succession.

Many writers who treated of the magical powers of stones regarded these
powers as having their origin in the substance of the stone itself.   Others
considered them to have been developed under favorable conditions in
the mineral or stone by the action of one or more of the Aristotelian
elements or by the influence of the planets or of the fixed stars (as ex-
planed in Chapter IV).   Others again state that in gems having these
supernatural qualities a certain symbol (or sigil) engraved upon the gem
will greatly intensify its power.   This belief goes back to the most remote
antiquity.

Albertus Magnus in his *De Mineralibus et Rebus Metallicis* points out
that it is important that sigils should be engraved on gems at a time
when the constellations influencing the operation of the celestial influences are favorably situated in the heavens.   Owing to the fact that the
rays from the planets fall directly upon the equatorial regions and only
obliquely upon the more temperate zones, the gems found in India and
the East were believed to possess greater power and virtue than those
found in more northerly or more southerly latitudes.

Lists and descriptions of such sigils are given by many of the ancient
writers.   Leonardus in the third book of his Speculum Lapidum reproduces those attributed to Rhagael, Chael, Hermes, Tethel and Solomon.
Tethel was a Jewish astrologer presumably identical with Sahl ibn
Bishr[31] (first half of the ninth century).

Conrad von Megenberg in Book VI of his *Buch der Natur* which appeared some one hundred and fifty years earlier than the work of Leonardus, treats of these engraved stones at some length and gives a German
translation of a Latin edition of Tethel's book.   This book, he says,
"Written by a great Jewish man of learning named Tethel, claims to have
been compiled by the Children of Israel during the time when they were
journeying through the desert on their way to the Holy Land."   The
author of the Latin edition of Tethel is quoted as saying that in his
opinion the book is not to be implicitly trusted and that the engraved
figures were intended more for purposes of adornment than for any
expectation that they possessed occult powers, since men should look to
God alone for sure mercies.   "I am," continues von Megenberg, "of

[31] George Sarton, *Introduction to the History of Science*, Washington, 1931, Vol. II, p. 593.

the same opinion, although the author states that the figures which are seen on these stones have been engraved solely by the art of man and never result from the powers of Nature. Herein he is in error, for many figures are to be seen in stones while they are still growing in the earth's crust." This same opinion is expressed by Albertus Magnus, who states that these sigils must for the most part at least have had their origin in the influence of the stars, since no material occurs in nature which is sufficiently hard to be used in engraving gems. Megenberg's book (quoting from Tethel) presents a long enumeration of gems and their sigils with the occult powers which they respectively possess. A few sentences will give an adequate picture of the contents of Tethel's treatise:

If one finds the stone called Jasper with the form of a man with a shield hanging from his neck or carried in one hand while he holds a spear in his other, with a snake at his feet, this is a protection from all enemies. A chrysolite with the figure of a woman holding a bird in one hand and a fish in the other has the power of furthering the wearer in all undertakings. A turtledove with an olive branch awakens love for all mankind. A white stone having a figure upon it which is half woman and half fish, holding a mirror in one hand and an olive branch in the other, makes its wearer invisible. A cross on a green jasper will prevent its wearer from drowning.

4. LATER LAPIDARIES, 1550–1650

Other well known lapidaries, all of which followed the same traditional treatment of the subject, appeared during the sixteenth and seventeenth centuries and may here be mentioned in the order of their succession.

Christophorus Encelius: De Re Metallica, 1551. The author was a physician in Thuringia, a pupil of Luther and a friend of Agricola and Mathesius. The book is prefaced by a letter from Melanchthon.

Andrea Cesalpino—De Metallicis, 1583.

Andrea Baccio—De Gemmis et Lapidibus,[32] *1603.*

Anselm De Boodt—Gemmarum et Lapidum Historia, 1609. This book which appeared in the same year as that in which Kepler discovered the laws of planetary motion, is in many respects the most important lapidary of the seventeenth century and exerted a widespread influence. After De Boodt's death two further editions in Latin and two in French, edited with annotations by Andreas Toll. were published.

De Boodt was a citizen of Bruges and a man who occupied a high social position, being physician to the Emperor Rudolph II in Prague; and who

[32] This is translated from an earlier Italian edition.

also acted as his advisor in all matters relating to gems and precious stones, a subject in which the emperor was much interested. He was moreover not only a mineralogist but a philosopher and a theologian. In his views concerning the nature and origin of gems and other stones De Boodt follows the stream of medieval teaching but in several respects his work shows the influence of the newer age in which he lived. While refuting the assertions of those who deny that gems possess peculiar virtues, calling forth our wonder and admiration, he says that these gems, being themselves produced by nature, cannot in and of themselves produce supernatural effects. Such powers as they possess are not exerted by them but through them, the stones being merely the media or instruments through which the powers of good or evil act. These powers may come directly from God Himself or from Him through some good or evil spirit used by Him as an instrument, the former acting at His will and the latter by His permission. He however contemptuously dismisses the usual stories concerning the magical and medicinal properties of gems put forward by those who sell and trade in these stones.

He was evidently impressed with the necessity for the introduction of some adequate method of classifying minerals and the recognition of this fact is in itself evidence that he was moving to break new ground for an advance in the science of geology.

The system of classification which he suggests and follows is however quite worthless, but one which might present itself to a person who, like De Boodt, was interested in the study of a class of objects of whose real composition, character, structure and origin he was ignorant. He presents a scheme in which minerals are first divided into two classes "Gems" and "Stones" according to their size, that is to say, according to whether they are minerals that are found in small individuals as for instance diamonds, or in larger individuals as in amethyst, agate or basalt. Each of these classes is again divided and sub-divided according to "rarity," "hardness," "beauty" and "transparity," with the result that minerals of the most diverse and unrelated kinds are thrown together in one and the same group. For those who do not approve of this classification, he presents another which while somewhat different embodies many of the same features as the first and is open to similar objections. The book however marks a forward step in the development of the science. It is subdivided into chapters each treating of a division of the subject or of a mineral or group of minerals and is also illustrated by a number of woodcuts.

Two other important lapidaries which were written under the influence of De Boodt appeared shortly after the publication of his work. The first was by Joannes de Laet[33] of Antwerp. De Laet, in his introduction, refers to De Boodt's book but expresses it as his opinion that much still remains to be said which is not to be found in the works of De Boodt and other authors, and these deficiencies he hopes to supply in his own volume. It is a much smaller book than that of De Boodt and contains first an annotated translation of the work of Theophrastus "On Stones" from the original Greek into Latin, and then passes on to describe the more important "fossils."

The second lapidary is that by Thomas Nicols, "Sometimes of Jesus College Cambridge," the son of Thomas Nicols, M.D., a Cambridge physician. A certain amount of confusion has arisen concerning this book owing to the fact that three issues of it appeared during Nicol's lifetime, each bearing a separate title.[34] These are as follows:

1. *A Lapidary, or the history of the pretious stones* by Thomas Nicols, Cambridge, 1652.
2. *Arcula Gemmea, or a cabinet of jewels* by Thomas Nicols, London, 1653.
3. *Gemmarius Fidelis, or the Faithful Lapidary* by Thomas Nicols of Jesus College in Cambridge, London, 1659.

All three are identical with the exception of the title pages.

Nicols follows De Boodt closely and quotes him frequently. He adopts his classification of gems and his views concerning the occult powers which certain of them possess, as derived from the divine Being, acting through good or evil spirits, the stones being intermediaries, and thus they are "Oft times the habitacles of *daemones* and intelligences which Johannes Langius in his epistle calleth *syderum & orbium motores*." The introduction to his chapter entitled *Of the Emerauld or Smaragde* is an interesting example of the quaint phraseology of the time:

The Emerauld is a precious gemme or stone of so excellent a viridity, or spring-colour, as that if a man shall look upon the Emerauld by a pleasant green meadow, it will be more amiable than the meadow, and overcome the meadows glorie, by the glory of that spring of viriditie which it hath in itself: The largeness of the meadow it will overcome with the amplitude of its glory, wherewith farre above its greatnesse it doth feed the eie: and the virescencie of the meadow it will overcome with the

[33] Joannes de Laet, *De Gemmis et Lapidibus libri duo, quibus praemittitur Theophrasti liber de Lapidibus*, Leyden, 1647.
[34] See the *Gentleman's Magazine*, December 1842, p. 594. Also the Catalogue of the Library of the British Museum and that of the Bodleian Library.

brightnesse of its glory, which in it self seemeth to embrace the glorious viridity of many springs. This stone is known by its apparent coldness in the mouth, by its gravity being weighed; and in this, that being cast into a fire, it will not burn, nor send forth any flame, and that in the brightnesse of the Sunne, it will keep its excellent viridity and greenness.

And in referring to pearls he puts forward a quaint and charming theory of their origin, which however did not originate with him:

The Margarites and Unions differ in the manner of their generation, from the generating of other Gemms or pretious stones, for these are generated of the pearly drops of chrystall morning dew, drunk in by the shell fish called Scallops and Cheripo ... and are increased by the new addition of fresh draughts of purest chrystall dew, even by fresh supplies of that purest restorative liquor taken in as morning draughts to serene and chearfull days.

5. LARGER WORKS OF THE SEVENTEENTH CENTURY

Finally, in bringing this chapter to a close, two other books should be mentioned, those by Caesius of Modena[35] and by Aldrovandus of Bologna. These differ in many ways from the lapidaries which have just been considered and of which they are the successors. The lapidaries are all small volumes, these are great folio tomes. Appearing toward the middle of the seventeenth century after the lapidaries had had their day and when the "New Learning" had already won its way in Europe, these encyclopedic works aim at presenting in its entirety the whole body of knowledge (or opinion) concerning the mineral kingdom which had been accumulated in times preceding their publication.

Caesius

With regard to the work of Caesius, the estimate of it given by Webster[36] in his usual blunt though forceful manner presents it in its true character:

The Jesuite Bernardus Caesius writ a voluminous Piece of Mineralogie, or Natural Philosophy; wherein, though he expatiated too far to fetch in all things that might seem any way of kinred to that kind of knowledge; and that it was but a meer Collection and heap stoln from other Authors, and hardly any thing except notions; yet is there something in it (especially concerning the signs of discovering Mines and Ores)

[35] *Cesi or Cesio*—or in its Latin form *Caesius (Bernardus)*. *Mineralogia sive Naturalis Philosophiae Thesauri in quibus Metallicicae concretionis medicatorumque fossilium miracula, terrarum pretium continentur.* Lyons, 1636.

[36] *Metallographa, or an History of Metals*, London, 1671, p. 30.

PLATE II

PORTRAIT OF ALDROVANDUS

(From Fantuzzi: *Memorie della Vita di Ulisse Androvandi*, Bologna, 1774)

that may advantage such a Reader, as hath the skill, or will take the pains to sever the tares from the wheat, and separate the gold from the dross.

The book is printed in double columns of rather small type but is excellently indexed so that while, as Webster says, it contains but little that is new, such detailed references are given to the original works from which its statements are derived that it is of great value as a guide to those who desire to explore the jungle of the earlier literature and find their way to the sources of the "notions" to which Webster refers.

Aldrovandus

Aldrovandus was one of the most renowned naturalists of the sixteenth century and occupied the chair of Natural History in the University of Bologna. He was a man of very wide learning who had travelled extensively and prosecuted his studies with great diligence. Being suspected at one time of coming under the influence of the teachings of Luther, "A heresy," his biographer Fantuzzi[37] says, "whose poison was then permeating ever more widely through Italy," he was arrested by the Inquisition and imprisoned at Rome, but was subsequently liberated. He was the author of a number of great volumes on insects, fishes, birds and quadrupeds respectively, as well as one on serpents and dragons and left on his decease a large collection of unpublished papers. He refers to these in his will, which is reproduced in Fantuzzi's book and in which it is stated that one volume or package of these papers is labelled *Geologia ovvero de Fossilibus*.

It may be of interest to mention here that this is the first instance in which the word *Geologia or Geology* appears in literature when used approximately in its present sense.[38] So far as is known at the present time, Lovell's work, *Pammineralogicon or an Universal History of Minerals*, etc. published in 1661, is the first work in the English language in which the word "geologia" appears.

The work of Escholt, written in the Danish language but published at Christiania in 1657, is the first printed work in which the word occurs.

But Aldrovandus employed the word, in some manuscript notes and

[37] *Memorie della Vita di Ulisse Aldrovandi medico e filosofo Bolognese*, Bologna, 1774, p. 12.
[38] See Frank D. Adams, *Earliest Use of the Term Geology*, Bull. Geol. Soc. of America, Vol. 43, 1932, p. 121, and
Further Note on the Use of the Term Geology, Bull. Geol. Soc. of America, Vol. 44, 1933, p. 821.

in his will, essentially in the modern sense, at least as early as 1605, the year in which he died.

The word "geologia" was used at a much earlier time, but in an entirely different sense, by Richard de Bury, Bishop of Durham, in his interesting and entertaining little book, entitled *Philobiblon, or the Love of Books.* In this work, first printed at Cologne in 1473, and which was first translated into English by J. B. Inglis (London) in 1832 and by E. C. Thomas in 1888 (an issue of this translation was subsequently published by the De la Mare Press of London in 1903), *the word "geologia," it is believed, made its first appearance in literature.* It was apparently invented, if this term may be used, by de Bury, who used it in an entirely different sense from that in which it subsequently came to be employed. By it, he designated the study of *Law*, which faculty, he says, "We may call by a special term *Geologia* or the earthly science," in antithesis to the sciences which aid in the understanding of divine things, comprehensively speaking, *Theologia.*

Aldrovandus himself published nothing about "Geologia," but some forty years after his death, Bartholomeus Ambrosinus compiled from the material left by Aldrovandus the great folio volume bearing the title: Ulyssis Aldrovandi Patricii Bononiensis Musaeum Metallicum in Libros IIII distributum Bartholomaeus Ambrosinus. . . Labore et Studio compDosuit. This was published at Bologna in 1648. While the materials used in its preparation were in all probability those mentioned in the will of Aldrovandus under the title "Geologia Ovvero de Fossilibus," the word "Geologia" does not appear in the volume itself. The four books referred to in the title of the work are comprised in a single volume. Each book is divided into a number of chapters, each chapter dealing with the consideration of a separate "fossil" or with some special topic. In a paper by Lodovico Foresti,[39] this author gives the modern names and discusses the recent opinions concerning the identity of the various organic "fossils" described in this work. The four books treat respectively of metals, earths, succi concreti and stones. Under this latter designation are included what are now known as minerals, rocks and fossils. Within each book there is no definite order in which the subjects are treated. The work, however, is provided with a good index. In each chapter the object or subject is treated of under a regular series of headings. Some of these, as for instance *Synonyms, Definition, Origin,*

---

[39] *Sopra alcuni Fossili illustrati e descritti nel Musaeum Metallicum di Ulisse Aldrovandi,* Soc. Geol. Boll., Italia, 1887, p. 81.

Nature and Properties, Varieties, Mode and Place of Occurrence, Uses, Historical References, are such as might be found in a modern textbook. Others however, as for instance, *Sympathia et Antipathia, Mystica, Miracula, Moralia, Mythologica, Somnia, Symbola* and *Lapidati* remind us that the book was written in an age far removed from the present.

As instances of sympathy and antipathy he gives the following: The diamond is said to have an antipathy to the lodestone or magnet, since when it is present iron is no longer attracted by the magnet, and if a piece of iron has already been drawn to itself by a magnet, a diamond brought near it will cause the magnet to repel the iron. Such was the opinion of Pliny, Marbodus and many other authors. Sympathy and antipathy are also seen among metals. Thus lead is wooed by gold and silver for these when fused together unite readily, on the other hand bronze shrinks away from lead, as also does tin from gold and silver.

Under the head of *Mystica* he refers to various mystical references to stones in the Bible.

Under the heading of *Somnia* many portents indicated by stones or falls of meteorites seen in dreams are mentioned, drawn from the works of various ancient authors.

Under *symbola* he mentions symbolic teachings set forth in the rocks, as for instance in that great stone mentioned and figured by Costalius, which so long as it remained intact floated freely on the waters of the sea, but when broken at once sank to the bottom, symbolizing the importance of harmony and concord among peoples, acting for their protection when maintained but when broken leading to their ruin. Other examples of mystical and moral lessons drawn from amber will be found in Chapter XIII.

An outstanding feature of the great work of Aldrovandus is that it is copiously illustrated by hundreds of woodcuts. These are rather roughly executed but are full of interest, many of them occupying the whole folio page; most of them, however, are smaller. They represent all the varied objects which are referred to in the text. Since specimens of rocks and those of many minerals do not lend themselves easily to pictorial representation, some of these cuts can scarcely be said to illustrate the text, they rather require the text to explain them. The fossil shells however are much more clearly reproduced. Interspersed with these are drawings of imitative forms of plants and of all manner of animals, also of various members of the human body, hands, feet and the internal organs, whose spontaneous growth in the form of stones, suggests the action of some hidden and mysterious force in nature. Also strange pictures (undoubtedly much elaborated by the imagination of the artist) representing various ecclesiastical subjects, as well as the saints and even the Savior himself, revealed on breaking open blocks of marble. Belemnites, teeth of mammoths, bezoar stones, the "Precious Jewel" which the toad bears in its head, gems engraved with mystical devices, the Tartars with their sheep and camels all turned to stone, as well as many implements, weapons, domestic vessels, and idols of barbaric peoples (some of them of Neolithic age) but to him equally mysterious in origin are also depicted.

In reading Aldrovandus, however, one gathers the impression that he was writing at the opening of a new age, and that the authority of the past did not exert the same binding influence upon him as it had done upon the writers of the old lapidaries. He sets forth nevertheless the medicinal properties of various rocks, minerals and gems of which he treats, when taken internally or applied externally, but in almost all cases cites certain ancient writers as responsible for the statements which he makes. He says very little concerning the occult properties of these bodies, and their value when used as amulets and charms, and unhesitatingly rejects (with Pliny) the power assigned by Marbodus and most of the medieval writers to the gem Heliotrope of rendering its wearer invisible. He, however, still retains a belief in the efficacy of certain gems as amulets, as in the case of the golden topaz which, when attached to the left arm or suspended from the neck so that it is in contact with the skin, will strengthen the vital powers and ward off bad dreams and melancholy, and also the stone jasper which, when carved into the figure of a dragon, may be used "ad dolorem stomachi."

6. CONCLUSION

Looking back, then, it will be seen that medieval mineralogy had its origin in classical sources and especially in Pliny, who in his turn gathered his material from earlier Greek writers, whose works, except those of Theophrastus, have been lost long since, as well as from a body of local gossip and tradition, which in its turn was probably derived from still more ancient and occult sources in Chaldea and elsewhere in the far east.

The minerals, gems and stones both in classical and medieval times were held in esteem and considered to be of value chiefly because of the medicinal and magical properties which they were believed to possess. With the infusion of Arabian learning in the Middle Ages, belief in the magical powers of minerals and especially of gems received an additional stimulus, for notwithstanding all its wonderful achievements, Arabic science belongs to the same world as the Arabian Nights—a world of magic and mystery—and the man of science was the man who could control these mysterious forces by the power of secret knowledge.[40]

Each writer on medieval mineralogy based his statements on the authority of previous writers, and was in his turn quoted as an authority by those who followed him. Finally the later encyclopedists like Caesius and Aldrovandus (or rather Ambrosinus) made it their aim to bring together every fact and fancy which had been propounded by anyone anywhere and adding but little of their own, to incorporate it into one portentous whole. Osler says essentially the same thing about the history of medicine, "From Hippocrates to Hunter the treatment of diseases was a long traffic in hypotheses."

Medieval mineralogy in fact was not a science. It was not a solid tower of learning, as Gregory Reisch[41] pictures the knowledge of his day, but a fairy castle, the insubstantial fabric of a dream, often quaint and even beautiful, but destined to crumble away because it had no foundation in reality. It was now to be succeeded by a true science of mineralogy built upon the basis of close observation and diligent study of the materials of the earth's crust.

[40] Christopher Dawson: *Mediaeval Religion and other Essays* (The Forwood Lectures, 1934), London, 1934, p. 88.
[41] *Margarita Philosophica*, Strassburg, 1504 (frontispiece).

CHAPTER VI

THE BIRTH OF MODERN MINERALOGY AND ITS DEVELOPMENT FROM AGRICOLA TO WERNER AND BERZELIUS

I

As has been seen writers in Classical and Medieval Times, and indeed down to the beginning of the fifteenth century, with the exception of Theophrastus and Vitruvius, considered minerals as bodies which were of interest and importance because they were supposed to possess certain mystical or medicinal properties or, in the case of the metallic minerals, because metals could be smelted from them.

It is now recognized that chemical composition and physical characters, especially crystallographic form, are the true bases for the classification of minerals. But the ancients knew little or nothing concerning the composition of minerals and very little about any of their physical characters. That character on which they laid particular emphasis was color, which is really one of the least important characters of a mineral for purposes of classification, since the same mineral often has many different colors. This frequently led the ancients to believe that different colored varieties of the same mineral were different species, while on the other hand they often classed several entirely different minerals together as one species because they had the same color.

They had therefore no satisfactory basis for the classification of minerals, and from the time of Pliny onwards almost without exception those who wrote about these members of the Mineral Kingdom arranged them alphabetically according to the initial letter of their names quite irrespective of their character or composition. With the opening of the fifteenth century these old ideas concerning the mystical properties of minerals faded away and minerals commenced to be made the objects of careful investigation and research.

This change in the general attitude of human thought with reference to nature, so far as the mineral kingdom was concerned, received a powerful impetus through the discovery and development of great mineral deposits in the Saxon Erzgebirge, not only in Saxony itself, but on the other side of this mountain range in Bohemia, as well as in the Harz Mountains situated immediately to the west and in Hungary to the east of this region.

170

Mining was begun at Schemnitz in Hungary as early as 745 A.D.[1] This was followed by the development of the mines at Goslar in the Harz in 970, at Freiberg in Saxony in 1170, in the Schneeberg district of Saxony in 1420, at Annaberg in Saxony in 1495, in Joachimsthal in Bohemia shortly before 1520, and at Andreasberg in the Harz in 1570.

Mining was inveighed against and held in disrepute by Pliny and many of the classical writers, including Seneca, Horace and Ovid, on the ground that the precious metals pandered to men's luxuries, while iron was used to promote war with all its concomitant miseries. Furthermore, the mining of these metals entailed upon those engaged in this work untold hardships and even loss of life. Writers in later times also enlarged upon this same theme.[2]

A suggestion that there was a certain feeling of opposition to the invasion of the agricultural and farm lands of this portion of Northern Europe by the miners at this time, on the part of some of its inhabitants at least, may be gathered from a little book entitled *Judicium Jovis in valle amenitatis habitum ad quod mortalis homo a terra tractus propter montifodinas in monte Niveo aliisque multis perfectas ac demum parricidii accusatus.*

This is a very rare and interesting little work which is undated and bears no author's name. It is known from other sources,[3] however, that it was written by Dr. Paul Schneevogel, or, in the latinized form of the name, Niavis, between the year 1490 and 1495, and that it was printed by Conrad Kachelhofen in Leipzig. Niavis was a professor in the University of Leipzig. As a matter of fact, he says in the introduction, that the document was originally written in the "vulgar tongue" by some one whose name he does not mention and that it was brought to him with the request that he would translate it into Latin in order that it might secure wider publicity. It tells with humorous touches here and there, of a certain hermit who lived in a lonely cell in the mountains near Lichtenstat (a little Bohemian town on the border of Saxony). Near his cell there was a shrine to which he was accustomed to resort for prayer. One day, "Following the festival of the apostles Philip and James about

---

[1] These dates are taken from the *Meissnische Berg Chronica* by Albinus (Dresden 1590). Zittel and others are of the opinion that in Bohemia and Moravia mining commenced even earlier.

[2] Kirchmaier, G. C., *Institutiones Metallicae*, Wittenberg, 1687.

[3] See Gründig, G. S. and Klotzsch, J. F., *Sammulung Vermischter Nachrichten zu Sachsischen Geschichte*, vol. 1, Chemnitz, 1767.

the year 1475," being weak in body he lay down on a sunny bank to rest himself. He fell asleep and on waking set forth for home, but lost his way in the forest and came upon a beautiful little valley in which, to his great surprise, he saw an enclosure with several gates within which was a royal throne on which sat a king, with all his royal insignia, grave and distinguished in appearance, on whose head was a golden crown gleaming with pearls, on the front of which appeared the words, "The Father helper of all." The hermit approached quietly and sat down by one of the gates to watch what might happen. He saw that the king was Jupiter on his seat of judgment. There came forth and stood at Jupiter's right hand a woman—Earth by name—"noble and freeborn, clad in a green robe, who walked like a person rather mature in years." Her clothing was torn and the hermit saw that her body was pierced in many places. She was followed by Mercury, Bacchus, Ceres, a Naiad, Minerva, Pluto, Charon and others, who gathered around her. There then came forth a *miner* accompanied by three strange little forms who were his *Penates*. These stood on the left hand of Jupiter. The miner wore his pointed cap and his "Kaputze" and carried his tools, a hammer and a miner's pick. This is probably the earliest picture of a miner that we have. The "Glib-tongued Mercury" first addressed Jupiter, and in true legal style set forth at length the complaints of these persons, charging the miner with *parricide*. Then several of the complainants enlarged upon the individual injuries which they had sustained. Bacchus complained that his vines were uprooted and fed to the flames and his most sacred places desecrated. Ceres stated that her fields were devastated; Pluto that the blows of the miners resound like thunder through the depths of the earth, so that he could hardly reside in his own kingdom; the Naiad, that the subterranean waters were diverted and her fountains dried up; Charon that the volume of the underground waters had been so diminished that he was unable to float his boat on Acheron and carry the souls across to Pluto's realm, and the Fauns protested that the charcoal burners had destroyed whole forests to obtain fuel to smelt the miner's ores.

The miner refutes these charges at length. The various districts of the earth yield different products, one abounds in crops, another in wine, another in timber, and so on. Furthermore, it had been ordered by Jupiter himself that the deficiencies of one region should be supplied from the abundance of another. No more convenient medium of exchange exists than gold transformed into money. Earth—which takes

the name of Mother, and proclaims her love for man—hides and conceals
this metal in such a way in her inward parts that she fills the rôle of a
step-mother rather than a true parent. Metals provide the human race
with the means of helping one another. It is an act of piety to help the
poor and a holy act to decorate the temples and adorn the statues of the
Gods. Who will practise virtue if it finds no reward, and what rewards
entice more powerfully than money, for it can purchase everything that
the human race lacks? "Certain high philosophers have said that a
man cannot find freedom for philosophy if he is poor" and it is he who
has no money that is poor. He can nowhere find a teacher or a school.
What is it that cures all illness and is called "the universal cure" except
gold in its liquid form (i.e. "potable gold"). The miner states, furthermore, that he is carrying on his work so that mortals may cultivate the
earth and carry out Jupiter's commands.

Jupiter, "In order that he might not appear to lean to either side"—
that is to say, to interfere with the free course of economic development—
writes a letter passing the whole matter on to "Fortune" for decision.
Fortune, the "Queen of Men" as she calls herself, replied at once.
"What thou dost write concerning men's labour surprises me not at all.
For such is their condition by nature, that if they knew they must perish
in the evening, thou wouldst observe their mind to be elated even in the
morning. But their eloquence they derive from me when they are so
voluble and audacious. But, to be brief, my opinion, O Highest, is
this: Men ought to mine and dig in mountains, to tend the fields, to
engage in trade, to injure the Earth, to throw away Knowledge, to
disturb Pluto and finally to search for veins of metal in the sources of
rivers, their bodies ought to be swallowed up by the earth, suffocated by
its vapours and intoxicated by wine, and afflicted with hunger and remain
ignorant of what is best. These and many other dangers are proper for
men. Farewell."

"This the hermit heard and after he had reached the road he ascended
the mountain and soon saw his own dwelling and returned to it."

The little book has, as a frontispiece, a quaint, medieval woodcut
depicting the incidents of the story (see figure 30). Three scenes are
represented in the single picture. Above, the hermit is seen, first at
prayer, then descending to the valley where he saw the vision. Below,
Jupiter is seen enthroned with the various actors in the scene and the
hermit seated at the gate intently listening.

The great development of mining which took place in this relatively

small area about the end of the fifteenth and the beginning of the sixteenth century made it one of the richest and most prosperous districts

FIG. 30. Frontispiece in the *Judicium Jovis* by Niavis.

in Europe. Populous cities sprang up in all directions and the large
number of men from the mining centers who rose to distinction in all
walks of life showed that the development was not one of wealth and

material prosperity only, but also of high civilization and culture. Furthermore, as time went on, through this great mining industry not only were great advances made in the technology of Mining, Ore Dressing and Metallurgy, but owing to the close observation and practical study of the earth's crust which accompanied the development there was accumulated a great body of information and experience concerning all branches of knowledge connected with the geological sciences. It was here that the science of Mineralogy in the modern acceptance of the term, took its rise.

While many men, especially during the latter part of the period under review, contributed to the development of the new science of mineralogy, five were outstanding and will be selected for consideration here. Each of them put forward a "System of Mineralogy," all of which Systems are imperfect and inadequate in the light of present knowledge, but they display five successive advances in the development of the science of mineralogy, and are of especial interest on that account.

These men are Agricola, Gesner, Werner, Haüy and Berzelius.

The first—*Agricola*—who has been called the "Father of Mineralogy" wrote his book *De Natura Fossilium* in 1546.

The second—*Gesner*—who on account of his encyclopedic contributions to knowledge, has been termed by Desalliers d' Argenville[4] "The German Pliny," was a contemporary of Agricola, and wrote a few years later, in 1565.

The other three lived some two centuries after. *Werner* (1750-1817) and Haüy (1742-1822) were contemporaries.

Berzelius was a little later although he was still alive when Werner published his *Oryktognosie*, (1792) and his *Letztes Mineral System* (1817).

Of the first two of these renowned mineralogists the work of Agricola is indisputably the more important, but it will be well to consider that of Gesner first, for two reasons—first, because, although his book appeared nineteen years after that of Agricola, he stood in closer relation to the medieval writers of the old school which was passing away; and secondly, because Agricola was a fellow countryman of Werner both being residents of Saxony, and the former being the first mineralogist of what has been called the Freiberg School,[5] while Werner was the last, at least during that period of which the present chapter treats.

---

[4] *L'Oryctologie*, Paris 1755. This title was also given to Agricola by Gesner and others. See Hoffmann's *Georg Agricola*. Gotha, 1905, p. 2.

[5] See Chapter on the *Origin of Ore Deposits*.

II

Conrad Gesner was born at Zurich in 1516.[6] He belonged to a family that was poor and of no outstanding position in the community. When he was 15 years old his father, a furrier, fell fighting for the Protestant cause in the battle of Kappee, and throughout his life he had a hard struggle to compass his material needs. In one of his letters he says that throughout his life while working for his daily bread—he was ever hurried onward by "Zwei hart Gebietende Göttinnen—die Durftigkeit und die Notwendigkeit." This however seems to have been due in part to the fact that he was, as his biographer observes, "Ebenso schlechter Kaufmann wie grosser Forscher." He received a good classical education, taking his degree at the University of Basel, and while intending at first to enter holy orders, later turned his attention to medicine and finally devoted his life chiefly to the study of natural science. He was a man of deep religious convictions, following Calvin and Zwingli, and having among his friends Beza, Peter Martyr and Hooper the Bishop of Gloucester, who in a letter to Bullinger says that he is sending to Gesner "A Welsh dictionary and some writings in the language of Cornubia, commonly called Cornwall, and should Master Gesner wish at any time to come over to us I will provide him with suitable companions, who will show him the rivers and fishes and animals of this country."[7]

In order to prepare himself for his work Gesner studied many languages. He spoke Latin, Greek, Hebrew, German, French, Italian, Dutch and probably a little English, and had a reading knowledge of Arabic. He wrote no fewer than 72 books which were published during his lifetime and about 18 others in manuscript at the time of his death. These embraced a wide range of subjects and many important contributions to medicine and theology, but those on natural history are the only ones which are of especial interest to us here.

In the study of nature it was botany that especially attracted Gesner, but he is known chiefly as a zoologist. On this subject he wrote five great folio volumes, two of them treating of Quadrupeds, and the others of Birds, Fishes and Serpents respectively, and in addition, other folio volumes of pictures of various classes of animals with notes concerning them. On his favorite subject, botany, he wrote three great volumes, while a fourth, in which he was to present a much more complete and

---

[6] Willy Ley, *Conrad Gesner, Leben und Werk*. Munich, 1929.

[7] *Reformation Contacts with Zurich (1530–1575)* in *The Modern Churchman*, June 1933.

PLATE III

PORTRAIT OF CONRAD GESNER

comprehensive treatment of the subject remained unfinished at the time of his death.

These works were of necessity largely compilations. Gesner's aim was to embody the work of all previous writers and then to complete the treatment of the several subjects by information drawn from his own observations, or from those of the great host of naturalists in all countries with whom he maintained a continuous correspondence. Most of his works on natural history furthermore were elaborately illustrated, so that he was obliged to keep a draughtsman and an engraver continuously at work in this house preparing illustrations for his works. The labor entailed in the writing of these works was so great that he was obliged, year in and year out, to rise early and to continue his labors late into the night, so that he was continuously taxed to the extreme limit of his endurance.

At this earnest request he was, in 1554, appointed to the position of Stadtarzt in Zurich, from which position he hoped to receive an income which would supplement that which he derived from the publication of his works, and thus relieve him of financial worry. An interesting side light is thrown upon the troubles of a harassed author at this time by Gesner's complaint that the printers will accept only large books which require an immense outlay of time and care in their preparation, and will not print small books even if the author is willing to furnish them with the manuscript gratis.

After a life of intense toil and anxiety Gesner died of the plague in Zurich in the year 1565 at the age of forty-nine, beloved by all who knew him. It is recorded that when he felt the hour of death had come, he caused himself to be taken to his study that there, where the greater part of his life had been spent, he might pass away, surrounded by the books which had been his constant companions for so many years.

His book on *Fossils, Stones and Gems* was written in the year of his death, while his great work on plants still remained unfinished. It was the last of his books which appeared before his death. It is a work of small octavo size, consisting of 169 leaves with a dedication and a preface and is stated by Gesner to be the first book on the subject of which it treats to be illustrated by cuts and figures. This is substantially correct, although the *De Re Metallica* by Encelius, which was published at Frankfort in 1551 contains four small wood cuts, two of shells and two which represent varieties of aetites, all of which are reproduced in Gesner's work.

Gesner's book bears the following title: *De Rerum Fossilium, Lapid um et Gemmarum maxime, figuris et similitudinibus Liber: non solum Medicis, sed omnibus rerum Naturae ac Philologiae studiosis, utilis et jucundus futurus,* Zurich, 1565. It is one of eight short works on various mineralogical subjects by different writers assembled in one stout volume of small octavo size having apparently been brought together by Gesner. This composite volume was published in Zurich in 1566.

This book Gesner states, was written offhand and rapidly as a pleasure and recreation, and was intended to stimulate and encourage all students who were interested in minerals, fossils and stones to continue their studies of these objects and to communicate with him concerning any points on which he was in error and to indicate any directions in which this book should be extended and elaborated, since he intended later to write another and larger work in which the whole subject would be treated more fully and completely. He hoped that the illustrations in the book would be of assistance to such students and refers to the difficulty which he experienced in giving adequate pictorial representation of minerals and rocks, as compared with animals and plants, especially without the aid of color, which in the present case he was unable to employ.

One of the first requirements in the presentation of a satisfactory System of Mineralogy was, of course, the selection of some adequate basis of classification. The writers of old lapidaries, as has been stated, arranged the minerals in alphabetical order according to the initial letter of their Latin names.

Gesner throughout his life had been feeling his way toward a true system of classification in dealing with the Kingdoms of Nature as they were then called. He treated first of the Animal Kingdom in his great work entitled *Historia Animalium* in five large volumes, dealing respectively with "land animals" (2 volumes), "birds," "fishes and water animals," and "snakes" (reptiles in general). While a general classification of the animal kingdom was thus adopted, the arrangement of the animals in each volume was merely alphabetical. He did not consider the alphabetical arrangement to be altogether satisfactory, but he says that it at least enabled the reader to find with ease any animal concerning which he desired information.

In his great work on the vegetable kingdom on which he was at work in the last years of his life, and which was still unfinished when he died, but was completed by Schmiedel and published in 1753 under the title

Opera Botanica Conradi Gesneri, it was found that Gesner had devised and adopted a true system of classification in which plants were arranged in Orders, Families, Species and Subspecies based on certain mutual resemblances, a system which is the same as that which was worked out, probably independently, by Linnaeus and which now bears his name.

When, however Gesner came to describe the third kingdom of Nature as it was called, that is to say to write his little book on minerals and rocks, he was faced with a difficult problem. These had no well defined and easily recognized forms which could be used for purposes of classification and little or nothing was known of their composition. Furthermore what we now know to be the remains of animals or plants embedded in the rocks were then grouped together with minerals and rocks under the general term "fossils" and this miscellaneous lot of objects, which had in common only the fact, from which their name was derived, that they had been dug (fossus) out of the earth's crust, had to be embraced in any system of classification which was selected.

The method which he adopted was quaint and interesting—he explains it in the dedicatory epistle to his book *De Rerum Fossilium. . .Figuris.* He says that he is unwilling to adopt an alphabetical arrangement as most of the earlier writers had done, since he considers this method to be commonplace and trivial. Nor will he take substance as a basis of classification. He states that he intends later, if an opportunity presents itself, to write a second book in which he will describe all stones and metallic substances, explaining at the same time the derivation of their names. In the present work Gesner says he wishes to draw especial attention to the figures with which the book is illustrated and goes on to say that he always experiences a feeling of delight in the contemplation of the forms and shapes in which the Creatress Nature has expressed herself pictorially as it were with the brush of a painter. And he intends in the discussion of these fossils to follow in the steps of Nature herself and adopt a classification based on the forms displayed by her.

Taking first those fossils whose forms are of the simplest character resembling lines, angles, circles or allied forms, as seen in the heavens or in the elements, he will pass on to consider those which bear a resemblance to more complex forms, descending by degrees from the heavens and celestial things to the earth and things which live upon or are nourished by it, and consider those fossils which in their form resemble plants or animals or parts of these.

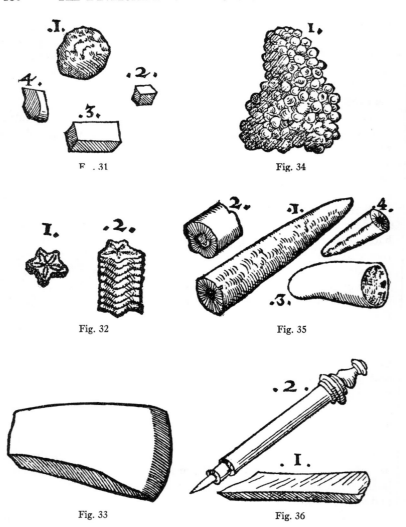

F . 31

Fig. 34

Fig. 32

Fig. 35

Fig. 33

Fig. 36

FIGS. 31 TO 36. Gesner's classes of fossils.   (From his *De Rerum Fossilium, Lapidum et Gemmarum . . . Figuris.*)   Natural size.
FIG. 31. Class 1.   Geometrical forms—Pyrites.
FIG. 32. Class 2.   Forms like heavenly bodies (stars)—Crinoids.
FIG. 33. Class 3.   Fallen from the sky—Stoneaxe.
FIG. 34. Class 4.   Like some terrestrial object—Peastone.
FIG. 35. Class 5.   Like an artificial object (dart)—Belemnite.
FIG. 36. Class 6.   Made from metal or stone—*Lead* Pencil.

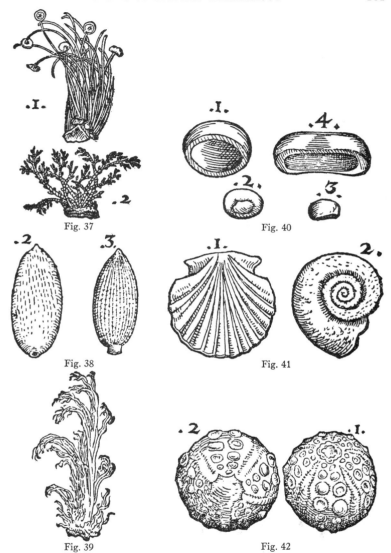

Fig. 37

Fig. 40

Fig. 38

Fig. 41

Fig. 39

Fig. 42

Figs. 37 to 42. Gesner's classes of fossils. (From his *De Rerum Fossilium, Lapidum et Gemmarum ... Figuris.*) Natural size.

Fig. 37. Class 7. Like plants or herbs.
Fig. 38. Class 9. Like parts of trees (fruit)—Judenstein.
Fig. 39. Class 12. Like parts of animals (hair)—native silver.
Fig. 40. Class 13. Name derived from birds or batrachians (in toad's head)—crapodina.
Fig. 41. Class 14. Like things in the sea—Pecten.
Fig. 42. Class 15. From insects or serpents—"Serpent's Eggs."

Following this method of classification the members of the mineral kingdom are brought into one or other of the following 15 classes.

Class 1 (fig. 31). Those whose forms are based upon, have some relation to, or suggest the geometrical conception of points, lines or angles. This includes all transparent or translucent minerals such as sunstone, within which there are minute and more or less *rounded* inclusions; also minerals upon whose surface little *dots* are seen, together with rounded concretions of all sorts and fossils having a more or less *rounded* shape. Also bodies like pumice stone, whose surface shows depressions often rounded in shape. It also includes asbestos, natrolite and all *fibrous* or *acicular* minerals also all minerals whose forms display *angles*, such as cubes of pyrite, "Judenstein," hexagonal prisms of quartz, pologonal columns of basalt and so on.

Class 2 (fig. 32). Those which resemble or derive their name from some heavenly body or from one of the Aristotelian elements. This class includes moonstone, selenite, carbuncle, aquamarine, opal and agate which often display little pictures of rivers or landscapes, also diamond which is clear and transparent like air, and Asterias (the separated elements of the stalk of a Mesozoic Coral) which in transverse section displays the form of a star.

Class 3 (fig. 33). Those which take their name from something in the sky. This embraces the fossil Echinoderms, as well as the Palaeolithic and Neolithic stone axes and other weapons, all of which were believed to have fallen from the sky. Also certain minerals and stones which display smoke-like markings or appearances which resemble clouds.

Class 4 (fig. 34). Those which are named after inanimate terrestrial objects. In this class are pisolite (peastone), chrysolite (after gold), iron ores, lead and other metallic ores.

Class 5 (fig. 35). Those which bear a resemblance to certain artificial things— belemnites (like darts) stalks of Palaeozoic crinoids (round like wheels).

Class 6 (fig. 36). Things made artificially out of metals, stones or gems. Rings, whetstones, seals, vases, the Mariner's Compass (made from magnetic iron ore), writing tables, lead pencils. He gives here the earliest picture of a "lead pencil." This was apparently metallic lead not "black lead" or graphite.

Class 7 (fig. 37). Those which resemble plants or herbs—plants turned into stone, (Encrinus 'the stone Lily') stones which have the form of apples and other fruits.

Class 8. Those which have the form of shrubs—Rhodites (which bear a resemblance to a rose), siliceous minerals and rocks which have branching coral-like forms.

Class 9 (fig. 38). Those which resemble trees or portions of trees. Trees which have been petrified and coal plants.

Class 10. Corals.

Class 11. Other sea plants which have a stony nature. Lithophyton and others.

Class 12 (fig. 39). Those which have some resemblance to men or to four-footed animals, or are found within these. Native silver in hair like forms, carnelian (blood stone), hematite, calculi, bezoar stones, ocellus, glossoptera (like tongues) and others.

Class 13 (fig. 40). Stones which derive their names from birds.

Class 14 (fig. 41). Those which have a resemblance to things which live in the sea. Fossil fish, glossoptera (sharks teeth), orthoceras, pecten and other sea shells. *Class 15* (fig. 42). Those which resemble insects or serpents. Ammonites, ophites, and others. Here he describes the first nummulites from the Paris Basin and refers them to the ammonites.

It is interesting to note that some fossils which resemble the remains of living animals he considers to be such, while others in which the resemblance is less distinct he thinks may not be such, but may have originated through some forces at work within the earth itself.

It will be noted that in each class or division there are brought together a most miscellaneous group of bodies which have no relation to one another other than that of a supposed resemblance in shape. The grouping is interesting however, as presenting one of the very earliest attempts to classify the members of the mineral kingdom.

As Gesner died immediately after this book was printed he never wrote the second work which, as above stated, he intended to prepare later.

This little book is of especial interest in that it presents a picture in miniature of the mineral kingdom as seen through the eyes of the greatest naturalist of his time. It marks a departure from the medieval lapidaries in that it says little or nothing concerning the miraculous properties and "virtues" which had been attributed to minerals but endeavours rather to describe the characters by which they may be recognized. It is in fact a more or less popular handbook for amateurs and others, written to arouse an interest in "fossils" and to encourage students to collect and study them.

With regard to the sources from which it is drawn, it is seen to be a compilation of the observations and opinions of other men, supplemented in a number of instances by those of Gesner himself. He owes much to Kentman and he borrows largely from Agricola, although he does not acknowledge his indebtedness to the latter writer as fully as he ought to have done.

III

Turning now to Agricola (1494–1555), we come to one of the most outstanding figures in the history of the geological sciences, not only of his own times but of all time. Goethe compares him to Lord Bacon.[8]

[8] See *Farbenlehre*.

Quid Medici poffent manibus? quas iungere plagas
Vlceribus fordes, figna mouere loco?
Extitit hic folus qui pondera, vifcera Terræ
Rimatus, nobis bella metalla fodit.

FIG. 43. Portrait and signature of Agricola.   It is probably in a collection of portraits made by Sambucus who died in 1584 shortly after Agricola's demise.   (From Hofmann: *Dr. Georg Agricola*, Gotha, 1905.)

Werner, himself a great mineralogist, calls him the Father of Mineralogy,[9] and Vogelsang[10] remarks that since the science of geology developed out of the older science of mineralogy he might also be called the Forefather of Geology. He was born in Glaucau in 1494 and was thus not only a contemporary of Gesner but of Leonardo da Vinci, Copernicus, Luther, Erasmus and Paracelsus—a great humanist and a son of the Renaissance. His true name was Georg Bauer, but as was the custom of many learned men of the time, he adopted a latinized form of the name and was always known as Georgius Agricola. Unlike Paracelsus who generally wrote in German, Agricola invariably wrote in Latin.

Little or nothing is known of his boyhood. At the age of twenty years he entered the University of Leipzig and subsequently went to Italy and continued his studies at the universities of Bologna and Padua. He began his career as a philologist and thus acquired a thorough knowledge of the classical languages and of the works of the writers of classical times, to which, like all the humanists, his mind continually reverted.

During the sixteenth century Italy stood out from among all the countries of Europe as foremost in the cultivation of the natural sciences, medicine and mathematics. Erasmus speaking of Italy at this period says that it was a country "Whose very walls are more learned and eloquent than men are in our regions."[11]

Students from all lands flocked to the Italian universities to prosecute their studies. Agricola from his classical studies turned his attention to medicine, took his degree in this subject at the university of Ferrara, and adopted medicine as a profession.

During his stay in Italy, when in Venice, he for two years acted as a collaborator in the renowned printing and publishing house founded by Aldus Manutius. Upon his return from Italy, Agricola was appointed to the position of City Physician ("Stadtarzt") in the flourishing mining town of Joachimsthal in Bohemia, where he remained from 1527 to 1533. He took up his residence in Chemnitz, another celebrated mining town, in 1534, where he remained during the rest of his life, practised as a physician, occupied the position of "Stadtarzt" and in 1545 was appointed Burgermeister.

Agricola was thus brought into intimate touch with the mining industry in all its ramifications, in what was at that time the greatest

[9] *Neue Theorie von der Entstehung der Gänge*, Freiberg 1791, p. 10.

[10] *Philosophie der Geologie*, Bonn, 1867, p. 58.

[11] See R. W. Chambers, *Life of Thomas More*, London 1935, p. 74.

mining region in Europe. He continually associated with mining men, and in their company spent much time in the mines and smelters. Thus while practising his profession of medicine he also acquired an intimate and practical knowledge of mining, mineralogy and the allied sciences, so that when writing on these subjects he brought to his treatment of them a wealth of new knowledge derived from direct and long continued personal observation and research.

Agricola, while at first strongly drawn towards the Reformation on account of its moral teachings did not, like Gesner and so many others, leave the Church of Rome, being attracted to it on account of his aesthetic and humanistic tastes and by his love of the ceremonial of that church. The democratic attitude of many of the Reformers was also distasteful to him, and on this account his adherence to the Church of Rome became more pronounced in his later years.

While most of his writings deal with the geological sciences, there was scarcely a field of human endeavor in which he did not take a part and to which he did not make important contributions, and but few subjects worthy of investigation escaped his keen attention.[12]   On this account Gesner speaks of him as the Saxon Pliny.

The many editions of Agricola's books which were printed shows the wide demand that there was for them when they first appeared, yet strange to say by the second half of the seventeenth century Agricola was almost forgotten, although Alstedius[13] writing in 1649 refers to him as *the* great authority on rocks and minerals.   It was only toward the end of the eighteenth century that the interest in Agricola revived, when Werner in his lectures drew attention to his work.   Even Von Kobell writing as late as 1864 barely mentions his name.[14]   Raspe[15] in the introduction to his translation of Born's letters to Ferber, writing in 1777 says: "Father George Agricola is undoubtedly the first and I dare say, till very late, unparalleled in respect to some scientific knowledge

[12] See Reinhold Hofmann, *Dr. Georg Agricola*, Gotha 1905, pp. 4 and 74.   This little volume gives an excellent account of Agricola's life and work.

Also Ernst Darmstaedter, *Georg Agricola Leben und werk*, Munich, 1926.   This gives a fuller account of his writings.

[13] *Scientium omnium Encyclopaediae*, Lyons, 1649, Vol. II.

[14] *Geschichte der Mineralogie*, Munich, 1864.

See also—Adalbert Wrany, *Die Pflege der Mineralogie in Böhmen*, Prag, 1896.

[15] Travels through the Banat of Temeswar, Transylvania, and Hungary in the year 1770 described in a series of letters to Prof. Ferber on the Mines and Mountains of these different countries by Baron Inigo Born &c, by R. E. Raspe, London, 1777, p. xxxiii.

of the veins, their run and their rules. What he knew and drew of it, he knew from the miners; but as ever since they have scarce been consulted at all by philosophers who attempted to create and to dream mountains and worlds, and systems of mountains and worlds, it is no wonder that hitherto the learned should have so little added to that stock of science which he has left us." Löhneyss in his *Bericht vom Bergwerck* which appeared in 1617, drew largely on Agricola's *De Metallica*, as well as incorporating verbatim almost the whole of Ercker's celebrated treatise on assaying which was published in 1574, changing only the order of the text and this without a word of acknowledgment to either author.

It may be of interest here to digress briefly in order to make a further reference to Rudolph Eric Raspe just mentioned, who had an eventful, checkered and unenviable career. He was born in Hanover of poor parents and studied at the universities of Göttingen and Leipzig. For a time he held an appointment at the latter seat of learning and translated the philosophical works of Leibnitz and the poems of Ossian. He was then appointed to a professorship in Cassel and became the custodian of the collection of gems and curios of the landgrave of Hesse, he however purloined some of these and when threatened with arrest escaped to Great Britain where he spent the rest of his life. He was described by the authorities of Cassel in the advertisement for his arrest as "Councillor Raspe, a long faced man with small eyes, a crooked nose, red hair under his stumpy periwig and a jerky gait."

In 1769 he communicated a paper to the Royal Society of London on the former existence of mammals in northern regions and was elected an honorary member of the Society, but on account of later occurrences his name was expunged from the roll of membership.

He wrote on a very wide range of subjects and is the author of several books on geological topics, among these the best known is his "Account of some German Volcanos and their productions with a new hypothesis of the Prismatical Basaltes," which was written in English and was published in London in 1776. From 1782 till 1788 he was assay-master and storekeeper at the Dolcoath Mine in Cornwall. Thence he went to Scotland and ingratiated himself with Sir John Sinclair who was interested in mineralogy. He pretended to have discovered some ore bodies on the latter's estates, but when the fact that he had "salted" them was on the verge of discovery, he absconded to Ireland where in a remote part of Donegal, still masquerading as a mining expert, he died in 1794. He is believed to have been the prototype of Dousterswivel in Walter Scott's novel *The Antiquary*. Raspe's chief title to fame is that he was the author of that well known work *The Surprising Adventures of the Renowned Baron Munchausen* which has probably been translated into more languages than any English book except *The Pilgrim's Progress, Robinson Crusoe* and *Gulliver's Travels*.

It is only in more recent times that the full significance of Agricola's work and the great influence which he exerted on his contemporaries, on his immediate followers and through them on the development of science has been fully recognized. This neglect is possibly due in part at least to the form in which his works were published.

He wrote seven books dealing with geological subjects, which have come down to us.[16] These bear the following titles:[17] *Bermannus* (1530), *De Ortu et Causis Subterraneorum* (1546), *De Natura eorum Quae effluent ex Terra* (1546), *De Natura Fossilium* (1546), *De Veteribus et Novis Metallis* (1546), *De Animantibus Subterraneis* (1549), *De Re Metallica*, (1556).

In the treatise *De Veteribus et Novis Metallis*, Agricola gives a sketch of the history and geographical distribution of the various metals so far as these were known to the ancients and in his time, as well as an account of the development of mining in Germany and Austria. The *De Animantibus Subterraneis* is the least important of these works and deals chiefly with the creatures which live in underground places.

The first editions of all these works were from the celebrated press of Froben in Basel. The *Bermannus* was issued in small octavo form, and the others in folio. *Bermannus*, his first contribution to geological science, appeared in 1530, while his latest work *De Re Metallica* was published in 1556, the year after his death. This latter is the best known of his books and is a treatise chiefly on the arts of mining and smelting.[18] Its popularity was undoubtedly enhanced by the fact that it is illustrated by a large number of very interesting and attractive woodcuts, while his other works are devoid of illustrations.

[16] An interesting statement is made by Hofmann in his *Georg Agricola* p. 63, to the effect that an otherwise unknown work by Agricola entitled *De Metallis et Machinis*, Basel, 1543, was sold to someone in America in 1896 by the bookseller J. St.Goar of Frankfurt. Who the purchaser was is not known. It may therefore happen that this lost work of Agricola's will sometime in the future be discovered somewhere in America.

[17] See Ernst Darmstaedter, *Georg Agricola, Leben und Werk*.

[18] Reference is made in Chapter IX to Agricola's contributions to the subject of the origin and mode of occurrence of ore deposits.

In the second book of the *De Re Metallica*, Agricola treats at length of the forked twig or "mineral rod" and is of the opinion that the miner who uses it wastes his time and effort.

A lover of old books gazes with admiration at the opened pages of one of these great folios—the type is beautiful and the paper excellent. It is an admirable example of the printer's art. But a reader who wishes to ascertain Agricola's opinion on any subject views the volume from a

FIG. 44. Miners employing the mineral rod.   (From Agricola: *De Re Metallica*.)

somewhat different standpoint. The text runs on continuously in solid pages of type, with no headings, subdivisions or paragraphs, in long sentences which sometimes occupy half a page. In *De Natura Fossilium*, which has been called the "First Text Book of Mineralogy," and which deserved the title, the minerals are described one after the other in a continuous succession, without separate headings and without a break or breathing space between them. It needs some little study to ascertain which particular mineral he is describing in any particular portion of the text. A good index is however provided.

Fortunately for the student, the two works in which Agricola's chief contributions to the science of mineralogy are set forth—his *Bermannus* and his *De Natura Fossilium*, have been admirably translated into German—the former by F. Aug. Schmid[19] and the latter by Ernst Lehmann,[20] a Saxon mining official. Though these translations are but little known they are of great value to any student who wishes without undue labor to gain a clear and accurate knowledge of Agricola's mineralogical work.

Lehmann not only translated the *De Natura Fossilium* but subdivided the text into numbered paragraphs, supplied them with appropriate headings, and provided an extended series of tables and annotations, as well as a number of excursuses, thus greatly facilitating the study of the book.

Bermannus, his first mineralogical work, was apparently written in 1528 but was first published in 1530. It was named after Agricola's friend Lorenz Bermannus, who was like Agricola a classical scholar and a man versed in mining lore.   It is arranged in the form of a conversation, which was supposed to have taken place between Bermannus and two learned physicians—Naevius and Ancon. A brief introduction to this little book[21] is written by no less a person than Erasmus, in which he says that he hardly knew whether he was more pleased or instructed by his perusal of the work, and refers to Agricola's Attic style, which is so vivid that as he read he saw before him the hills and valleys, the mines and machinery, which Agricola describes, while the references to so many veins of gold and silver almost aroused a feeling of avarice in his own bosom.

Agricola himself elsewhere speaks of the difficulty which he experienced in describing certain of these things in classical Latin.

The book commences with an historical review of the rise of the mining industry in Germany and the early development of the great

---

[19] *Georg Agricola's Bermannus, oder Gespräche über den Bergbau aus den Lateinischen übersetzt und mit erläuterungen, ammerkungen und excursionen begleitet,* von F. A. Schmid, Freiberg, 1806.

[20] *Georg Agricola's Mineralogische Schriften übersetzt und mit erläuternden ammerkungen begleitet von Ernst Lehmann,* Dritter Theil, Oryktognosie (*De Natura Fossilium*), Freyberg, Ester Band 1809, Zweiter Band 1810.

[21] Dr. Naevius was a physician in Annaberg.

mining centers in the region of the Erzgebirge. The discussion then passes on to a variety of subjects connected with mineralogy and mining, the demons which were supposed to haunt many of the mines, the opinion of the ancients concerning Pyrite and Galena and of the various "Kiese" and "Glanze," and the exact sense in which the Classical and Arabian writers used the various terms which they employed in describing these and other minerals. The ores of silver, copper and other metals are then referred to and their mode of occurrence described, together with other matters. It is a short treatise and serves as an introduction to the larger work which followed it.

In the *De Natura Fossilium* he reviews the systems of classification proposed by Aristotle, Avicenna, Albertus Magnus and other of the older writers, and, rejecting these, puts forward an entirely new and original classification of "fossils" based on the physical properties of these bodies. These physical properties he enumerates as follows:— Color, Weight, Transparency, Luster, Taste, Odor, Shape and Texture. He also refers to "Fat" or "Unctuous" minerals such as sulphur, bitumen and marl, and to "Lean" ("Mager") minerals such as salt, ochre, sand, as well as to the Form, Hardness, Friability, Smoothness, Solubility, Fusibility, Brittleness, Cleavage, Combustibility of minerals. All these terms he defines and carefully explains.

He also mentions the medicinal properties of certain minerals and then striking a modern note goes on to say: "Of the powers which the Persian magicians and the Arabians attribute to certain stones and gems I will say nothing. Dignity and propriety obliges a man of science to reject them entirely." But it is strange to find that, notwithstanding Agricola's modern outlook, he states that the magnet loses its powers of attraction if a diamond is brought in contact with it and that he repeats (in the *De Animantibus*) Aristotle's fable of the "pyrigonoi" or flies that live in the fires of the Cyprus copper furnaces and which die if they are removed from these.

Referring to the form of minerals, Agricola notes especially that some occur in the form of balls, others have cylindrical, turbinate, cone, wedge or plate-like forms. Still others he says display angular shapes, such as cubes, triangular forms, or hexagonal forms as in rock crystal or the basalts of Saxony, the latter, however, sometimes having five and in other cases seven sides. He then passes on to present his classification of "Inanimate Subterranean Bodies" based on their physical characters—this is as follows:

1. *Fluids and Vapours.*
2. *Fossils.*

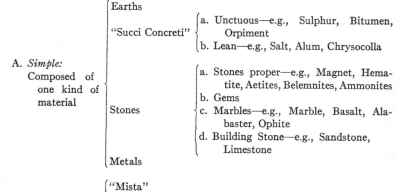

A. *Simple:*
Composed of one kind of material

Earths

"Succi Concreti" {
a. Unctuous—e.g., Sulphur, Bitumen, Orpiment
b. Lean—e.g., Salt, Alum, Chrysocolla
}

Stones {
a. Stones proper—e.g., Magnet, Hematite, Aetites, Belemnites, Ammonites
b. Gems
c. Marbles—e.g., Marble, Basalt, Alabaster, Ophite
d. Building Stone—e.g., Sandstone, Limestone
}

Metals

B. *Compounds:*
Composed of two or more kinds of material

"Mista"
In these the constituents are so intimately mixed that they cannot be separated except perhaps by fire. e.g., Galena, Siderite, Arsenopyrite, the bituminous copper schist of Mansfield.
"Composita"
In these the different materials are generally visible, they can often be separated by water or even by hand. e.g., Ores composed of galena associated with pyrite—Quartz and native gold, conglomerates of various kinds

He explains certain of the terms which he employs:

Earths are materials which when moistened with water can be rubbed between the fingers into a clay-like material. Some of them have special names assigned to them, others are as yet unnamed.

The *Succi* or juices have been referred to in another chapter (see page 90). Most of these would now be designated as mineral solutions. When these become dried and hardened they form "*Succi Concreti.*" These are rather hard bodies which do not become soft, like the earths, when sprinkled with water, but dissolve in it; or if they do become soft, either on account of their unctuous character or owing to the nature of the material of which they are composed, differ in a very marked manner from the Earths.

The *Stones* are hard dry "fossils" which either soften very slowly when immersed in water and fall to powder in a fire, or are not softened at all by water and melt in a very hot fire.

Divison (c) of the Stones consists of the *Marbles*. These are defined by Agricola as stones which when cut to a flat surface will take a high polish. He therefore includes Basalt among the Marbles. In fact this division includes a series of rocks which are now known to be utterly diverse in character, origin and composition, but which have in common only the property of taking a high polish. Marbles sometimes differ from the opaque gems only in the fact that they occur in larger masses.

The *Metals* are "fossils" which are either fluid by nature or become fluid when subjected to fire and when allowed to cool resume their original character. The latter part of this definition excludes fusible stones and minerals. It is the generally accepted view that there are six metals, Gold, Silver, Copper, Tin, Iron and Lead, but there are others. Quicksilver and Bismuth are metals, although the alchemists will not allow that such is the case, and here Agricola interjects a remarkable observation for the time, in the form of a quotation from one or other of the several writers bearing this name, "Ammonius says correctly that there are many metals as well as many plants and animals which are as yet unknown."

Passing to those bodies which are composed of two or more kinds of materials he says the *Mista* are those bodies which are composed of two or more "fossils" which are so intimately intermingled that in every little particle of the "Mista" all the constituents are present but so intimately associated that they cannot be separated except by fire, indeed by their union a new body is formed, one in which the properties of the original materials can no longer be recognized.

After having given examples of a number of the "Mista" and "composita" Agricola concludes his book with the words: "There are an infinite number of other combinations, which however I will not attempt to mention, because in the first place it would take too long and secondly because anyone who will apply himself diligently to the task can discover them for himself."

In considering the system of classification presented by Agricola several facts must be borne in mind. In the first place, as has already been mentioned, the word "fossil" in Agricola's time and in fact down to the first half of the nineteenth century had a different meaning from that which it now bears. A "fossil," as the derivation of the word indicates, was any body which had been dug up out of the earth's crust. It included those bodies which we now designate as Minerals, Rocks and Fossils. Agricola's *De Natura Fossilium* while, therefore, a treatise on mineralogy, touches also upon petrography and palaeontology.

This system of Mineralogy set forth by Agricola marks, as will be seen, a great advance on all former classifications and serves as a basis for many subsequent ones. From the standpoint of the present time however it leaves much to be desired.

He can make no sharp distinction between simple and compound bodies. Many of his "stones" are really compound and not simple bodies. Nor could he distinguish chemical compounds from fine grained and intimate mechanical mixtures. His "Mista" include both.

This was owing in part to the fact that in Agricola's time alchemy was just commencing to change into chemistry and he knew little or nothing of the latter science, in fact there was little or nothing to be known: and in part to the fact that he lived some three centuries before the microscope was introduced in the study of minerals and rocks. Furthermore the science of crystallography had not as yet come into being and chemical analysis was unknown, so that the most important bases for the classification of minerals, form and composition, were not available in Agricola's time. The various remains of animals and plants which he was obliged to include in his System, such as Judenstein (the club-shaped spine of an echinus), Trochit (elements of a Crinoid Stalk) Glossoptera (shark's tooth) Belemnite and others, presented another and further difficulty because while he believed that they grew within the rocky crust of the earth, like the minerals with which they were associated, he was not certain that this was the case but thought that they might be of organic origin. He mentions that many of the stones used in building contain sea-shells.

Agricola sets aside as unworthy of consideration the many quaint stories and idle tales which were commonly believed concerning certain minerals and fossils such as Thunder Stones, Fraueneis and Rock Crystal. The magnet, he says, is, on account of its extraordinary power of drawing to itself fragments of iron, the best known and most celebrated of all stones. This power it possesses is one which "Theologians hold to be supernatural but which medical men consider to be natural, although one which cannot be explained." He employs this term because in his time most of the men who made a study of nature were members of the medical profession.

The name "Schistus" he says, is given to a variety of hematite, that which occurs in nodular or concretionary masses, not because it is easily cleaved, for such is not the case, but because it looks as if it had been broken, the mineral having a grain like wood. Pumice is a stone found in places where subterranean fires are burning or have burned in times past.   It is burned earth or stone.   Ammonite is the name which he gives to the rock now known as Pisolite.   Magnetes is a white mica.

He considers the economic value of minerals and rocks, treating at length of the various kinds of soils and mentioning various stones as suitable for building purposes, paving stones, or for other purposes.   He also describes a number of furnace products which result from the smelting of various mineral substances.   Very extensive references are also made to many localities where certain minerals or rocks occur, not only in Europe but in other parts of the world.

His system of classification, however, was not by any means Agricola's chief contribution to mineralogy.   This lay rather in the description which he gives of many new minerals, especially the metallic ores, their mode of occurrence and mutual relations, drawn from his wide and intimate knowledge of the products of the mines in the region of the Erzgebirge.   Agricola certainly deserves the title which is commonly bestowed upon him—*The Father of Mineralogy.*

III

Among tne contemporaries of Agricola and Gesner there are two men who also wrote on mineralogy and who are worthy of especial mention.

The first is Johann Kentmann who was a doctor of medicine and practised his profession in Torgau.   He was a man of wide and varied learning, a very diligent student of nature, and an intimate friend of Gesner.   Kentmann is stated to have been the first man in Europe to make a collection of minerals.   His little book on "Fossils"[22] appeared in the small octavo volume, which also contains Gesner's *De Rerum Fossilium* and six other short works, and to which reference has already been made.   In this book is a plate showing Kentmann's cabinet of fossils, and the work itself presents an annotated list of some sixteen hundred specimens, apparently those ccntained in his collection, the list being given in the Latin language, frequently followed by a German translation.   These comprise minerals, several varieties of the same mineral being often included, with some remains of animals and plants and some furnace products.   The minerals are arranged in classes— Earths, Efflorescences, Stones, Sands, Marbles, Petrified Woods, Pyrites,

[22] *Nomenclaturae Rerum Fossilium quae in Misnia praecipuè et in aliis quoque regionibus inveniuntur,* Zurich, 1565.

etc. It is interesting as giving a conspectus of most of the minerals known at that time, with the localities from which they were derived as well as an exact equivalent in German of the various names by which they were known in Latin.

The second is Johann Mathesius, who wrote a celebrated work entitled *Sarepta oder Bergpostill*.[23] Mathesius was a scholar and intimate friend of Luther, as well as a close associate of Agricola and a man beloved by all who knew him. It was the time of the Reformation, and the movement took a strong hold in these mining centers of the Erzgebirge. Mathesius was a Protestant minister in the great mining community of St. Joachimsthal, and was led to the study of mining and metallurgy by reading Agricola's *Bermannus*. *Sarepta* was published as a large folio volume and went through several editions. It consists of twenty long and learned sermons preached to his congregation of miners, in the course of which every statement concerning minerals, metals, mining or smelting in Holy Writ is referred to, explained and frequently made the basis of a moral or spiritual lesson. The analogies and comparisons which he makes between the mining of ores from the hard and stubborn rocks of the earth's crust and the work of the Spirit in the human heart undoubtedly appealed to and deeply impressed his great congregation of miners, but are in some cases rather startling to the modern reader.

"It is," says Webster, whose comments upon the works[24] of ancient writers are always quaint and to the point, "an exposition of all those places of the Holy Scripture where any mention is made of Metallick matters. A work of that worth, for declaring experimentally the nature, generation, increase and decrease, ascension and descension, perfection and decay, and other properties of metals that I know few can equalize it." It is furthermore not the work of a man who had merely a general acquaintance with technical matters, but of one whose life had been spent in continuous and intimate association with the residents of a great mining community. His sermons not only strike a high spiritual note but also afford an excellent presentation of the knowledge of the time on the many objects connected with mining and metallurgy of which he treats.

The name *Sarepta* which he gives to his book, is taken from that old city to which, during the great famine in Israel, the prophet Elijah was

---

[23] *Sarepta oder Bergpostill. Sampt der Jochimsthalischen kurtzen Chronicken*, Nürnberg, 1564.

[24] *Metallographa* or *An History of Metals*, London, 1671.

sent—"Unto Sarepta a City of Sidon, unto a woman that was a widow."
In the opening sermon he tells his congregation that Sarepta was a renowned center for mining and smelting in the ancient world. It must
be confessed that some of the evidence which he proceeds to advance in
support of his story—to our cold and critical gaze in modern times—
seems scarcely conclusive. The passages in Holy Writ which he cites
seem scarcely convincing and the adumbrations of the Church Fathers
on the question even less so. The haze which often obscures the historical landscape in these very remote times cannot always be dispelled by
Mathesius, but his intentions are good and his narrative highly entertaining.

He narrates the story, told by Pliny,[25] concerning the origin of glass
making, how as "The common voice and fame runneth" some merchants,
sailing up the eastern coast of the Mediterranean in a ship which carried
a cargo of "Nitre" landed at a place on the Phoenician shore near that
on which the city of Sidon now stands, for the purpose of cooking a meal,
but could find no stones on which to place their pots and cooking utensils,
the beach being of pure white sand. They therefore propped up the
pots on some blocks of nitre which they brought from the ship. After
the fires had been burning for a long time they noticed, to their great
surprise, streams of a clear white fluid running out from beneath their
vessels. This was glass; and through this discovery there arose a great
glass industry at Sarepta, a great city grew up on this site, and the
Sidonians became famous artificers in glass. Sidon and Sarepta, adjacent cities, were in the territory assigned to the tribe of Asher, of whom
Moses prophesied "Thy shoes shall be of iron and brass." Tubal Cain,
Malachi, Jeremiah and many others make their contributions to the
story, and finally Strabo mentions Sarepta as a city where smelting is
carried on. And so Sarepta, this ancient city of mines, was adopted as
a prototype of Joachimsthal, and Mathesius named his volume of mining
sermons after it.

The writer has recently come into possession of a very interesting
manuscript letter written by Mathesius to his friend Paul Eberus, Professor of Philosophy in the University of Wittenberg, on December first,
1547, in which he makes an interesting reference to Agricola, who was
one of his contemporaries. It is written in a small but clear hand on a
large double sheet of folio (sized) paper. After giving some news con-

[25] *Natural History*, Book 36, Chapter 26.

cerning his family he says: "I am running through the work of Agricola and I am looking for certain things that may be of interest to you; the work itself is brilliant and attractive and can deservedly be compared with the works of all the Gentile philosophers. It contains too little reference to Christ and some remarks upon the sun and the planets, which these people consider in their prayers. I am surprised indeed that not even a mention is made of holy water. In the enumeration and review of the sources of water mention could have been made quite easily of these waters which were sweetened by Elisha when from that spring of salt water, the holy water gushed forth, but I shall cease and I shall draw upon the book as the Israelites drew upon the storehouses of the Egyptians. I may reasonably wonder why he wrote nothing about the nature of Magnetis, how it does not cut the meridian line, unless perhaps all that comes later in the work. For I have not yet perused the fifth book on the Nature of Fossils."

Mathesius is evidently referring to the large folio edition of Agricola's geological works which was published by Froben in 1546. He had read the first two books, *De Ortu et Causis Subterraneorum* and *De Natura eorum, quae effluent ex Terra*, the latter of which is devoted chiefly to a discussion of waters of various kinds. The third book, which he erroneously refers to as the fifth, is that entitled *De Natura Fossilium* which Mathesius says he had not read. It *does* as he conjectures deal with magnetic iron ore but curiously enough, while he discusses at length its various properties, he does not refer to its use in the construction of the mariner's compass.

Mathesius was evidently struck by Agricola's method of dealing with the objects and phenomena of nature, describing only the immediate or proximate causes which give rise to them and making no reference to the great First Cause as his predecessors were accustomed to do, but taking this as understood and requiring no further mention. It may be observed however that while he speaks of the sun and the planets in the first of the books referred to above, he does so only to refute the views which had been put forward by the alchemists concerning the marvellous powers which they supposed them to possess.

Somewhat over 100 years later Johann Frederick Suchland,[26] Deacon of Clausthal, then a great mining center in the Harz Mountains, wrote a work which closely resembles *Sarepta* although it is smaller in size. The

[26] *Allegorische und Historiche Beschreibung des gantzes Berg-Wercks, darinnen enthalten alles was dem Berg-Werck nützlich zu betrachten vorfället*, Clausthal, 1687.

book consists of four long sermons which were preached to the Clausthal miners, filled with references and lengthy illustrations drawn from the operations of mining and smelting. Appended to the book there is, very appropriately, an extended glossary of contemporary mining terms explanatory of words used in the text. Similar sermons to miners are to be found in Christian Melzern's *Berglaufige Beschreibung der Bergstadt Schneeberg*, Schneeberg, 1684.

In addition to these admirable sermons a number of hymns[27] sung by these old miners have come down to us and are also of the greatest interest, showing as they do the hold which the principles of the Reformation had taken in these great mining centers. Like the sermons, they are full of mining and metallurgical terms and striking comparisons. Agricola in his *De Re Metallica* refers to the fact that the miners commonly sang when at work. The following is one of these hymns:

Ein Berg Furst ausserkohren, der man Jehova nennt
Ein Berg Fürst ausserkohren, der man Jehova nennt, aus
David's stamm geboren, ein Herr in ganzen Land, der thät
sich offenbaren, die Zeit war nun dahin, er wolt eine
Grube thun befahren, zu bauen stund sein sinn.

Einen Durchschlag thät er machen, in dem kleinen Bethlehem,
thät einen Gang nachtrachten, fuhr nach Jerusalem, bald
schlug er ein in den Tempel, und offenbahrt sich wol, gab
darmit gute Exempel, wie man recht schirffen soll.

Der Gang thät mit sich führen, ein unartig Gestein, wenig
Erz thät man da spüren, dass reichhältig thät seyn, viel feste
Stämm und Knauren, schossen gewaltig für, seine Arbeit
wird ihm sauer, im ganzen Leben hier.

Herr Christ hilff dass wir alle, gute Goldstuffen seyn,
und führ uns bald mit Schalle in deine Hütten ein.
thu uns mit Freuden bringen, in dein schönes Paradeis,
so wollen wir dir singen, ewig Lob, Ehr und Preiss.

IV

From the time of Agricola onward increased attention was directed to mineralogy, due chiefly to the rapid development of the mining in-

[27] *Drey schöne neue Berg Reyen—Der erste, Das Bergwerck ist doch lobens werth, mir thuts gefallen. Das ander Ein Berg Lied will ich heben an. Der Dritte, Ein Berg Furst ausserkohren, den man Jehova nennt. Gedruckt im Jahr, 1661.*

dustry in Europe.   In this connection the science had a direct economic value.   Its study was therefore carried on chiefly by men associated with the industry in the various great mining centers.   But its growing importance came to be widely recognized and professorships of mineralogy were founded, not only in the mining academies, but in many universities; and the outstanding men who were appointed to these positions contributed very largely both to the development of the science itself and to its economic applications.   The rise of the great learned societies in the various countries of Europe, publishing proceedings and transactions, was also an important factor in the development of mineralogical studies and the literature of the subject commenced to grow rapidly.

Gmelin in his translation of the *Natursystem des Mineralreichs* by Linnaeus,[28] enumerates no less than twenty-seven "Systems of Mineralogy" written by various authors in different countries of Europe, during the one hundred and twenty-eight years which followed the appearance of De Laet's book, that is, between 1647 and 1775, when Gmelin's work was published.

An earnest search was being made for an ideal system.

In the last quarter of the eighteenth century a new principle presented itself for consideration in connection with the classification of minerals: this was their *Chemical Composition.*

As a result, in as far as methods of classification are concerned, mineralogists became divided into two groups—the first of these advocated what came to be known as the *Natural History Method* of classifying minerals, while the second adopted a *Chemical Method.*[29]

The first school having before them the threefold division of nature into the animal, vegetable and mineral kingdoms, followed the method which was used by naturalists in classifying the members of the two former kingdoms, namely, that of arrangement according to external form, and accordingly based their classification of the mineral kingdom upon the external characters of minerals, that is to say, on their form, color, hardness, lustre, cleavage, etc.

The second group considered that the true nature and character of a

[28] *Des Ritters Carl von Linné vollstandiges Natursystem des Mineralreichs nach der Zwölften lateinischen Augsabe, in einer freyen und vermehrten Ubersetzung* von Johann Freiderich Gmelin, Nürnberg, 1771, p. 84.

[29] See William Whewell, *History of the Inductive Sciences,* vol. 3 (Parker & Son), London, 1857.

mineral was determined by the substance or substances of which it was composed. They therefore believed the chemical composition of the mineral to be its most important character, one which was internal rather than external, while the external physical characters were of secondary importance for purposes of classification.

It must be borne in mind that the science of chemistry was just taking form at this time, and it was only as chemical knowledge developed that the mineralogist could make use of it. Furthermore, even after the existence of the various elemental bodies and their combinations came to be known, the art of chemical analysis had still to be developed. It was not until about the year 1800 that approximately accurate chemical analyses of minerals were generally available. Robert Townson, writing in 1798 says:

> Chemistry of late years has made a most rapid progress and every branch of human knowledge within its reach has been advanced by it. Mineralogy should be the first to speak its eulogium as the small tribute of gratitude for great favours. Chemistry has done much for Mineralogy; it has raised it from a frivolous amusement to a sublime science and will in time with the progress of science, bring to light many things which now lie concealed and unveil many of the hidden mysteries of nature.[30]

Among the writers who favored the Natural History Method, that is to say, who employed this method chiefly, are Linnaeus, Mohs, Jamieson and Hill. While foremost among those who give prominence to the chemical composition of the minerals are Cronsted, Berzelius, Bergman and Kirwan.

In spite of the strenuous advocacy of each of these two systems by their respective supporters, neither was found to be satisfactory. Too little was known at the time of either the chemical composition or the physical characters of minerals. Much further study and research was acquired before a true conception of the mineral kingdom could be obtained.

Mineralogists therefore adopted "Mixed Systems" in which what they judged to be the best points of the Natural History System and the Chemical System respectively were combined. As an example of these "Mixed Systems," that adopted by *Werner* might be selected, it being a classification which was very widely employed in the opening years of the nineteenth century, owing to the great number of Werner's students who were teaching mineralogy in all the countries of Europe.

---

[30] *Philosophy of Mineralogy*, London, 1798, p. 114.

A somewhat extended reference to Werner's important contributions
to the science of geology is made in chapter VII, but Werner was before
all things, a mineralogist. He lectured on this subject at Frieberg
throughout his entire academic career. During this time his views

Von den

äußerlichen

Kennzeichen der Foßilien,

abgefaßt

von

Abraham Gottlob Werner,
der Bergwerks = Wissenschaften und Rechte Be-
flißenen, auch der leipziger öconomischen Gesellschaft Ehren = Mitglied.

Leipzig,

bey Siegfried Lebrecht Crusius,

1774.

FIG. 45. Title page of book on fossils.

naturally underwent a progressive growth and expansion, and each year
from 1780 until the time of his death, which took place in the year 1817,
he made additions and alterations to his system of mineralogy.

In the first edition of his celebrated work *Von den ausserlichen Ken-*

nzeichen der Fossilien, which was published in Leipzig in 1774, Werner refers to these two schools into which mineralogists were divided, and gives his own views on the question of classification. He first points out that mineralogists have two distinct problems to solve: first, to discover an ideal system of classification, and secondly, to ascertain how the individual minerals can be recognized and distinguished from one another. He then goes on to say that the method of classification which is adopted for the animal and vegetable kingdoms cannot be employed in the mineral kingdom. A plant, for instance, has many parts, its roots, branches, leaves, flowers, and these differ from one another. In every species each of them has a certain form and these various parts are arranged in a definite manner with reference to one another. It is this form and arrangement of the parts ("Zusammensetzung") which afford the basis for classification in the vegetable and also in the animal kingdoms.

With minerals the case is quite different. If a mineral is broken into minute fragments each one is identical with all the others. It has not "parts" as plants and animals have. The fundamental difference between different minerals lies in their composition ("Mischung"). He cites the example of "spröde glaserz," which is composed of the union of silver, sulphur and arsenic, and this passes over into other species when changes take place in its composition. Thus the "spröde glaserz" by the addition of iron passes into "rothgültigerz" or by the addition of copper into "weissgultigeserz."

"My opinion is," he writes, "that fossils must be classified according to their composition" ("mischung"). He felt that a true and final classification could be arrived at only after much more had been learned concerning the chemical composition of minerals, and that it would be possible to recognize the various minerals and to determine them with certainty only after much more was known of their external characters and physical properties.

In the meantime, Werner goes on to say, that what is required is a clear and adequate description of the external character of each and every mineral which will make it possible to recognize it and distinguish it from other minerals. "Ich will lieber ein Fossile schlecht geordnet und gut beschrieben, als gut geordnet und schlecht beschrieben haben." And to this study of what would now be termed "Practical Mineralogy" Werner cheifly devoted his attention. This was but natural, his chair was in a mining academy founded for the purpose of training men for the profession of mining, it was therefore of especial importance that his students should be taught to recognize readily the individual minerals and know their econimic value as ores, fluxes or for the various uses in the arts and this without the necessity of resorting to chemical examination.

After his death there was found among his papers a copy of his classification of minerals embodying the latest changes and alterations which he had made in his Mineral System.   It was a manuscript which he had used for the course of lectures which he gave immediately before his decease.   This was his last word on his beloved science of mineralogy, and was published in 1817 in Freiberg and Vienna as a small pamphlet under the title *Abraham Gottlob Werner's Letztes Mineral System*.   This classification in its main outline is as follows:

Class I. Earthy Minerals—subdivided into the following "Geschlechter"
1. Diamond
2. Zircon
3. Silica—This includes most of the silicated minerals, also quartz, opal, obsidian, pitchstone, spinel, corundum, etc.
4. Clays—This includes Clays of all sorts, also Clay Slate, Mica, Trap and Lava, Lithomarge, etc.
5. Talcs—including Serpentine, Asbestos, Tremolite, etc.
6. Lime Minerals, including Carbonates, Phosphates, Sulphates, Borates, etc.
7. Barite
8. Strontium
9. Halite, Kryolite

Class II. Saline Minerals—including Alkaline Carbonates, Nitrates, Chlorides and Sulphates.

Class III. Combustible Minerals—Sulphur, Coals and Bitumen, Graphite, Resins.

Class IV. Metallic Minerals—including in each case the native metal, and all its ores and compounds:

1. Platinum	8. Tin	15. Cobalt
2. Gold	9. Bismuth	16. Arsenic
3. Quicksilver	10. Zinc	17. Molybdenum
4. Silver	11. Antimony	18. Wolfram (Tungsten)
5. Copper	12. Tellurium	19. Titanium
6. Iron	13. Manganese	20. Uranium
7. Lead	14. Nickel	21. Chromium
		22. Cerium

There are 317 "minerals" included in his list.

V

There are two other men whose researches were of the most far reaching importance in the further development of the science of mineralogy, and who on this account should now be mentioned. These are Haüy and Berzelius. The results of the work of both of these men were published during Werner's lifetime, but he apparently took little notice of the work of the former, while that of the latter appeared only three years before Werner's death—too late possibly to have been seriously considered by him in all its implications.

It is to Haüy that we are indebted for the first clear exposition of the crystallographic forms of minerals, while Berzelius first demonstrated the outstanding importance of the chemical composition in every mineral species.

René-Just Haüy (1743–1821) was born in the little town of Saint Just[31] in Northern France. He was of humble origin, simple, modest and unassuming in character, but a lover of learning, whose attention was, one might almost say by chance, directed to the study of mineralogy, but who by close observation, diligent study and indefatigable patience in following out a restricted and at that time practically unknown branch of knowledge, founded and developed the science of crystallography, and became one of the greatest mineralogists of his time. Berzelius dedicates his book to him as the one "Dont le génie a élevé la Mineralogie au rang des sciences."[32] Haüy's discoveries indeed constitute an epoch in the history of the science.

His father was a cloth worker who lived in poverty and his son would, in all probability, have followed in the same lowly occupation, had not certain charitably disposed persons come to his aid. From the time that he was a child the young Haüy displayed a marked interest in religion and the services of the church, and had a taste for music and especially religious music. His constant attendance at divine service attracted the attention of the Prior of the Abbey in his native town, who found the boy to be possessed of vivacity and intelligence, and through his interest and that of others who befriended him at a later date, he received a good education at Navarre and studied afterwards in Paris. Among other accomplishments he became a good botanist and one day when walking through the Royal Gardens in Paris, which he visited frequently in the prosecution of his studies in this subject, he saw a large number of people crowding into a lecture hall where the aged Daubenton was to

---

[31] Cuvier, *Eloges Historiques dans les Séances Publiques de L'Institut Royal de France,* T. III, p. 123.

Grothe, *Entwicklungsgeschichte der Mineralogischen Wissenchaften,* Berlin, 1926.

[32] *Nouveau Système de Mineralogie,* Paris, 1819. (Translated from the Swedish.)

give one of his charming popular lectures on mineralogy. This lecture turned Haüy's attention from botany to mineralogy.

The constancy of the complicated forms displayed by the various flowers and fruits had always impressed him and he felt sure that a similar law of constancy must prevail in the various species of the mineral kingdom, in spite of the almost bewildering variety of forms which their crystals exhibit. His mind was filled with this idea and one day when examining some minerals in the collection of one of his friends, by what may be termed a happy misfortune, a fine group of prismatic calcite crystals slipped from his hand, fell to the floor and one of the crystals was broken. On examining this broken prism Haüy noticed that it had assumed a new form bounded by faces as smooth and shining as those of the original prism, and to his great surprise he found that this form was identical with that presented by the rhombic prisms of Iceland Spar. A new light broke upon the subject. He returned home, and, selecting from his collection a number of calcite crystals, some prisms, some scalenohedrons and others of the diverse forms in which this mineral crystallizes, he broke them all and found that the resulting cleavage form was in every case that of a rhombohedron. "Tout est trouvé," he exclaimed, all the molecules of calcite have one and the same form, and it is merely the manner in which they are grouped that gives rise to the various external forms which this mineral displays. If this be the true principle of crystallization other minerals should also show analogous cleavage forms revealing their internal structure. In the pursuit of knowledge he did not hesitate to break up all the crystals in his own collection and all which he could obtain from his confiding friends. In all species, at least in those which displayed a cleavage, he believed that he could find some nuclear form about which its crystals had been built up and from which their angles might be predicted before they were actually measured.

While Steno,[33] Romé de l'Isle,[34] Bergman and others share with Haüy the honor of having founded the science of Crystallography, and while Haüy's theory of crystal structure has been modified by more recent investigations, he is, says Cuvier, "Le seul véritable auteur de la science mathématique des cristaux" and his great work[35] will ever remain a

[33] De Solido intra Solidum Naturaliter Contento, Florence, 1649.

[34] Essai de Cristallographie, 1772.

[35] Traité de Minéralogie, 4 Volumes with Atlas, Paris, 1801. Traité de Cristallographie suivi d'une application de cette science à la détermination des espèces minérales et d'une nouvelle méthode pour mettre les formes crystallines en projection, 2 Vols. with Atlas, Paris, 1822.

monument in the history of the development of this science, to the study of which he devoted his life.

VI

While Haüy demonstrated the great importance of crystallographic form in the mineral kingdom, it is the Swedish chemist Berzelius to whom the honor belongs of having been the leader of that succession of mineralogists whose work finally demonstrated what Werner believed to be true, namely, that it was the chemical constitution of minerals which afforded the basis for a satisfactory classification of the members of the mineral kingdom.

The chemists of the eighteenth century who busied themselves with the study of the composition of minerals for the most part confined their attention to qualitative analysis. Klaproth (1743–1887) was the founder of quantitative mineral analysis. Delamétherie, writing in 1797 complains that the mineral analyses[36] in his time were imperfect, those made by Klaproth were different from those made by Bergman, but that the methods of analysis were being rapidly improved, and that it would be impossible to secure a good classification of minerals until accurate analyses were available.

It was Berzelius however who, by improving the methods of chemical analysis through his own investigations and by inspiring and directing the work of other Swedish chemists, led the way to a more widespread recognition of the paramount place which chemistry must occupy in the study of mineralogy. It was he who first presented a classification of minerals on the basis of their chemical composition and who, recognizing silica as an acid, introduced into mineralogy that great group of minerals which he first named the *Silicates*, and brought order[37] out of the chaos which formerly reigned in that division of the mineral kingdom. He showed by his studies of "Multiple Proportions" that a chemical or atomic formula for the mineral could be deduced, and through the discovery in his laboratory of the principle of isomorphism demonstrated the existence of great groups of minerals differing indeed from each other in chemical composition, but whose differences were due to the replacement of certain elements by other allied elements, a substitution which took place with little or no change in the crystallographic form.

---

[36] *Théorie de la Terre*, Paris, 1797, Vol. I, p. XIII.
[37] Berzelius J. T., *Nouveau Système de Minéralogie*, traduit du Suédois, Paris, 1819, p. 23. The original Swedish edition of this work was published in Stockholm in 1814.

Further researches carried out by many able investigators during the nineteenth and twentieth centuries, not only into the chemical composition of minerals but, by the aid of polarized light and x-rays, into their inmost structure, have immensely extended our knowledge of minerals, which now in the light of modern knowledge display a beautiful and orderly system based primarily on chemical composition, which may fitly take its place beside those exhibited by the animate Kingdoms of Nature.[38]

[38] See Graham, R. P. D., *The Development of Mineralogical Science*, Trans. Royal Society of Canada, 3rd Series, Section IV, Vol. XXVIII, 1934.

Chapter VII

THE BIRTH OF HISTORICAL GEOLOGY WITH THE RISE
AND FALL OF THE NEPTUNIAN THEORY

1. WERNER AND THE NEPTUNIAN THEORY

The rise and fall of the Neptunian Theory of the Earth forms one of
the most important, interesting and even romantic chapters in the history
of geology.   It embraces a period of only a little over half a century
(1775–1825).   But while his theory has had its day and ceased to be,
it was Werner who, as Fitton[1] has said, first elevated geology to the rank
of a real science.

Many before Werner had written on the origin of the earth, from the
time of the early Greeks down to that of the Cosmogonists of England
and the Continent.   These writers, especially the later ones, presented
theories spun out of their own vivid and brilliant imaginations, in which,
to use the words of Lord Bacon, each "Builds a ship with materials
insufficient for the rowing pins of a boat."   They are, in fact, veritable
"wonder stories" containing here and there a passage of fine writing,
quite as interesting as most of the stories written about other things
during their respective periods, for the delectation of the public.

Perhaps the most celebrated of these treatises was that of Thomas
Burnet, which bears the modest title *The Sacred Theory of the Earth—
containing an account of the Original of the Earth and of all the General
Changes which it hath already undergone or is to undergo till the Consumma-
tion of all Things.*   This was published in London in 1681, under the
patronage of Charles II, and enjoyed a widespread popularity not only
in England but on the continent of Europe.   In none of them, however,
was the story narrated with more pomp and circumstance than by Buffon
in his *Histoire Naturelle.*[2]   This brilliant picture, however, marked the
close of this long period of imaginative effort and ushered in a new era.

---

[1] *On the Geological System of Werner,* Nicholson's Philosophical Journal, Vol. XXXVI
1813.

[2] *Théorie de la Terre,* Paris, 1749, Vol. I, also *Epoques de la Nature,* Paris, 1778.

These early fables of geological science should be read by all who are in need of mental recreation and who possess the required leisure and a certain sense of humor, although many of them make a further demand upon the seeker after amusement and recreation in this fairyland of science, namely, that he shall seek this relaxation in the somewhat unaccustomed field of medieval Latin.

But the time had now come when those who looked for an explanation of the origin of the earth should turn from mere speculation to a study of the earth itself and seek in it for an answer to this riddle which had presented itself to the mind of man from the earliest time at which mankind contemplated the aspects of nature.   We will therefore turn to the labours of certain men connected with the great mining industries of the Saxon Erzgebirge and its adjacent regions, whose daily work brought them into direct and continuous contact with the earth's crust and its manifold problems, and a great part of whose life was lived in the bowels of the earth itself.   Through successive generations they gradually accumulated a great body of actual knowledge and indisputable fact, and on this sought to base a true explanation of the earth's origin.

The advice which Severinus[3] gave to his students was laid to heart and adopted:

> Go my Sons, buy stout shoes, climb the mountains, search the valleys, the deserts, the sea shores, and the deep recesses of the earth.   Look for the various kinds of minerals, note their characters and mark their origin.   Lastly, buy coal, build furnaces, observe and experiment without ceasing, for in this way and in no other will you arrive at a knowledge of the nature and properties of things.

Thus Field Geology had its birth.

The new movement, as has been said, had its home in the great mining region of Northern Europe situated in the Saxon Erzgebirge and the adjacent regions, it took its rise with Agricola and reached its culmination in Werner, or rather in the men trained by him.

Reference has been made elsewhere in this volume to the outstanding contributions which Agricola made to the sciences of geology and mineralogy, as well as to the work of Werner in the field of mineralogy, in the restricted sense in which that word is now employed.   Here it is purposed first to make some reference to Werner's character and career and then to consider in some detail his geological work, for he was indeed an out-

---

[3] Quoted by J. G. Wallerius, *Systema Mineralogicum*, Preface, Vienna, 1778.

PLATE IV
PORTRAIT OF WERNER
(From Beck: *Abraham Gottlob Werner*)

standing figure in the history of geology and one of the greatest teachers of the science who ever lived, attracting as he did to his lecture hall at Freiberg multitudes of eager students from every land, among them many who, taught and inspired by Werner, became the foremost geologists of Europe and by their researches greatly advanced geological knowledge.

Abraham Gottlob Werner was born in 1749 at Wehrau in Prussian Silesia, not far from Görlitz and near the border of the Kingdom of Saxony. Even at that time mining had been carried on in the Saxon Erzgebirge for close upon 1,000 years and Werner's forefathers had occupied important positions in the mining and metallurgical industry of this old mining region, at least as far back as the opening of the 16th century. His father held the official position of Inspector of Iron Works at Wehrau. The boy was thus from his very childhood brought up among people whose daily work for generations had been in mines and furnaces, and from the first he took a keen interest in all these things. It was about them that his father taked to him daily, and the boy delighted in collecting minerals which his father named for him, and to pore over the pages of Hübner's[4] *Berg-Gewerk und Handlungs Lexicon* and Minerophil's[5] *Bergswerks Lexicon*, two books which were in constant use at that time and which, although now out of date, retain their interest to the present day.

When ten years old he was sent to school at Bunzlau in Silesia and five years later at the age of fifteen received his first appointment as assistant "Hüttensschreiber" to his father in the management of the Wehrau blast furnaces. He held this position for five years and then entered the Mining Academy at Freiberg, which had been established only three years previously. Here he pursued his studies with diligence and by repeated visits to all the mines and furnaces of the district, enlarged and extended his knowledge of the practice of mining and metallurgy, and accumulated a body of technical experience which was to stand him in good stead later. By frequent field excursions also, he laid the foundation of his mineral collection which was later to become so renowned. Most fortunately for Werner, the well-known Berghauptmann Pabst von Ohain, the most renowned mineralogist of his time, was

[4] Johann Hübner's *Curiöses und reales Natur-Kunst-Berg-Gewerk und Handlungs-Lexicon*, Leipzig, 1792.
[5] *Minerophili: Neues und Wohleingerichtetes Mineral und Bergwecks Lexicon*, Chemnitz, 1743.

then "Curator" of the Academy, and gave him free access to the great mineral collection under his charge.

Werner studied at the Freiberg Academy from 1769 to 1771, when he entered the University of Leipzig. Here he gave his attention chiefly to the study of law, languages, history and philosophy, while devoting some time to mineralogy as well, and thus enlarged the borders of his knowledge and prepared himself for that wide range of teaching which he was subsequently to undertake. While still a student at Leipzig he wrote his first work, entitled *Von den äusserlichen Kennzeichen der Fossilien*. This was published in 1774 and introduced entirely new methods in the description of minerals. The next year he was caught up in that "Tide in the affairs of men which taken at the flood leads on to fortune." The authorities of the Freiberg Mining Academy being impressed with the evidences of marked talent on the part of the young man, to his great surprise called Werner to the position of lecturer on mining and metallurgy in that institution. He accepted the position and entered upon his duties with zeal and enthusiasm, being at the time twenty-five years of age.

His lectures on mining and ore dressing followed the general lines laid down in Oppel's *Bericht von Bergbau* published in Leipzig in 1772, but as time went on he extended the scope of his instruction and gave courses of lectures on mining machinery, mining finance, mining law and other subjects related to his work, mentioned in Oppel's book as those which, according to the prospectus issued at the time of its foundation, were to form part of the curriculum of the newly established Freiberg Mining Academy.

Werner, his biographers tell us,[6] was a man with flashing blue eyes, friendly dispostion, of medium size, always closely shaven and wearing a peruke. He took a fatherly interest in his students, had them frequently to his house and invited certain of them to dine with him when he was entertaining distinguished guests. He insisted, however, on their paying strict attention to their work and always behaving in a seemly manner. He was untiring in his instruction and taught by lectur-

[6] *Lebensbeschreibung Abraham Gottlob Werners* von D. Samuel Gottlob Frisch, *nebst zwei Abhandlungen über Werners Verdienste um Oryktognosie und Geognosie* von Christian Samuel Weiss, Leipzig, 1825.

Richard Beck, *Abraham Gottlob Werner, Eine kritische Würdigung des Begrunders der modernen Geologie, zu seinem hundertjährigen Todestage*, Berlin, 1918.

Christian Keferstein, *Geschichte und Literatur der Geognosie*, Halle, 1840.

ing and demonstrations. He was the first to introduce regular courses of practical instruction as part of the training in the natural sciences—as had always been done in the teaching of anatomy. When his classes in practical mineralogy, on account of the great number of students who flocked to hear him, became so crowded that it was impossible for him to give efficient instruction, he divided the class into several sections and gave separate and personal instruction to each. In these classes he questioned the students continuously, which led Napione to refer to him as the "nuovo Socrate nella mineralogia." He also occasionally took his students on geological excursions to various parts of Saxony. His lectures were not confined to a mere exposition of the nature of the minerals themselves, but he endeavored to lead the thoughts of his students to the wider aspects and implications of the subject. From a couple of mineral specimens which lay before him on the table as a text, he would pass to the consideration of the influence which the character of the rock underlying a district or country exerted upon its surface features, and then to the influence of these upon the history of the country and the development of the character of its people.

Some idea of the charm of his lectures and of their influence upon his hearers can be obtained from the account of them given by one of his most distinguished students, D'Aubuisson de Voisins. He was a Frenchman from Toulouse who studied at Freiberg under Werner in 1800 and 1801:

À mon arrivée a Freyberg, je fus reçu chez un conseiller des mines, et un miné-ralogiste (M. de Charpentier) entièrement opposé à Werner sous tous les rapports: il m'inspira son éloignement, et je restait deux ans a l'école des mines, sans vouloir même suivre le cours de minéralogie de ce professeur. Enfin, j'y allai, mais avec les préventions les plus défavourables: les commencemens les confirmèrent, outres les choses communes que j'entendois (et qu'on étoit obligé de dire aux jeunes éleves) le vague et le peu de précision qui regnoient dans les descriptions et les détermina-tions ne pouvoient que rebuter une personne élevée dans les études des sciences exactes. Mais je vis bientôt que ce vague étoit ici inhérent à la nature des choses, que dans la minéralogie, tout n'est pas rigoureux et précis comme en geométrie; et que si je voulois la savoir, que si je voulois connoître les minéraux (et j'en avois besoin, pour l'exploitation de mines), il falloit bien écouter Werner; car enfin, ce qu'il enseignoit étoit la vraie minéralogie: et il n'y avoit pas de meilleur moyen de l'en-seigner. Son cours de Géognosie produisit sur moi le même effet. Les préliminaires renfermoient des détails, des lieux communs sur la cosmologie et la géographie phy-sique, connus de toutes les personnes instruites (qui cependant ne l'étoient pas de la majeure partie de ses auditeurs). Mais lorsqu'il vint a traîter de la structure des masses minérales, de leur stratification, de leur superposition, de l'histoire de chacune d'elles, tous ces objets me parurent extrêmement intéressans; et je dois le dire, il est impossible de la présenter d'une manière plus séduisante et plus propre à en pénétrer que ne le fait Werner. Ici, sa manière de professer est au comble de la perfection: il n'a jamais écrit un mot sur les objets dont il a à parler; ainsi ce n'est ni une froide lecture, ni un froid débit d'une leçon apprise par coeur: avant d'entrer dans son auditoire, il se recueille un instant dans son cabinet, il s'y rappelle la matière dont il va traiter: ensuite, il l'expose d'abondance, et avec un tel enchâinement et une telle gradation dans idées, qu'il fixe singulièrement l'attention de ses auditeurs: et qu'il les pénètre complètement de son objet. Il les en pénètre souvent jusqu'a l'enthousiame; et c'est au sortir de ses leçons que tant de personnes sont allées courir les montagnes de la Saxe, de toute l'Allemagne, de l'Hongrie, de la Suisse, de la France, etc.[7]

He lectured without manuscript, and Cuvier says: "He treated his subjects in such an admirable manner that he roused the enthusiasm of his hearers and inspired them not only with a taste but with a passion for his science." Cuvier continues:

At the little Academy of Freiberg, founded for the purpose of training mining engineers and mine captains for the mines of Saxony, there was renewed the spectacle presented by the universities of the Middle Ages, students flocked thither from every civilized country. One saw men from the most remote countries, already well advanced in years, men of education holding important positions, engaged with all diligence in the study of the German language, that they might fit themselves to sit at the feet of this "Great oracle of the sciences of the earth."

Werner was a most diligent collector, devoting much care to the proper classification and arrangement of his minerals and rocks. He was also an inveterate collector of all contributions to the literature of his subject, and thus built up at Freiberg an extremely valuable mineral collection and a very large and excellent library of books and papers dealing with the subjects upon which he lectured. Both of these now form part of the equipment of the Mining Academy at Freiberg.

It should here be mentioned that while accomplishing so much and such excellent work, Werner, throughout his life and especially in his later years, suffered from the great handicap of continuous poor health.

While Werner was one of the most attractive and inspiring lecturers, he wrote but very little. During his long career he made but twenty-six contributions to the literature of the science, and most of these are merely short magazine articles. Even his few longer contributions are

books of very moderate size.   He repeatedly announced his intention of
writing extended works embodying his views and discoveries and treating
of the subjects upon which he lectured, but these books never appeared.
In one of his short publications,[8] which Beck says was in great part
printed as early as 1783, but which was not published until 1811, he
gives the Table of Contents of a great work which he planned to write,
covering the whole field of mining, mineralogy and geology (as the words
are now understood), as well as the concentration and smelting of ores,
mine surveying, mining law, etc., a work which never saw the light.

So great and widespread was the demand for books by Werner treating
of the subjects on which he lectured, that when after a series of years
these books did not present themselves, certain persons spread abroad
the statement that Werner did not intend to write such books and proceeded themselves to issue what purported to be reports of his lectures,
without obtaining his authorization.   Werner complained of this in the
introduction to one of his books,[9] and stated that he was engaged in the
preparation of such works embodying his most recent views and conclusions and that they would soon appear.   Again these books were
never published.   Frisch and Karsten are of the opinion that Werner
did not write these various and much desired works owing to his inability
to find time to do so.   It is known, however, that Werner disliked writing
even the simplest note, and he probably found it difficult to prepare such
books, a difficulty which was accentuated by the rapid advance of knowledge in the subjects of which they were to treat.   This dislike became
an obsession toward the close of his life, so much so that when in 1812
he received the high honor of election to Foreign Membership in the
French Academy—an honor which undoubtedly he highly appreciated,
the letter to him announcing the coveted honor was actually left un-
answered.

Turning now from Werner as a man to the consideration of his contributions to science, it will be noted that at Freiberg he covered a very
wide range of subjects in his courses of instruction.   In the table of
contents of the great work which he contemplated writing and to which
reference has already been made, one subdivision or chapter bears the
title, the "Mineralogical Part or Mineralogy."   This is subdivided as

[8] *Kleine Sammlung Mineralogischer Berg—und Huttenmannischer Schriften, Herausgegeben
von Abraham Gottlob Werner*, Leipzig, 1811.

[9] *Neue Theorie von der Entstehung der Gänge*, Freiberg, 1791, p. xxvi.

follows:—(1) Oryktognosie, (2) Mineralogical Chemistry, (3) Geognosie, (4) Mineral Geography, (5) Economic Mineralogy. Oryktognosie, a word apparently coined by Werner, was what might be called "Determinative" or "Practical" mineralogy and was of especial value to persons engaged in mining, but it did not include theoretical or mathematical aspects of mineralogy as that term is now understood. The word is a specimen of polysyllabic cacophony which fortunately gradually disappeared.

Werner, as has been said, was first and foremost a mineralogist and his contributions to that science (sections 1, 2 and 5) will be referred to in another chapter, here it is more particularly his "Geognosie" (subdivision 3 of the above classification) that is to be considered. This word Geognosy which Werner is commonly supposed to have introduced, he really borrowed from Füchsel, who 24 years earlier in a paper written in Latin, employs the term "Scientia Geognostica."[10]

Werner defines Geognosy as the science "Which treats of the solid body of the earth as a whole and of the different occurrences of minerals and rocks of which it is composed and of the origin of these and their relations to one another." That is to say, it differed from geology and geogony in that it treats of what was *known* concerning the earth's crust and not of speculations concerning its origin with which the Cosmogonists were especially concerned. In one of his lectures delivered in 1816–18 he said:

It will be the aim of teachers of Geognosy, so to train their students in the practice of true methods of geognostical research, that we shall be led to a wide and at the same time certain knowledge of the solid framework of our earth.

It is interesting to note that while Werner was in revolt against the speculations which characterize the geological writings of his predecessors and even went so far as to adopt the new term geognosy to designate a purely observational science which he intended should replace the old speculative geology and geogeny,[11] he himself later on put forward a theory of the origin of the earth which was as fantastic as many of those which had preceded it. The explanation of the paradox is that Werner

---

[10] *Historia Terrae et Maris ex historia Thuringiae per montium descriptionem erecta*, in *Acta Academiae Electoralis Morguntinae*, Erfurt, 1761.

See also T. E. Gumprecht, *Einige Beitrage zur Geschichte der Geognosie*, Berlin, 1851.

[11] See Christian Weiss in the Lebensbeschreibung Abraham Gottlob Werners von Samuel Gottlob Frisch, Leipzig, 1825, p. 147.

believed that his theory was based upon fact and confirmed by actual observation at every point, and that if anyone wished for conclusive evidence of this, all that he had to do was to go and examine the rocks for himself. It is now evident, from the standpoint of our greatly increased knowledge, that in a number of instances he misunderstood what he saw, so that his supposed facts were not facts, and that furthermore his sphere of observation was far too limited.

Werner in his lectures on geognosy at Freiberg pointed out to his students that the earth did not consist of an unorderly assemblage of rocky materials, but that the earth's crust at least, which is the only part of the earth accessible to human observation, in many places presented a succession of "formations" each consisting of a series of beds or layers, reposing one upon each other. This very important fact Werner was not the first to discover—it had already been made known by Strachey[12] in 1719, by Swedenborg[13] in 1722, by Lehman[14] in 1756, by the Rev. John Mitchell[15] in 1760 and by Füchsel in 1762,[16] by Linnaeus in 1768[17] and by von Justi in 1771.[18] The last mentioned writer gives a list of the successive strata—49 in number—which are to be seen in the Mansfeld copper district. Werner, however, emphasized the importance of this very significant fact, described the petrographical character of the beds with greater accuracy and through his lectures obtained for this discovery a world-wide recognition.

Werner's earliest statement in print setting forth his views of the geological succession is contained in a little pamphlet, entitled "Kurze Klassifikation und Beschreibung der verschiedenen Gebirgsarten." This was employed by Werner as a "Leitfaden" in his lectures on geognosy and was, he states, sent to the press in 1777. It was not however actually printed for distribution until ten years later, by one of his friends in

[12] *Philosophical Transactions of the Royal Society*, Vol. XXX, p. 968, 1719, and another paper in the same, 1725.

[13] *Miscellanea Observata circa Res Naturales*, Leipzig, 1722.

[14] *Versuch einer Geschichte von Flötzbürgen.* See also *Essai d'une Historie Naturelle de Couches de la Terre*, in Vol. III of his *Traité de Physique, de Histoire Naturelle, de Minéralogie et de Metallurgie*, Paris, 1759 (Berlin, 1756).

[15] Phil. Trans. Vol. LI, 1760, part II, p. 582. See also A. C. Ramsay, *Passages in the History of Geology*, London, 1849, p. 33.

[16] *Historia Terrae et Maris ex historia Thuringiae* (to which reference has already been made).

[17] See A. G. Nathorst, *Carl von Linné als Geolog*, Jena, 1909, pp. 61, 62.

[18] *Die Geschichte der Erdkörpers*, Berlin, 1771.

Dresden.   In this Werner describes the geological column as consisting of four chief subdivisions as follows:

> (1) Aufgeschwemmte Gebirge—Gravel, sand, clay.   These in places contain fossils.
> (2) Vulkanische Gebirge—Ashes, pumice, &c.   "True volcanic" & "Pseudo-volcanic" rocks.   In the latter fossils are occasionally found.
> (3) Flötz Gebirge—Limestone, sandstone, grauwacke, puddingstone, coal formation, chalk, salt, gypsum, clay rocks.   Often holds fossils.

FIG. 46. The Scheibenberg in the summer of 1873.   (From a sketch by Beck.)

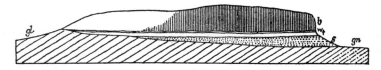

FIG. 47. Section through the Scheibenberg by Beck. *gn* is gneiss. *gl* is mica schist *s*, sand and gravel. *t*, clay. *w*, "wacke." *b*, basalt with columnar structure.
     The "wacke" is a decomposed basal portion of the basalt sheet.   Werner held that there was a transition from the clay into the "wacke" and from the "wacke" into the basalt and that the whole series, including the basalt, was therefore of sedimentary origin.   (From Richard Beck: *Abraham Gottlob Werner*, Berlin, 1918.)

> (4) Uranfangliche Gebirge—Granite, gneiss, mica schist, clay slate, porphyry schist, porphyry, basalt, amygdaloid serpentine, primitive limestone, quartz rock, topaz stone.   No fossils are present.

The "Flötz Gebirge" of this table may be translated as "stratified rocks," the German word "Flötz" meaning a bed or layer.

It is not always possible, he says, to separate these four divisions sharply from one another as they often seem to show transitional phases. It will be noted that basalt is here placed among the "Uranfangliche Gebirge" of Class 4.   In 1787 Werner visited a little hill near Scheiben-

berg in the Erzgebirge where he found basalt overlying sand, clay and the
"wacke," and from the examination of this occurrence became firmly
convinced that all these rocks belonged to one and the same group, and
that they were all of aqueous origin, forming part of the upper Flötzgebirge. Figures 46 and 47 show a sketch of this hill and a section through
it showing its structure, as given by Beck.

Later, about 1796, Werner in his lectures introduced into his system
another subdivision, that of the "Uebergangsgebirge" or transitional
rocks which were inserted between the Uranfängliche and the Flötzgebirge. His geological column eventually took the following form, as
given by Beck:—[19]

I. The Urgebirge containing the formations: Granit, Gneiss, Glimmerschiefer,
Thonschiefer, Urkalk, Urtrapp, Serpentin, Porphyr, Syenit, Urquartz,
Topasfels, Kieselschiefergebirge, Urgyps, &c. Fossils are absent.

II. The Übergangsgebirge, subdivided into:—
   1. Übergangskalkstein (e.g., the Devonian of Planitz near Zwickau, the
   Carboniferous Limestone of Derbyshire).
   2. Übergangstrapp (e.g., the Diabase and Diabase amygdaloid in Vogtland).
   3. Grauwackengebirge, much thicker than 1 and 2, the first well-marked
   mechanical deposits from the primitive ocean (e.g., in the Harz where it
   carries ores).

III. The Flötzgebirge, which is subdivided into:—
   1. The old red sandstone or Todtliegende (now known the Rothliegende).
   2. Flözkalkgebirge or first old Kalkgebirge, with the Kupferschiefer and the
   Rauchwacke (e.g., at Mansfeld = Zechstein).
   3. First Flötzgypsformation, gypsum with Stinkstein and clay with many
   caves (= Zechstein)
   4. Salzgebirge, salt beds with clay and gypsum (= upper Zechstein)
   5. Second Sandsteingebirge or Buntsandstein with Rogenstein (e.g., at Bebra)
   6. Second Flötzgypsformation with clay, often red (= Röt)
   7. Muschelkalkgebirge or second Kalkgebirge (= the Muschelkalk + Jura-
   kalk + Kreidekalk).
   8. Third Sandsteingebirge with calcareous beds (now known as the Quader-
   sandsteingebirge) e.g., in Elbtalgebirge.
   9. Third Flötzgypsformation, e.g., at Montmartre near Paris (= Tertiary).
   10. The Steinkohlengebirge, consisting of Schwarz- und Glanzkohle, Schief-
   erletten, soft Sandstones, Conglomerat, Marl, Limestone, also Porphyry
   and Clay Ironstone (The stratigraphical position of this important subdivision was still very much in doubt).
   11. The Flötztrappgebirge, consisting of Basalt, Wacke, Porphyrschiefer
   (= Phonolite), Sand, Grus, Clay, Stone and much brown Coal
   (= Tertiary).
   12. (?) The Kreidegebirge (= Upper Chalk) (According to Werner the position
   of this subdivision was doubtful, but it was very young).

[19] See his paper entitled *Abraham Gottlob Werner*, Berlin 1918, p. 29.

IV. Das aufgeschwemmte Gebirge, embracing as formations: Nagelflue, Sand, Clay,
Loam, Gravel, Grus, Fuller's earth, Calcareous Tufa, Bitumenous Wood,
Alum Shale.

V. Volcanic Rocks, of which there are two subdivisions: 1) True volcanic rocks
(Lava, Ejectamena, Lapilli, Ashes, Peperino, Pumice, Tufa; 2) Pseudovolcanic Rocks (Burned Clay, "Porcelain Jasper," Hornstone, "Erd-
schlacken."

Werner, however, still remained in doubt concerning the relative position
of several members of the Flötzgebirge, as for instance the Jura, Chalk
and Tertiary.

Only once—and this in the last paper which he wrote—did Werner
commit to paper a presentation of his views concerning the *structure* of
the earth's crust. This was in a popular lecture which he delivered
before the Mineralogical Society of Dresden in 1817 under the title
Allgemeine Betrachtungen über die festen Erdkörper. It was printed as
the first paper in Volume I of the Publications of this Society, and the
latter portion of it, giving a résumé of Werner's views on the subject, is
reprinted in Frisch and Weiss' *Lebensbeschreibung Abraham Gottlob
Werners*, Leipzig, 1825, pp. 108–114. This paper aims, Werner says,
at giving an account of the "Innern Verhältnisse des festen Erdkörpers",
so far as these can be determined by a close study of the earth's surface.
He speaks of the soft, incoherent and earthy materials which mantle the
lower levels of the earth's surface and of the more solid rocks which form
the higher land and the mountains and which also underlie the loose
material of the lower levels. The evidence goes to show that the former
have been derived from the waste of the latter. The harder rocks of
the mountains are very diverse in character and usually occur in the
form of beds, sometimes vertical but usually having a horizontal attitude
or dipping at an angle to the horizon. When there is a series of these
inclined beds overlying one another and rising into mountain masses,
it often comes about that the lowest beds of the stratigraphic series form
the highest mountain peaks. In such a series it will be found that the
lowest rocks of the series are almost invariably crystalline in character,
consist of siliceous and aluminous materials and are evidently of chemical
origin, while the higher and newer members of the series are largely of
mechanical origin. In the upward succession calcareous beds are found
intercalated with the siliceous and aluminous materials, while in the
highest parts of the stratified series limestones preponderate.

Where the beds of mechanical origin begin to come in, the first remains of living things are found in them. Indeed, it appears that different kinds of "fossils" are found in the different beds, so that a certain order in their succession is indicated. And so we must conclude, he says, that *"Our earth is a child of time and has been built up gradually."* The rocks abounding in organic remains, and still more the uncompacted materials which mantle the lower portions of the present surface of the earth and which are clearly derived exclusively from the disintegration of the older rocks, indicate that they belong to a recent age.

And again the very uneven character of the earth's surface with its deep valleys and high precipitous cliffs testifies to the action of great floods and other powerful influences, to which they owe their origin. And finally we recognize (he says) some few occurrences, limited in their extent, whose origin must be attributed to fire. He concludes by stating that the study of the action of all these forces will lead us to a clear understanding of the character and conditions of the exterior and interior of the earth, and of its origin and gradual development, through the many changes which it has undergone.

This is, strange as it may seem, all that Werner ever *wrote* with reference to his *Theory of the Origin of the Earth*, a theory which has played such an important rôle in the history of geology. For a detailed statement of his theory, developed and elaborated as it was by him as years went by, and set forth in his lectures at Freiberg, it is necessary to turn to the accounts of these lectures which have been left by certain of his students.

Werner's theory of the origin of the earth may be set forth as follows:— Of the inner portion or *nucleus* of the earth, although it constitutes the greater part of the globe, little or nothing is known, for mankind has no access to it. This hidden nucleus of the earth however has not a smooth and even surface, but one accentuated by high elevations and deep depressions. "In the beginning" it was completely enveloped by and submerged beneath the waters of a primeval ocean, which rose above the tops of its highest mountains. These waters were not clear and transparent like those of the present ocean, but were thick and turbid, holding in suspension or in solution (these two terms, apparently, not being sharply defined and separated in the minds of most of the Neptunists) all the materials which now form the earth's crust. These as time went on, were deposited in succession upon the surface of the earth's nucleus. The first of these deposits, forming the basement of the succession, consisted of granite. Upon this and often more or less intermingled with it, were laid down a thick series of gneisses, schists, porphyries and other crystalline rocks, which from their character were believed to be *chemical precipitates*. These Werner calls the Primitive Rocks.

The waters of this primitive ocean then began to subside, and while the subsidence was taking place another series of "formations" were laid down. They were partly chemical precipitates and partly mechanical sediments. These form what Werner terms the Transitional Strata. They consist as already mentioned, chiefly of various varieties of slate, schist, grauwacke, and greenstone with occasional limestones. They sometimes, though rarely, contain fossils, which when present consist chiefly of the remains of marine animals.

The abatement of the waters still continued and a third series of strata was laid down. These were the Floetz Strata, consisting of sandstones, limestones, puddingstones, chalk, gypsum, trap, rock salt, coal and other materials. In this series mechanical sediments are relatively more abundant and organic remains are very frequently present, often in large numbers.

Finally there were laid down the Alluvial Deposits which consist of sand, clay, gravel, shingle, tufa and similar uncompacted or partially compacted materials. While these were in part deposits from the waters of the primitive ocean, they originated chiefly from the disintegration and waste of the underlying and older strata. They lie chiefly on the lowlands of the globe and over wide areas form its present surface. The deposits laid down from this primitive ocean before the waters had subsided sufficiently to allow the first mountain tops to appear, were *"Universal Formations,"*[20] since they enveloped the entire surface of the globe. So soon, however, as the mountain tops appeared above the waters, the continuity of the deposits was interrupted owing to the small areas of dry land thus formed, but since the areas of dry land which appeared in these very early stages of the earth's history were very limited in area, Werner considered that all the Primary and a large part of the Transitional Formations might for practical purposes also be considered as Universal Formations.

The other and later formations were "Formations partielles," covering as they did only portions of the earth's surface. But each formation had a definite position in the geognostical column and therefore when it was exposed would always be found at its own proper horizon.

But the strata which make up a large portion of the earth's crust and which, it might be supposed, would be deposited from the ocean in flat-

---

[20] See D'Aubuisson de Voisin: *Traité de Géognosie*, Tome I, p. 327.

lying beds are often especially in the Primitive and Transitional Formations, seen to have a steeply inclined or even vertical attitude.
These Primitive and Transitional formations consisted almost entirely
of chemical precipitates from the primeval ocean. Now when any
substance is crystallized out of solution, it will be deposited not only on
the bottom but *on the sides* of the containing vessel, and so in any great
mountain chain, such as the Alps, the primitive granite rocks, which
often display a more or less distinct foliation, owe the steep inclination
and often undulating attitude of this foliation to the fact that it is conforming to the contours of the surface of the nucleus of the earth on which
it was deposited.

Werner believed that in other cases the steeply inclined attitude of
these rocks was due to the irregular manner in which the deposited mass,
while still uncompacted, had settled down; in still other cases to great
bodies of the precipitated materials having slipped down the steeply
inclined surface on which they had been deposited. Such explanations,
to the minds of the latter Wernerians, did not seem to be altogether
satisfactory, and D'Aubuisson concludes his discussion of this subject
with the words: "La détermination de la stratification, de ses circom-
stances et de ses lois, est encore un problème à résoudre; et c'est peutêtre
le plus important de la Géognosie."[21]

Furthermore, the primitive ocean did not subside slowly and quietly
or at a uniform rate, but was swept and driven by the furious winds and
great storms which characterized those times of universal chaos, and
also from time to time for reasons that are unknown, it experienced
oscillations in level. Powerful and shifting currents set up by the winds
and by the draining away of the subsiding waters, cut deep channels
through the sediments in all directions and by their erosive power gave
rise to deep valleys separated by high mountains. As time went on,
with the gradual approach to modern conditions, the ocean waters became more quiet and the strata deposited from them were laid down as
horizontal beds. With the final withdrawal of the ocean into its present
basin, still mantling more than one half of the earth's surface, still depositing mechanical sediments and some chemical precipitates as well,
the world as we see it came into being.

Nevertheless there were certain matters which the Wernerians found
it very difficult to explain. Where, for instance, did the waters of the
primeval ocean go, when this ocean subsided and the continents ap-

---

[21] *Traité de Géognosie*, Vol. I, p. 352.

peared? Some said they retreated into the interior of the earth, but it was difficult to believe that empty spaces large enough to receive such an enormous volume of water ever existed in the earth's interior. Werner[22] himself was inclined to believe that the waters might have vanished into space, drawn away by the attraction of some celestial body which by chance passed near the earth. All that he knew definitely was that the universal ocean had now and somehow vanished.

Then there was the question as to why the volume of materials precipitated from the primeval ocean, that is to say the thickness of the formations, was so much greater in some places than in others.

But the greatest difficulty which presented itself in connection with Werner's theory was the existence of volcanoes. There were, in all the divisions of Werner's geological column, certain fine-grained dark or black rocks which were allied to basalt in character or in composition. Intercalated in the succession of the Primitive and Transition Rocks were the "Primitive Greenstones." In the Floetzgebirge there were beds of Basalt, dark Tufas and allied rocks, and on the surface of the earth were volcanoes with their lava streams, lapilli, ashes tufas and pumice stone. These latter undoubtedly owed their origin to fire. Furthermore in all the formations there were to be found dykes of Trap and "Toadstone" cutting through the sedimentary deposits. The question therefore at once presented itself:—Were not all these rocks, occurring as they did in all the groups of formations constituting the earth's crust, really of igneous origin like the modern lavas, and therefore did not igneous action play an important rôle from the beginning in building up the earth's crust? Werner answered this question in the negative. He was obliged to admit that volcanoes existed, but he held that they were few in number and played a very subordinate and unimportant rôle in nature, and furthermore that they came into existence only in the latest period of the earth's history.

In treating of the "Vulkanische Gebirgsarten" in his *Kurze Klassifica-tion und Beschreibung der verschiedenen Gebirgsarten*, Werner says that these either owe their existence to fire or have been changed to their present form by fire. The first group which have been actually produced by fire and have come into existence through violent volcanic eruption he calls True Volcanic Rocks. The second group which have been merely given their present form and character by subterranean fire, he calls Pseudovolcanic Rocks. He then describes briefly the character of the lavas, volcanic ashes and volcanic tufas which are true volcanic rocks. These are generally, he says, very irregular in their mode of occurrence, but not infrequently have a character which simulates bedding owing to the products of successive eruptions having been laid down one upon the other. These true volcanic rocks are grey, red or black in color and more or less vesicular in structure, often containing small eight-sided crystals which, he says, most mineralogists call schorl but which are really crystals of a variety of hornblende.

The pseudovolcanic rocks include lava-like slags often vesicular or scoriaceous in character, "porcelain-jasper" and all manner of half burned clays. These are largely carbonaceous Floetz rocks which have been partially or completely melted by the combustion of coal beds occurring in the vicinity and they retain to a large extent their original bedded character. The porcelain-jasper often displayed the impressions of plants, owing we now know to the fact that it was really a shale indurated and altered by the action of heat.

Werner held that basalt was a rock which was quite distinct and different from the *lava* of volcanoes. It was found in the form of beds alternating with beds of sandstone, shale and other rocks which were indisputably of aqueous origin and like them had been deposited from the waters of the primitive ocean, but as shown by its crystalline character it was a chemical precipitate. In other cases it was found in the form of dykes cutting across such strata but these dykes, like other "mineral veins," were also precipitates from the ocean waters deposited in fissures. When however he proceeds to define the exact difference between basalt and a basic lava, his explanations are indefinite and obscure, on account of the fact that he is endeavoring to describe the difference between two substances which are identical. His students at this time, however, were firmly persuaded that he was quite right in making this distinction, and many of them continued to hold this view for many years, in spite of all that was urged against it.

And here a short digression may be permitted for the purpose of noting an interesting incident which relates the work of Werner to that of Darwin. When Darwin in 1825 entered the University of Edinburgh as a student, among other courses of lectures he attended those of Jamieson who had been a pupil of Werner's and was one of the last and most outstanding of his followers. Darwin[23] tells us that he found Jamieson's lectures "incredibly dull"[24] but also states that all the other

[23] *Life and Letters of Charles Darwin* edited by his son, Francis Darwin, Vol. I, p. 41, London, 1887.
[24] Ibid., p. 36.

courses of instruction which he followed at this seat of learning, with the exception of those in chemistry by Hope, were "intolerably dull," so that it is difficult to know whether in Darwin's opinion Jamieson's lectures suffered by comparison with those of his fellow professors at the time. He goes on to say, however, that "The sole effect produced on me was the determination never as long as I lived to read a book on geology or in any way study the science." He further states that in one of the geological excursions on which Jamieson conducted his students:

I heard the Professor at Salisbury Craigs discoursing on a trap dyke with amygdaloidal margins and the strata indurated on either side, with volcanic rocks all around us, say that it was a fissure filled with sediments from above, adding with a sneer that there were men who maintained that it had been injected from beneath in a molten condition. When I think of this lecture, I do not wonder that I determined never to attend to Geology.

And yet only five years later, Darwin, writing from the "Beagle,"[25] when in South America, says, "Geology carries the day. I find in geology a never failing interest, it creates the same grand ideas respecting this world which astronomy does for the universe." It was a case of sudden conversion due to the influence of Lyell. "The very first place which I examined," Darwin writes in 1876, "namely St. Jago in the Cape Verde Islands, showed me clearly the wonderful superiority of Lyell's manner of treating geology compared with that of any other author whose works I had with me or ever afterwards read. The science of geology is enormously indebted to Lyell—more so, I believe, than to any other man who ever lived."

One of the most difficult problems with which Werner was confronted in his explanation of the origin of the volcanic and pseudovolcanic rocks, was that of explaining the source of the heat by which they were produced. His theory of the origin of the globe postulated no "Fire at the centre of the earth." All that portion of the earth's crust to which man had access was aqueous in origin with the exception of a few volcanoes to which reference has just been made, and these he regarded merely as sporadic phenomena confined to the highest and latest portions of the geological column. When in 1777 he made a visit to the Mittelgebirge of Bohemia, he saw a certain locality where in former times some of the great coal seams in that district had taken fire and by the intense heat

---

[25] On the Structure and Distribution of Coral Reefs, also Geological Observations on the Volcanic Islands and parts of South America (Ward, Lock & Co.), 1890, p. 159.

which had thus been developed, certain of the adjacent rocks had been baked and partially fused, giving rise to products which bore a close resemblance to certain of his pseudovolcanic rocks. So impressed was he with this resemblance, suggesting as it did an explanation, although of course an entirely inadequate one, for the origin of his pseudovolcanic and volcanic rocks, that he adopted it as the true solution of the problem; an explanation which, he states, receives further support from the fact that so far as he could ascertain deposits of coal or other combustible materials were invariably present in the vicinity of volcanoes.

Such then in brief was Werner's theory. One of the best and most complete presentations of it is to be found in a work entitled *A Comparative View of the Huttonian and Neptunian Systems of Geology in Answer to the Illustrations of the Huttonian Theory of the Earth by Professor Playfair.* This book was published in Edinburgh anonymously but was written by John Murray. It appeared in 1802, that is to say, while Werner was still lecturing in Freiberg, and sets forth very impartially the explanations put forward by the advocates of each of these theories to account for the observed facts, and is most interesting as showing how the associates and followers of Werner explained with a certain degree of plausibility geological occurrences which would at first sight seem to be irreconcilable with the thory in question.

2. LEOPOLD VON BUCH AND HIS "INTERROGATIONS" OR THE TESTING OF WERNER'S THEORY

Werner's students, a goodly fellowship indeed, having completed their courses of study at Freiberg, inspired by their teacher, now the most outstanding geologist in Europe, and firmly convinced that his views were sound and true, returned to the various lands from which they came, eager to apply their newly acquired knowledge of that most fascinating subject geognosy, to the solution of the problems presented by the geological succession in their several countries.

"One can say of Werner," writes D'Aubuisson,[26] "what has been said of Linnaeus, that his disciples have covered the earth and that from one pole to another nature has been *interrogated* in the name of one individual man."

It will therefore now be of interest to inquire and ascertain what answer nature gave to these "interrogations" concerning the truth of the Neptunian Theory.

[26] *Traité de Géognosie,* Strasburg & Paris, 1819, Vol. I, p. xiv.

From among the multitude of Werner's disciples Leopold von Buch stands out as the most distinguished, and among the others Alexander von Humboldt and D'Aubuisson de Voisins were men who achieved great renown. Upon leaving Freiberg all three took up geological field work with great enthusiasm and carried out very important researches in widely separated portions of the world. A study of the results which they obtained will afford a full and satisfactory answer to our inquiry as to how Werner's theory responded to the acid test of further study and research, without examining the conclusions reached by other and less renowned disciples of the great master.

In this connection the work of Leopold von Buch is of especial interest and value, not only on account of its wide extent and the importance of the results which he achieved, but also since it has a certain psychological interest, displaying as it does his continued reluctance to relinquish, as he was forced to do step by step through the logic of stubborn facts, the most cherished teachings of his beloved teacher.

Leopold von Buch was born in 1774 at Schloss Stolpe in Angermünde, a town of Brandenburg not far from Berlin. He belonged to a wealthy and noble house, and was the sixth son in a family of thirteen children. From his boyhood he displayed a singular distinction of character and took a keen interest in the study of nature. When thirteen years of age he was struck by the following motto which he chanced to see and which might be taken as the maxim which governed his life:

Das Neue erweitert, das Grosse erhöht unsern Gesichtskreis, das eine wie andere verstärkt das Gefühl der innern Triebkraft und Vollkommenheit.

The "New" expands our intellectual horizon, the "Great" exhalts our outlook, both strengthen the sense of the power of the ideal within us.

Humboldt[27] who was his life-long friend, in a letter dated March 13, 1853, says of von Buch that he showed "A happy blending of the most noble philanthropic sentiment, momentary impulses and a little despotism of opinion."

When but fifteen years of age von Buch attended lectures on chemistry and mineralogy in Berlin and the following year, 1790, he entered the Mining Academy at Freiberg. He was here brought into very intimate relations with Werner, living in his house most of the time during the three years in which he remained at Freiberg. From Freiberg he went

---

[27] *Letters to Varnhagen von Enze*, New York, 1860.

Leopold Von Buch

PLATE V

PORTRAIT OF LEOPOLD VON BUCH

to the University of Halle to continue his studies, and while there wrote
his first little treatise, that on "Cross Stone."[28]   He also studied for a
time in Göttingen.

In 1796 he received an appointment in the Mining Service of Silesia,
entered with great zeal upon the study of the geognosy of that province
and wrote his treatise entitled *Versuch einer mineralogischen Beschreibung
von Landeck*, published at Breslau in 1797.   It is an excellent description
of the geology of a little area in Silesia, and in carrying out this field
work he found no facts which, in his opinion, would seriously conflict
with Werner's theory, although he states that, in an "amygdaloidal
Wacke of the basalt series" which formed the hill of Finkenhübel he
found specimens of Turbinites, adding, "This fact is very remarkable on
account of the small number of organic remains which have hitherto
been discovered in rocks of the trap formation."   He also shows that
the Silesian basalts lie upon strata of very diverse ages, some of which
are lower than the coal-bearing horizon and that, therefore, they could
not have been produced by the action of fire resulting from the combustion of coal seams, as Werner supposed.

In 1797 von Buch, being a man of independent means, resigned his
position in Silesia in order that he might devote his attention exclusively
to the prosecution of independent scientific research.   He made some
most important studies in the Alps (to which reference will be made in
the chapter dealing with the Origin of Mountains) and then passed on
into Italy.   The results of his work in the Alps and in Italy are set forth
in his *Geognostische Beobachtungen auf Reisen durch Deutschland und
Italien*, Berlin, 1802–1809.   This book he dedicated to Werner, and
commences a brief letter following the Dedication with the following
significant words:

In den wenigen Stunden gütiger Belehrung, die Sie mir, kurz vor meiner Abreise
nach Italien in ihrem Hause zuzubringen erlaubten, schienen sie mein verehrter
Lehrer, die Hofnung zu äussern, dass meine Reise veilleicht der Wissenschaft selbst
von Nutzen seyn könnte.   In wie weit diese Hoffnung erfüllt seyn mag, müssen
Ihnen diese Bogen, welche die Resultate meiner Beobachtungen enthalten, beweisen.
Sie werden oft die Worte und die Ideen—wie sehr wünschte ich hinzu fügen zu können—auch den Geist des Lehrers wiedererkennen.   Ich darf deswegen Ihre Misbil-
lung nicht fürchten.   Denn wie könnte der Schüler seine Dankbarkeit lebhafter
äussern, als durch das Bestreben, der Schöpfung des Lehrers weitere Verbreitung,
neue Ausdehnung, neue Festigkeit zu verschaffen.   Und wenn es, in diesem Falle
auch immer sein Schicksal seyn muss, seine Lehrsätze mit den Irthümern des Schülers

---

[28] *Beobachtungen über den Kreuzstein*, Leipzig, 1794.

durcheinandergeworfen zu sehen, so leitete ja von jeher der Weg zur Wahrheit über
Irthümer hin.

He went first to Rome. Through his studies of the volcanic rocks
there and especially those of the Alban Hills, his Neptunian views on
the origin of basalt received their first shock. In his book to which
reference has been made, von Buch states that while he does not wish to
contradict the opinions of all the naturalists who had formerly examined
the Capo di Bove and who had held that it was indisputably a lava flow,
an opinion in support of which, he says, much evidence could be adduced,
he wished to point out a number of facts opposed to this view, which he
then proceeds to do.[29] His absolute bewilderment of mind however
arising from the study of these Roman occurrences is graphically set
forth in a letter which he wrote from Rome on September 23, 1798, and
which appeared in von Moll's Jahrbuch. In this he says that the
exposures in this district are so remarkable that he can scarcely believe
his own eyes and that here nature actually seems to contradict herself.
This letter is of especial interest as showing the "growing pains" which
he endured while gradually abandoning the Neptunian views of his
teacher, Werner, and passing to the wider outlook of the Plutonists.

In February 1799 von Buch arrived in Naples and with great enthusiasm commenced his study of Vesuvius and the surrounding district. He
had with him Breislak's earlier description and map of the area, and
speaks in the highest terms of these and of their author:

> Der scharfsinnige Breislak, der einzige Geognost am Vesuv. . . . Die Vesuvischen
> Laven kennen wir in der That nur durch ihn, und wenn auch die Kenntniss nicht
> vollständig ist, so hat er doch seinen Nachfolgern nur einige Lücken auszufüllen
> gelassen.[30]

Von Buch gives a clear and altogether admirable description of
Vesuvius and the Phlegrean Fields. For the first time he was able to
examine the structure of a great active volcano. He recognized at once
that the whole area was of volcanic origin, and was greatly impressed
with the terrific power manifested by the volcanic forces and by the
great variety of the products of volcanic activity. He states that
certain of the lava flows consist of basalt but others, especially those
rich in leucite diverge greatly from this in character and differ much

---

[29] *Geognostische Beobachtungen auf Reisen durch Deutschland und Italien*, Bd. II, p. 64.
[30] *Geognostische Beobachtungen auf Reisen durch Deutschland und Italien*, Bd. II, p. 180.

among themselves. Even the ground mass of these rocks cannot be considered as a "simple" rock, that is to say, it is always composed of two or more different minerals. Until much more is known of the miscellaneous group of eruptive rocks it is better to designate them collectively as lava. His Neptunian views however, received a further shock when he found that certain rocks that he could not have believed to be of volcanic origin occurred in the form of indisputable lava flows. Among these was "Feldspar Porphyry." His bewilderment may be gathered from the following passage in his description of the Phlegrean Fields:—[31]

Also ein Feldspathporphyr, ein Gestein, dem mann nimmermehr ein vulkanisches Fliessen hätte zuschreiben mögen, wenn nicht alle Lagerungsverhältnisse so unmittelbar, beynahe so unwiderleglich darauf hinwiesen! Wie sollen wir uns die Erhaltung so grosser, so schöner Feldspathkrystalle, und in dieser Menge in einer feurigflüssigen Masse vorstellen?

Aber, wenn es nun Thatsache ist!—So müssen wir von der Zeit über die Möglichkeit Belehrung und Aufschlüsse erwarten. Es fehlt uns eine Beobachtungsreihe, zu deren Aufsuchung uns dieser Scheinbare Widerspnuch zwischen dem Wirklichen und dem Möglichen aufruft.

The question of the locus from which the terrific forces which manifested themselves in the formation of the volcano proceeded, thrust itself upon him at every turn:

Dass der Sitz des Vulkanischem Heerdes im Vesuv selbst wohl schwerlich seyn könne, ist einleuchtend. Im Conus nicht;—weil man schon oft die ganze innere Hölung des Conus gesehn hat;—und in der unteren Hälfte des Berges nicht, weil die Lavenströme, welche sich von je her über den Abhang ergossen, wahrscheinlich den grössten Theil des Innern ausfüllen wurden. Auch ist der ganze Conus selbst nur ausgeworfen, aus dem Innern heraufgebracht. Daher muss die Hebungsursach, das vulkanische Feuer, noch ungleich tiefer liegen, und also wahrscheinlich weit unter dem Fusse des Berges. Warum aber unmittelbar darunter? Dazu ist keine nothwendige Urasache. Denn es ist doch möglich, dass die Dämpfe in einiger Entfernung von Entstehungsort zufällig einen leichteren Ausweg fanden, als unmittelbar darüber; einen Weg, den sie sich dann immer offen erhielten. Und dürfen wir den Mofetten trauen, so müssen wir uns ehe gegen das Meer wenden, und diesen Sitz vielleicht unter dem Meere selbst suchen; um so mehr, da uns die Bergölquelle im Neapolitenischen Golf hinreichend beweist, dass vulkanische Wirkungen sich auch noch wirklich unter dem Grunde des Meeres zu äussern vermögen. Denn diese Quelle steigt fast allemal stärker und heftiger nach grossen Ausbrüchen.

[31] Op. cit., Bd. II, p. 206.

The seepage of petroleum into the sea water referred to in the passage just quoted took place, according to Breislak, about a mile off shore opposite to the castle of Pietra Bianca, the oil appearing as flecks on the surface of the water.

There was the further question also as to the origin of the intense heat which fused the immense masses of rock poured out of the crater of the volcano in all the successive periods of eruption. Could it be derived from the combusion of reservoirs of oil beneath the surface as suggested by the seepages of this material into the sea water at one point near the shore? Scarcely. Could it be produced by the burning of immense subterranean beds of coal, as Werner supposed? No evidence whatsoever could be found for the existence of such coal deposits in the region.

Wozu dienen auch die scharfsinnigsten Meynungen über die Ursache dieser Feuerwerkstatt, so lange unsere Erfahrung noch bis dahin nicht hat durchdringen können?—Wir haben kein Mittel, die Wahrheit dieser Theorien zu prüfen. Denn wir kennen von den Erscheinungen im Innern nur so wenig, dass zu ihrer scheinbaren Erklärung sich mit gleichem Recht eine Menge Ursachen angeben lassen. Wir wissen nichts mehr, als dass dort ein nie aufhörender Feuerquell sey, der Laven schmiltz und Dämpfe erzeugt.—Selbst die befriedigendste dieser Theorien, die *Wernersche*, der Steinkohlenentzündung, muss um so behutsamer angewandt werden, je einnehmender sie ist. Denn vergebens suchen wir am Vesuv und in der ganzen Gegend umher die Orte, wo diese Steinkohlenflötze könnten gelagert seyn.—Unter dem Grunde des Meeres? Es ist möglich; aber noch sind keine Erscheinungen gefunden, welche die wirkliche Existenz dieser Flötze verbürgen.—Die Bergölquelle wohl schwerlich; denn das Bergöl ist hier, wie im Elsass und Jura, in Gebirgsarten häufig, die mit den Steinkohlen wenig gemein haben.

And finally the dreadful question presents itself to his mind—What if it should be shown eventually that these volcanic forces really break their way upward through the underlying primitive rocks? Such a discovery would be a fatal blow to Werner's Theory! He was to receive a very clear and decisive answer to this very important question in his study of the Auvergne district to which he next directed his attention in the year 1802.

Here in the Auvergne district of Central France he found a series of extinct volcanoes, still retaining their typical form, from which streams of basalt had flowed forth. These volcanoes rose from a great plain underlain by granite. The volcanoes were set upon the granite, and in this primitive granite or beneath it was evidently the locus of the volcanic

forces.[32] Another indisputable fact was that the Puy de Dôme and
many other of these mountains consisted of "Trap Porphyry" which he
now called "Domite" and which was composed of the same constituents
as granite itself, although in the Domite the feldspar had a bright
vitreo-pearly luster instead of the duller lustre presented by the feldspar
of the underlying granite.

These Domite hills however had rounded tops and no craters on their
summits. From a close study of them he reached the conclusion that
they were formed at certain places where the granite had been softened

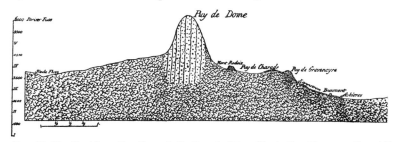

FIG. 48. Von Buch's section through the Puy de Dôme illustrating his conception of its
origin. The granite of the plateau from which it rises has been changed into domite and
pushed upward in the form of a dome-shaped mass. (From his *Geognostische Beobachtungen
auf Reisen durch Deutschland und Italien,* Vol. II, Berlin, 1809.)

by the action of the heated vapors and through the force exerted by these
had been pushed upward out of the granite mass in the form of mush-
room-like protuberances.

In a paper entitled *Observations sur les Volcans d'Auvergne,*[33] which he
published at this time, he says that he sees in the Pûy de Chopine,
which belongs to the same class as the Pûy de Dôme:

> La preuve finale que la roche du Puy de Dôme est un granite changé et soulévé.
> Par quoi a-t-il été changé? Ce n'est pas par fusion. Rien n'est fondu (contre l'opin-
> ion de Guettard et Legrand). C'est un changement, opéré par l'action d'une vapeur
> quelconque aqueuse ou acide qui en même-tems a suffi par sa force expansive à
> soulever ces masses.

The Pûys which have craters, on the other hand have been elevated
by eruption. In the Pûy de la Nugére one can trace distinctly this

[32] This fact does not seem to have become generally recognized until many years later,
for William Phillips, F.R.S., and a member of the Geological Society of London, in his
Outline of Mineralogy and Geology which was published in London in 1815, p. 173, says,
"No known volcano is seated in granite."

[33] *Journal des Mines,* Paris, Vol. 13, 1802–3, p. 249.

"porphyry", as he calls the Pûy de Dôme rock ("Domite"), passing gradually into a black and porous lava.   From his subsequent examination of the great composite mass of Mt. Doré he thought that it might almost be affirmed that there the basalt itself had resulted from the alteration of the granite and "Könnte von diesen Gestein ein geflossenes Product sein."   There was moreover no evidence of the existence anywhere in the region of coal beds or other fuel to feed the fires.   But even so he asks, is it necessary to abandon the belief in the geological succession so clearly established from the field evidence in Germany?   Is it necessary to abandon the clear evidence of the aqueous origin of the basalt there when it seems to form part and parcel of a group of undoubtedly sedimentary formations lying above the coal formation?   Surely not. These French occurrences are so strange and unusual as to suggest that they may be the result of some special and aberrant conditions.

In conclusion he says "The facts are really so bewildering, that we must await the results of further observation for the solution of the riddle which they propound."

Boué has said, in writing of von Buch, that "The learned geognost left Germany in 1798 as a Neptunist and returned in 1800 as a Vulcanist," and Geikie[34] also speaks of his complete and sensational conversion to the opinions of the latter school.[35]   As a matter of fact, however, as appears from what has just been stated, while von Buch was now convinced that the basalts of the Auvergne and Mt. Doré were of volcanic origin, he was still of the opinion that the German basalts and probably those of most other regions were the products of aqueous precipitation. Von Buch's studies in Germany, France and Italy were however only the beginning of his life work.

But a further stage of his history now developed.   He went from France to Norway, a part of Europe of whose geology very little indeed was known at that time, and spent two entire years in Scandinavia, from Norway crossing Lapland and thence continuing southward to Christiania.   In Norway he found that the geological succession in the main agreed with that of Werner's Formations.   He saw there, however, great sheets of granite overlain by a highly fossiliferous limestone into which the granite sent numerous apophyses and which was highly

---

[34] *The Founders of Geology*, pp. 246, 247.

[35] See Chapter X on the *Origin of Mountains*, where a further reference to this subject is made in considering von Buch's theory of Craters of Elevation.

metamorphosed along the contact, evidently owing to the heat emanating from the intrusive mass. Another striking fact which he observed was the distinct evidence that a slow differential elevation of the whole Scandinavian Peninsula had taken place, the land rising from the district about Frederiksall on going in an easterly direction toward Abo and probably on toward St. Petersburg.[36]

Our indefatigable geologist then turned his attention to the north-west coast of Africa, and made a prolonged study of Teneriffe, the Grand Canaries, Palma, and Lanzarote. Here he found a display of volcanic forces on a much grander scale than he had formerly seen, learned something of the manner in which volcanic islands are developed by the suboceanic fires, and expressed the opinion that most if not all islands in the ocean originated in this manner, a great generalization which subsequent investigations have proved to be in the main correct.[37] He then visited the Hebrides and other parts of northern Scotland, as well as northern Ireland, for the purpose of examining the celebrated occurrences of basalt in these districts.

Returning to Germany he there resumed his investigation of the geology of his native land, devoting his attention more especially to the geology of the Alps, and endeavoring to find in the elucidation of their structure an explanation of the origin of the mountain ranges of the earth. This study he prosecuted for several years continuously and it led him to conclude that the Alps owed their origin to a process of upheaval, due to a force exerted by bodies of volcanic rocks beneath the surface, such as those which led to the development of the oceanic islands, but which here could not find an actual passage to the exterior of the earth owing to the resistance offered by the great thickness of the overlying rocks.

Thus, the further von Buch prosecuted his studies, and the wider the field over which his studies extended, the more evident it became that volcanoes were not, as Werner believed, purely local and relatively unimportant factors in the history of the earth, but were of widespread and far-reaching importance and furthermore, that they had their origin deep down within the earth's crust. He also found that, quite contrary to the teachings of Werner, basalt and granite were not pre-

[36] *Reise durch Norwegen und Lappland*, Berlin, 1810.

[37] *Physikalische Beschreibung der Kanarischen Inseln*, Berlin 1825. See also *Ueber die Natur der Vulkanischen Erscheinungen auf den kanarischen Inseln und ihre Verbindung mit andern Vulkanen der Erdoberfläche*, Poggendorf's Annalen, X, 1827.

cipitates from a universal ocean but in all cases igneous rocks which played quite another rôle in the geological succession.

Von Buch brought his field work to a close in 1826 with the publication of his great geological map of Germany in 24 sheets, after which he turned his attention to palaeontology and the classification of the fossiliferous formations.

He was one of the very greatest geologists of his time, a clear thinker, an excellent observer and an untiring worker. Murchison met him in the Alps in 1847, and in his notes gives a picture of him, even then when he was in his seventy-fifth year, as a man of extreme determination and great perseverance with a contempt for physical privation.

Von Buch (writes Murchison)[38] still toddles along from ten to twelve leagues on foot. . . .

Seeing that at von Buch's slow pace we could not reach St. Cassian until far in the night, I walked on in a heavy drenching rain to beat up the Curé of St. Cassian and get supper ready. . . . Von Buch has just arrived wet through. He calls at once for his dinner, and will not change his clothes. He eats a hearty dinner with his shoes full of water, and all drying upon him. He was indignant when I begged him to do as I had done, and take the priest's great woollen stockings.

Next day we had not gone far on our road to Brunecken, distant twelve leagues, when it was evident that M. de Buch could never accomplish half that distance. He seemed to De Verneuil and myself to be staggering, and every now and then he sat down on a block of stone. He would, however, hear no reason; said he had often such megrims in the streets of Berlin, and would persist. At a snail's pace, and after a hundred halts, passing through some Alpine villages, he allowed that he was fairly beaten, and at last accepted my proposal to walk on fast to Brunecken, and bring or send back a light calêche to the point to which a practicable road came. . . .

Arrived at the priest's house, the old man in his dripping clothes (angry with us if we alluded to his state, his hands, which I touched, being icy cold), actually sat for two hours, fortunately in the hot *Stube* of the priest, the thermometer outside being nearly at freezing point. He joked, told stories, ate a good dinner, and was up at five o'clock next morning, ready to start in his still moist and damp garments. Hear this, you chamberlains of the Courts of Germany, and imitate, if you can, your brother!

He died after a short illness in the year 1852.

It is not necessary here to follow in such detail the studies of von Humboldt and D'Aubuisson which led them, at first ardent advocates of the Neptunian theory, to pass over into the camp of the Plutonists. Alexander von Humboldt who was born in 1769 in Berlin, studied at Göttingen and entered the Freiberg academy as a student of Werner's

[38] See also Chapter X.

in 1791, two years later than von Buch. He also lived with Werner for a time and was a life-long friend of von Buch. He was a strong adherent of the Neptunists even before he entered Freiberg, for in his first work, a little book published in Brunswick in 1790 when he was yet but 21 years of age, he strongly advocated the view that basalt was of aqueous origin.

Humboldt, like his companion von Buch, after leaving Freiberg, was appointed to a position as "Oberbergmeister," but after some years of mining experience resigned this position and devoted his life subsequently to a wide range of study in the natural and physical sciences, travelling very extensively for many years in various parts of Europe, Central and South America, Russia and Siberia. His life work may be said to have culminated in the appearance of his "Cosmos," a work of immense scope and erudition, which aimed at nothing less than the presentation of a picture of the whole world of nature. Von Humboldt's geological work which was of permanent value, lay chiefly in the study of volcanoes and earthquakes, in the pursuance of which he visited some of the most highly volcanic regions of the world. These demonstrated to him the great and widespread action of fire in the upbuilding of the earth's crust in all past ages as well as at the present time, and the truth of the contentions of the Plutonists.

D'Aubuisson de Voissins came from Southern France, having been born in Toulouse, and he studied at Freiberg from 1797 to 1802. On returning to France, he entered the service of the Government in the Department of Mines and speedily rose to the rank of "Ingénieur en chef au corps Impérial des Mines." Like the others, when he left Freiberg, he was an ardent Neptunist. In a paper read before the National Institute in Paris in 1803, he marshalls the evidence that these basaltic rocks were aqueous precipitates, stating that this evidence is conclusive. He goes on to say:

I think it needless to extend farther the enumeration of such arguments, any one of which, by itself, appears to me to be decisive of the question. If the hypothesis, which supposes the basaltic rocks of Saxony and some others to be volcanic products, is so completely at variance with the results to which correct observation leads, and with the principles of sound physics, it may naturally be enquired how this opinion has gained such currency. I would ascribe it in great measure to a love for the marvelous. It is well known how readily any thing that pleases by striking the imagination, is received, and with what difficulty the simple unaffected truth can procure a hearing. It delights the imagination to suppose the existence in former ages, of burning mountains, and vast streams of liquid fire,—nature, as it were in a conflagration,—at the very spot, perhaps, now occupied by our peaceful habitations. From our childhood, we are charmed with the pompous description of the terrible phenomena of volcanoes. The imagination pictures to itself the appearance of the fiery gulfs; and is wonderfully pleased with the employment of tracing the contrast between the present and former state of supposed volcanic mountains. The black and fuliginous aspect of basaltic masses, and the isolated and striking shape of the mountains which they constitute, greatly aid the illusion.[39]

Later he visited the Auvergne district—which was a veritable graveyard of the Neptunian opinions—in company with von Buch, and there made extended and careful studies of that most interesting and instructive region. As a result of his studies in the Auvergne, D'Aubuisson's views suffered an entire change and he became convinced that the basalt and rocks in that region were of volcanic origin, and after further years of study he reached the conclusion that the basalts of Saxony were of the same origin.

On trouve en Auvergne et dans le Vivarais des basaltes d'origine volcanique; c'est un fait évident; on retrouve en Saxe, et dans les terrains basaltiques en général, des masses qui ont une pâte exactement pareille, qui contiennent exactement et exclusivement les mêmes cristaux, qui ont exactement les mêmes circonstances de gissement: il y a un non-seulement analogie, mais parité complète, et l'on ne peut se dispenser de conclure à l'identité de formation et d'origine.[40]

3. HUTTON AND THE PLUTONISTS

The school of geologists which opposed the views of Werner and the Neptunists came to be known as the Plutonists. This name however is a rather misleading one. For while Werner regarded water as the sole agency which had been instrumental in building up the successive "formations" which constituted the earth's crust, the Plutonists held that both fire and water had played major parts in the development of the geological succession. More and more as a knowledge of the geological structure of the earth's crust in all parts of the world was revealed by the studies of an ever-increasing number of interested and eager students, the views of the Plutonists met with wider and wider acceptance. Most of Werner's students in time went over to the camp of the Plutonists.

[39] See *An Account of the Basalts of Saxony* (translated from the French by P. Neill), Edinburgh, 1814.
[40] *Traité de Géognosie*, Vol. II, p. 605, 1819.

It was in Scotland that the Plutonists found their leader. Here it was that Hutton (1726–1797), who may fairly be called one of the founders of modern geology was born, lived and wrote his epoch-making work, *The Theory of the Earth*.[41] It was here that Sir James Hall, the "Founder of Experimental Geology," carried out his celebrated researches, which gave such support and corroboration to Hutton's explanations of certain phenomena which he had observed in the field and it was here that Playfair, Professor of Natural Philosophy in the University of Edinburgh, wrote his *Illustrations of the Huttonian Theory*.

These were the three great leaders of the Scottish school of geology, but there were other men associated with them of whose work it is impossible here to make further mention. Hutton being the chief exponent of the views held by this school, his explanation of the origin and structure of the earth came to be known as the Huttonian theory.

James Hutton (whose portrait is the frontispiece of this book), was born and brought up in Edinburgh. After leaving the University he was for a time apprenticed to a lawyer. He abandoned law for medicine which he studied not only at Edinburgh but in Paris and at Leyden. He did not however practice his profession but turned to agriculture and settled on a farm in Berwickshire. In 1768 he gave up farming, took up his residence in Edinburgh and turned his attention more particularly to geology.

A circumstance which gave a marked impetus to the study of geology in Edinburgh was that about the same time that Playfair was appointed to the chair of Natural Philosophy at the University of Edinburgh, Robert Jamieson received the appointment of Regius Professor of Natural History at this same seat of learning. Jamieson had studied geology at Freiberg under Werner and, being an able and active exponent of Werner's opinions, promulgated them in Edinburgh and in 1808 founded there the Wernerian Natural History Society, in the publications of which Werner's views were set forth and strongly advocated. There thus arose in Edinburgh an active controversy between the supporters of the Huttonian and Wernerian Schools, which led to a close study and active scrutiny of the respective tenets of the two parties, especially as applied to the elucidation of the geology of Scotland. The views of Hutton prevailed over those of Werner in the end, and Jamieson is said

---

[41] *Theory of the Earth with Proofs and Illustrations* by James Hutton, M.D. and F.R.S.E. 2 Vols., Edinburgh, 1795.

to have frankly admitted his conversion to the views of his opponents.[42] And so in the words of Friederich Hoffman it may be truly said:[43]

Wir dürfen daher füglich Schottland als die Wiege der neueren Geologie betrachten, und wenn gleich die damals in unsern Vaterland herrschende Lehre die rasch folgende Verbreitung dieser neuen Ansichten hinderte, so bildeten sich dieselben doch an ihrer Geburtsstätte, trotz des Widerstrebenden Einflusses von *Jamieson*, bevor sie bei uns Eingang fanden, zu ansehnlicher Vollkommenheit aus.

One of Hutton's most important contributions to the science of geology was due to his recognition of the significance of the fact that in the geological column one series of beds often rested uncomfortably on the upturned edges of another series. Werner passed over this phenomenon as one of little or no importance, attributing the inclined attitude of the lower beds to their deposition on sloping surfaces of the nucleus of the earth when this was covered by the waters of the primitive ocean, or to the sliding of the sediments down such slopes after deposition; while the upper and newer series of beds were formed of later oceanic precipitates deposited horizontally upon the older series. Hutton saw that the beds of the lower series had also been originally deposited in a horizontal position but had been subsequently upheaved, folded and tilted by a force which he believed to have originated in the highly heated interior of the earth, and to have been exerted chiefly through the agency of great masses of fused rock intruded into the overlying strata.

Hutton in 1788 wrote his paper entitled "Theory of the Earth" which appeared in the first volume of the *Transactions of the Royal Society of Edinburgh*. A violent attack upon the conclusions reached in this paper was made by Kirwan, afterwards President of the Royal Irish Academy, who was a very ardent Wernerian. This led Hutton—after he had completed a number of long excursions in the Alps and various parts of Scotland and England in order to test his conclusions by more extended studies in the field—to reproduce his original paper, supplementing it with *Proofs and Illustrations*. It was published in two volumes at

[42] See Frank D. Adams, *Science*, October 26, 1934, and in the *Trans. Edin. Geol. Soc.*, Vol. XIII, 1935. Address given at Edinburgh on the occasion of the centenary celebration of the Geological Society of Edinburgh, September 3, 1934.

So far as is known, Jamieson never announced this in print. Geikie, in his *Life of Murchison* (Vol. I, p. 108), says that Sir Robert Christison and Prof. Balfour informed him that they were present at a meeting of the Geological Society of Edinburgh when the statement was made verbally.

[43] *Geschichte der Geognosie*, Berlin, 1838, p. 210.

Edinburgh in 1795 under the title *Theory of the Earth with Proofs &
Illustrations.* This was Hutton's great work. When Hutton died in
1797 his friend Playfair drew up an abstract of his work with a view to
preparing a biographical memoir. Hutton's literary style is very poor

THEORY

OF THE

E   A   R   T   H,

WITH

PROOFS AND ILLUSTRATIONS.

IN FOUR PARTS.

BY *JAMES HUTTON, M. D. & F. R. S. E.*

VOL. I.

EDINBURGH:
PRINTED FOR MESSRS CADELL, JUNIOR, AND DAVIES,
LONDON; AND WILLIAM CREECH, EDINBURGH.

1795.

FIG. 49. Title page of the first volume of Hutton: *Theory of the Earth.*

and Playfair decided to present the *Theory* in a more attractive form,
which he did in his work entitled *Illustrations of the Huttonian Theory,* a
work on which he bestowed much time and labor and in which he eluci-
dated Hutton's views and presented his theory in such a clear and

attractive form that it did much to make the Huttonian theory widely known and pave the way for its ultimate acceptance on the part of geologists.   The two printed volumes mentioned above, did not, however, contain the whole of the *Theory of the Earth* as Hutton wrote it. The second volume concludes with the words, "In pursuing this object I am next to examine facts with regard to the mineralogical part of the theory, from which, perhaps, light may be thrown upon the subject; and to endeavour to answer objections or to solve difficulties which may naturally occur from the consideration of particular appearances." Hutton evidently intended to issue a third volume but this never appeared.   For many years the manuscript of this third volume was sought for but without success.   It was discovered by the present writer in the library of the Geological Society of London in 1895.[44] It is a small quarto volume of manuscript comprising 208 pages, written by an amanuensis but with notes and corrections in Hutton's own handwriting, and on the fly-leaf is a note by Leonard Horner stating that he had presented it to the Library in 1856.   It was published by the Geological Society of London in 1899 as Volume III of *Hutton's Theory of the Earth*, edited by Sir Archibald Geikie.   It may be noted, however that this manuscript does not contain the whole of the missing portion of the work—there are still some additional chapters required to complete it and it will indeed be a treasure trove if anyone can find them.

After the publication of his *Theory of the Earth* which appeared in the *Transactions of the Royal Society of Edinburgh* in 1788, Hutton made a further communication to this Society in 1790, entitled *Observations on Granite* in which he says: "Granite hitherto considered by naturalists as being the original or primitive part of the earth, is now found to be posterior to the Alpine Schistus, which schistus being stratified is not in itself original."

The series of rocks thus contorted and penetrated by plutonic intrusions, had at a later time been wasted away by the continued action of the destructive forces of the atmosphere, or of the ocean, thus acquiring a more or less level surface on which the new deposits were laid down.   Each of these *unconformities*, according to Hutton, marks a *"revolution,"* and evidences of successive revolutions are found throughout the geological column.

[44] See *Science*, October 26, 1934, *Trans. Edinburgh Geological Society*, 1935, and *Nature*, October 10, 1895.

Instead, therefore, of a continuous and uniform development of the earth from the beginning down to the present time, the geological record reveals a repeated alternation of periods of quiet deposition and of violent upheavals or "revolution." Each series of deposits laid down in the periods of quiet deposition, even the oldest of them, is, according to Hutton, built up of materials derived from waste of older formations, the succession revealing in this way a long "*succession of former worlds.*"

The elevation of strata with accompanying folding and contortion Hutton held to be due to the expansive force of the internal fires giving rise to the Plutonic intrusion. To account for the retreat of the ocean Hutton considers the land to have been raised. Werner supposes the ocean waters to have diminished in volume. A volcano according to Hutton is a spiracle to the subterranean furnace, acting as a safety valve to prevent the unnecessary elevation of land and the fatal effects of earthquakes. Even the deposits on the low alluvial lands may in their turn be upheaved into mountain masses.

> The earth may thus be considered as an organized body which has a constitution in which the necessary decay is continually being repaired in the exertion of those productive powers by which it has been formed.[45]

In every landscape we see the results of the action of forces which are at work shaping the earth, and in contemplating the history of the earth, as recorded in the geological succession, to use the striking and oft-quoted statement of Hutton:[45] "We find no sign of a beginning—no prospect of an end."

The question as to how it came about that the incoherent sediments laid down in the sea became compacted into solid rocks, was one which presented itself to every observer, but to which the Neptunists and Plutonists gave entirely different answers. Werner and his followers held that the Primitive Rocks were chemical precipitates and owed their hardness to the fact that they were crystalline aggregates which had been deposited in their present forms from solution. The Transition and Floetz rocks which were composed largely of mechanical sediments, were compacted into solid rocks in part by the great pressure to which they were subjected by the weight of the materials deposited upon them, and in part owing to the constituent grains of these sediments being cemented together by a certain amount of material deposited from chemical solution. Hutton and his followers, on the other hand, contended that while the great weight of the overlying strata, added to that of the

[45] *Theory of the Earth. Trans. Royal Society of Edinburgh*, Vol. I, 1788, pp. 216, 304.

ocean waters, would result in intense compression of the subjacent beds, the fiery forces of the earth's interior had acted to fuse or partially fuse these sediments while they lay at the bottom of the sea, and thus give them that solidity and hardness which they are seen to possess. Some had been completely fused and had solidified into granite and allied rocks. In this way beds of sand are fused or sintered into sandstones and clays into shales. Even calcareous deposits, by fusion under intense compression, are transformed into marls, and cannot be changed into quicklime as they would be at the surface, owing to the pressure, to which reference has been made, which prevented the escape of the carbon dioxide.

Hutton denied that water could hold in solution many of the substances which are found cementing the elements of clastic rocks while "Foreign matter may be introduced in the form of steam or exhalation, as well as in the fluid state of fusion; consequently heat is an agent competent for the consolidation of strata, which water alone is not... we cannot from the natural appearances suppose any other cause as having actually produced the effects which are now examined... it may be, therefore, asserted that no siliceous body having the hardness of flint, nor any crystallization of that substance, has ever been formed, except by fusion."[46]

Hutton held that even the flints in chalk[47] and the silica of silicified wood had been brought into their present positions while in a state of fusion and that the beds of rock salt found in the earth's crust had been consolidated by perfect fusion at the bottom of the sea. Playfair in his *Illustrations of the Huttonian Theory* does not hesitate to attribute the origin of a "Siliceous Pudding Stone" to a "stream of melted flint forcibly injected among a mass of loose gravel." These fantastic ideas were later cast aside by the Plutonists and Hutton's great *Theory of the Earth* freed from such trivialities.

While, however, Hutton explained the upheaval of strata and the birth of mountains by the expansive action of fire, he was unable to offer any explanation as to the forces which maintained the strata in their elevated positions. "We know," he says, "that the land is raised by a power which has for its principle subterraneous heat, but how that land

---

[46] *Theory of the Earth. Trans. Royal Soc. of Edinburgh*, Vol. I, 1788, pp. 230, 232.

[47] A view actually put forward as an established fact in a popular magazine in this day of grace. (See *Chamber's Journal*, Sept. 1934, p. 662.)

is preserved in its elevated station, is a subject in which we have not even the means to form a conjecture."[48]

4. THE END OF THE NEPTUNISTS

The tide was setting against the Neptunists even before Werner's death. Not only did many of his own students as we have seen, when their work carried them to wider fields of observation and study, see that his theory was untenable, but a great number of the other leading geologists of the time were its vigorous opponents. Among these Boué, Desmarest, Faujas de St. Fond, Hutton, Montlosier, Macculloch, Playfair and Scrope may be especially mentioned. Desmarest held that the basalts of the Auvergne had been formed by the fusion of the underlying granite by volcanic fires.

Professor Sedgwick of Cambridge,[49] in a letter to Harkness on August 29th, 1856, advises the latter to examine certain rocks for fossils, adding:

I did not examine them in 1823–1824 because I thought that they were all below the region of animal life. At that time I had not quite learned to shake off the Wernerian nonsense I had been taught.

A footnote to the same page adds that Sedgwick in a letter to Lyell in 1845 speaks of himself as having been, in 1819, "Eaten up with the Wernerian notions, ready to sacrifice my senses to that creed—a Wernerian slave."

Shortly after von Buch examined the Auvergne district, another very intimate and trusted friend of Werner's, one in whose ability and critical judgment he had great confidence, also visited this classical district to see the occurrences which von Buch had described. This was Christian Samuel Weiss of Leipzig. Weiss upon his return visited Werner and gave him a detailed account of the remarkable phenomena which he had seen. Werner listened to him patiently and thanked him for the clear and detailed presentation of the evidence which was afforded by this volcanic region of Central France in opposition to his theory of the earth, but informed Weiss that he was not prepared to change his opinion without having himself seen with his own eyes the occurrences which Weiss had described.[50]

The authorities of the Freiberg Academy noted with great concern,

[48] *Theory of the Earth. Trans. Roy. Soc. of Edinburgh*, 1788, p. 285.
[49] *Life of Sedgwick*, Vol. I, p. 251.
[50] S. G. Frisch, *Lebensbeschreibung Abraham Gottlob Werners*, Leipzig, 1825, p. 131.

how immediately upon Werner's retirement, the number of students rapidly declined, so much so that they even feared the future of the institution itself might be seriously imperilled by the discrediting of Werner's views.   In order to ascertain whether some case might not be made out for the Neptunists, so that the courses in Geognosy at Freiberg might be given the support which they manifestly needed, they dispatched a brilliant young geologist, Ferdinand Reich, to the Auvergne to see whether *he* could not discover some interpretation of the geology of the region which might save the Neptunian cause, so that students might once more be attracted to the Academy which had given birth to it.   It was, however, another case of Balaam the son of Beor, for Reich having been called to denounce the enemy, was obliged to bless him.   He submitted upon his return a long report entirely corroborating the views of von Buch and Weiss.   Professor Beck of Freiberg long afterwards summed up the matter in the remark that the Scheibenberg had been vanquished by the Puy de Dôme.[51]   This report was never published.

The Neptunian Theory having taken its rise in Freiberg where Werner commenced to lecture in 1775, and having at once achieved a very wide acceptance, after a brief period of not much over half a century, may be said to have disappeared.   The names Neptunist and Plutonist now have a purely historical association and interest.

Werner had failed to reach a true interpretation of the structure and history of the earth because, as has been said, he misunderstood the true significance of certain phenomena which he saw in nature, but chiefly owing to the fact that his field work had been practically limited to Saxony.   One may be an excellent mineralogist—as he was—and make most valuable contributions to this science through researches carried out in his laboratory and museum, but if he is to reach a true conception of the structure and origin of the globe he must "Take the world for his parish."

Hutton himself on the other hand, had travelled rather widely and his followers had studied the geology of almost every country in Europe and some of those in other continents, and found the confirmation of the Plutonic Theory written in the records of the rocks.   The prophecy of Daniel was being fulfilled: "Many shall run to and fro and knowledge shall be increased."

These two theories—the Neptunist and the Plutonist—have, in the

[51] Abraham Gottlob Werner, *Eine kritische Würdigung.*   Berlin, 1918.

history of science, a certain analogy to the Ptolemaic and Copernican theories of the starry heavens respectively. The Ptolemaic theory of the heavens like the Wernerian theory of the earth, as has been stated in chapter III, presented a simple conception of a compact little world, approved of by the church and easily understood. The Copernican theory presented an infinitely vaster spectacle, taxing the powers of the human mind even to apprehend it, in many ways opposed to the teachings of the church, but having one important fact in its favor, namely that difficult as it might be to grasp, it was true.

Nevertheless Werner made a great contribution to the science of geology. As Fitton has said, "It was Werner who elevated geology to the rank of a real science."[52] And Cuvier has observed:[53]

C'est de lui seulement que datera la géologie positive, en ce qui concerne la nature minérale des couches

One hundred years after Werner's death a beautiful medal was struck in Germany to commemorate the event, on the reverse side it bears the following words concerning this very distinguished man: "Hundert Jahre nach seinem Tode ist sein Geist noch im Reiche der Steine lebendig."

5. GOETHE AND THE GREAT CONTROVERSY

This celebrated controversy between the Neptunists and Plutonists was one of the greatest and most bitter controversies in the history of geology, one which was prosecuted with a vigor and even a virulence which at first seems quite surprising and difficult to understand, for it passed beyond the field to which such scientific discussions are usually confined, and made itself widely felt in the literature of the time.

This was due chiefly to the fact that by some persons the controversy was thought to have a religious aspect. The Neptunian theory was believed to conform to the teaching of the Book of Genesis. It presented a finished world whose development might—with the exercise of some considerable imaginative effort—be compressed into the creative week; while the Plutonic theory, on the other hand, displaying an unfinished world, which showed so Hutton said, "No trace of a beginning and no prospect of an end;" a world, furthermore, which must in any

[52] *On the Geological System of Werner, Nicholson's Philosophical Journal,* Vol. XXXVI, London, 1813.
[53] *Récherches sur les Ossements Fossiles.* Discours préliminaire, Paris, 1821, p. xxvi.

event have required untold millions of years for its present development, incomplete as this might be, was an atheistic conception.   Thus it took on, in certain quarters, the vigor which frequently characterizes a theological controversy.

Perhaps the most striking emergence in the literature of the time is its appearance in Goethe's *Faust*.   In Act II of the Second Part of that great drama there is a dialogue between Seismos, personifying the earthquake and violence of the Plutonic forces, and the Sphinxes, who personify calmness, stability, permanence and the quiet and stately progress of the creative forces as Goethe saw them at work in the Neptunian conception of the origin of the world.   And in Act IV, Mephistopheles with his usual persiflage presents the Huttonian view, and Faust in more dignified phrases the Neptunian aspect of creation, as Goethe thought of them:

Mephistopheles:

When God the Lord—wherefore I also know,—
Banned us from air to darkness deep and central,
Where round and round, in fierce, intensest glow
Eternal fires were whirled in Earth's hot entrail;
We found ourselves too much illuminated,
Yet crowded and uneasily situated.
The Devils all set up a coughing, sneezing:
At every vent without cessation wheezing:
With sulphur stench and acids Hell dilated,
And such enormous gas was thence created,
That very soon Earth's level, far extended,
Thick as it was, was heaved, and split and rended!
The thing is plain, no theories overcome it:
What formerly is bottom, now is summit.
Hereon they base the law there's no disputing,
To give the undermost the topmost footing:
For we escaped from fiery dungeons there
To overplus the lordship of the air, . . .

Faust:

To me are mountain-masses grandly dumb:
I ask not Whence? and ask not Why? they come.
When nature in herself her being founded,
Complete and perfect when the globe she rounded,
Glad of the summits and the gorges deep,
Set rock to rock, and mountain steep to steep,
The hills with easy outlines downward moulded,

Till gently from their feet the vales unfolded!
They green and grow: with joy therein she ranges,
Requiring no insane, convulsive changes.[54]

Always seeking in nature for the evidences (or traces) of nature's plan manifested in perfect beauty, Goethe found in the Neptunian theory a magnificent picture of slow and stately progress in the development of the earth, while his indignant opposition to the Plutonic conception was due to the fact that it destroyed this fair picture, by introducing violent and sporadic upheavals and eruptions due to the igneous forces, which marred and spoiled the beauty and symmetry of the whole. He suffered his aesthetic instincts to mould his scientific opinions.

Goethe, who was a keen student of the natural sciences, and one who made many very interesting contributions to our knowledge of nature, was, as may be gathered from these quotations, an ardent Neptunist and bitterly hostile to the views of the opposing school. His contributions to geological science are collected in his *Naturwissenschaftliche Schriften*[55] and a very vigorous polemic by him against the Plutonian theory and all its exponents is to be found in a paper in this work entitled *Geological Problems and their Solution*.

[54] These extracts from *Faust* are taken from Bayard Taylor's translation.
[55] Published by W. Spemann, Berlin & Stuttgart, Vol. II.

Chapter VIII

"FIGURED STONES" AND THE BIRTH OF PALAEONTOLOGY

From the earliest times those who examined the strata of the earth's crust, even in a most cursory manner, were surprised to note that the rocks frequently displayed curious forms or markings, some of which resembled living things—plants or animals—while others bore a resemblance to certain geometrical forms, or even to castles, cities or landscapes. In the rocks of some localities, other forms still stranger and more mysterious were to be found.

In figures 50 to 54 some of these *"Figured Stones"* (Lapides figurati or Lapides idiomorphi) as they were called, have been reproduced from the works of ancient writers. These will show the character of and the great variety of the objects embraced under this term. The reader will note that the imagination of the copyist has in certain cases played a part in the delineation of the figure.

It has already been pointed out in Chapter II, which deals in some detail with the geological sciences in classical times, that some of the writers among the Greeks, whose works have come down to us in part at least, recognized certain of these objects to be the actual remains of plants or animals. It is not necessary here to make further reference to the views of these classical authors, but rather to pass on to medieval times when the nature and origin of these bodies again attracted much attention and was the subject of widespread and even violent controversy.

It is difficult to suppress a smile at the explanations put forward in the Middle Ages, even by men of the most undoubted distinction and ability, to account for these figures found in the rocks of the earth's crust, until it is remembered that they lived in an age when miracles were believed in as common occurrences, when the "Reign of Law" in nature was not recognized and when anything might happen.

These "figures" seemed to fascinate every writer who, throughout the Middle Ages and the Renaissance, referred to the rocks of the earth's crust and each gave his own explanation of their origin. Some regarded them as the works of an occult power or influence at work in nature, perhaps a Spirit or "Virtu divina" which intended to convey a hidden meaning or lesson, or as being, in the case of the picture of the Virgin and Child, or in that of the Crucifixion, a prophecy of great events to

250

FIG. 50. A portion of a slab of rock showing dendritic growths. The animal and the house probably represent indistinct forms elaborated by the active and formative imagination of the draughtsman. (From the *Rariora Naturae et Artis* of Kundmann, Breslau, 1737.)

FIG. 51. A dendritic growth. (From the *Recueil des Monumens des Catastrophes que le Globe terrestre a éssuiées*, Nuremberg, 1777.)

come, set forth in the very rocks of the earth's crust at the time of the creation, ages before the events themselves took place.   Others believed

FIG. 52. An agate showing a picture of the crucifixion.   Undoubtedly a suggestive form embellished by the artist.   (From the *Historia Lapidum Figuratorum Helvetiae Langius,* Venice, 1708.)

FIG. 53. A section of the "Ruin Marble" of Florence showing castellated forms due to a series of minute faults.   (From Knorr, *Recueil des Monumens des Catastrophes que le Globe terrestre a éssuiées,* Nuremberg, 1777.)

them to be due to operations of the forces of evil, intended to mislead or terrify mankind.

Still others[1] held that they had their origin in "Irradiations" from the heavenly bodies, the stars and planets, which had developed these

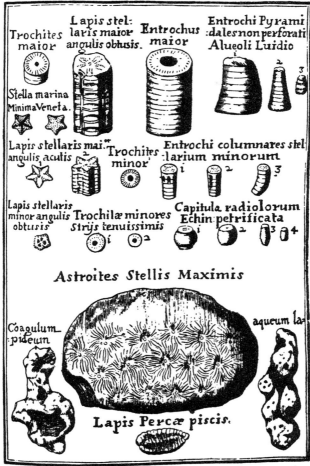

FIG. 54. Most of these "figured stones" are fossils in the modern acceptation of the term. (From the *Historia Lapidum Figuratorum Helvetiae* of Langius, Venice, 1708.)

curious forms, for in some of them, indeed, as for instance, Astroites (see figure 54), the stars had reproduced their own forms.   The opponents

[1] Mercati, *Metallotheca*, Rome, 1717, p. 219.

of this view, however, asked why the stars concentrated their powers in certain spots instead of acting everywhere throughout the earth's crust.

Others again suggested that they had come by mere chance, a view which, however, did not meet with general acceptance, but the opinion came to be widely held that the *Lapides Figurati* were "sports" or "jokes" developed in the rocks by "Nature," perhaps when in some humorous or capricious frame of mind. These curious forms, after all, they said,[2] were not so surprising for do not plants in their forms sometimes imitate the shape of certain animals, as for instance the mandrake which reproduces the human form? Do not certain animals furthermore, as for instance the parrot, imitate the voice of man? Why, then, is it particularly wonderful that sometimes in the "Third Kingdom" of nature also, forms and shapes seen elsewhere in nature appear. These *remembrances* which occur everywhere in nature serve to indicate the unity pervading all creation and to impress upon the thinking mind the majesty of the whole.

It was even suggested by some writers that before making the world and providing it with its inhabitants, the Creator engaged in several preliminary attempts or trials which proved unsatisfactory, and the plants and animals made in these experimental attempts at creation were discarded, their remains being seen scattered through the rocky strata of the earth, while the present fauna and flora of the earth represent the final and successful attempt which the divine being himself pronounced to be *good*.

In the thirteenth century Albertus Magnus[3] writes of the branch of a tree in which there was a bird's nest with the birds in it, the whole having been converted into stone by a certain *plastic* or *formative* force at work in nature which had already been called by Avicenna the *Mineral Virtue* and had still earlier been referred to by Aristotle as the *Vis Formativa*. The doctrine of equivocal (or spontaneous) generation which had been adopted by the disciples of Aristotle contributed to mislead many of these later writers who taught that there was at work in the earth such a hidden or occult power or virtue, which developed in the rocks figures which bear a resemblance to various vegetable or animal substances.

[2] See D. Giacinto Gimma, *Della Storia Naturale delle Gemme delle Pietre e di tutti i Minerali,* Naples, 1730, Vol. II, pp. 232 et seq.

[3] *Liber Mineralium,* Book I, Tract 1, Capt. VII, Oppenheim, 1518.

Falloppio[4] of Padua held that these Figured Stones were generated by vapors due to a process of fermentation set up in the rocks in those places where these were found, and even goes so far as to assert that it was from the circular movement of such vapors escaping from cavernous spaces that the buried urns found in many places, and even the earthenware pots and vessels which were heaped together to form the hill known as Monte Testaceo at Rome were formed, in fact many medieval writers held that the beautiful Etruscan vases and other earthenware vessels which had been dug up in Italy were not of human workmanship at all, but had grown within the earth itself from the same causes as other "figured stones."

Libavius,[5] referring more particularly to the forms of fishes, frogs, insects and other living things found in rocks, thought that "seed" of these might have been carried into the more or less porous or fissile rocks by means of percolating water, and have there developed into their adult forms.   In the case of some of the fossil fishes, such as those in the black shale at Mansfield, in Saxony, which are made of Pyrite, he suggests that the presence of this mineral was due to a metallic constituent in the water in which the "seed" was carried.

Athanasius Kircher,[6] a member of the Jesuit Order and a prolific writer to whose works reference is made elsewhere in this volume, gives an extended account with many illustrations, of wonderful markings and forms which are found in rocks.   Among them are the letters of the Greek and Latin alphabets, various geometrical figures (see figure 55), representation of the heavenly bodies, of trees, castles, animals of different kinds, the human form, as well as certain strange outlines which, as he presents them, seem to carry a suggestion of supernatural meanings. These latter, like those given by other writers in this period, represent the products of a glowing and highly imaginative fancy, inspired, in some cases at least, by an earnest desire to read into these obscure figures a deep religious significance, as the direct revelations of the Creator of the world.   Kircher held that most of these had been brought into being through the action of a *"Spiritus Architectonicus"* or *Spiritus Plasticus*, but thought the forms of leaves, mussels, fish and bones were the remains of living things.   He gives a picture of the *giants* to whom

[4] *De Medicatis Aquis atque Fossilibus*, Venice, 1564, pp. 109–110.

[5] See Gimma, Op. cit., Vol. II, pp. 235–236.

[6] *Mundus Subterraneus*, Amsterdam, 1678, Vol. II, p. 22.

belonged some of the larger fossil bones, now known to be the remains of various large mammals.

The renowned Barba[7] found similar perplexing objects in the rocks of the towering mountains of the South American continent and writes:

In the Highway as one goes from Potosi to Oronesta down the Hill, they gather Stones that have in them impressions of divers sorts of Figures, so much to the life, that nothing but the Author of Nature itself could possibly have produc'd such a

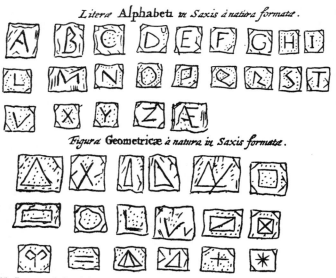

FIG. 55. Letters of the alphabet and geometrical forms.   (From the *Mundus Subterraneus* of Kircher, Amsterdam, 1678.)

piece of Workmanship.   I have some of these Stones by me in which you may see Cockles of all sorts, great, middle-siz'd, and small ones; some of them lying upwards, and some downwards, with the smallest Lineaments of those Shells drawn in great Perfection; and this Place is in the Heart of the Country, and the most double mountainous Land therein, where it were Madness to imagine that ever the Sea had prevailed, and left Cockles only in this one Part of it.   There be also among these Stones the perfect Resemblance of Toads and Butterflies, and strange Figures, which tho' I have heard from credible Witnesses, yet I forbear to mention, and not to overburden the belief of the Reader.

[7] *The Art of Metals*, written originally in Spanish by Albaro Alonzo Barba, Director of Mines at Potosi in the Spanish West Indies, and translated by the Earl of Sandwich in 1669. London, 1738, Part I, pp. 46 and 47.

Elie Bertrand[8] reached the conclusion that Figured Stones of all kinds were brought into existence at the same time as the rocks which enclose them and presents *in extenso* the arguments in support of this contention.

Even the great naturalist Gesner, who is the author of the earliest illustrated book on these objects, was of two views concerning their origin. Some, he thought, were animals turned into stone, but others were *sui generis*, the products of the earth itself. Agricola also, although he wrote but little on these figured stones, held essentially the same view as Gesner.

In another interesting old work which treats of the figured stones of Saxony, the author, who writes under the initials G. F. M.[9] describes representations of Noah's Ark, Moses and a crucifix, as well as those of many fishes and dendrites, found in the rocks in the vicinity of Ilmenau and gives it as his opinion that the best answer to the question as to how these fossils originate is to refer the inquirer to the words of "The wise Solomon," to be found in the eighth chapter of the Book of Ecclestiastes: "Then I beheld all the work of God, that a man cannot find out the work that is done under the sun; because though a man labour to seek it out, yet he shall not find it. Yea, farther: though a wise man think he know it, yet shall he not be able to find it."

It was not, however, only continental writers who experienced this sense of bewilderment, several outstanding English naturalists of the time were perplexed by the same doubts. J. Beaumont in an interesting paper on Crinoids (Trochites)[10] says that they have roots, stems and branches like plants, to which in fact they bear so close a resemblance that he refers to them as "stone plants." They possessed parts to assimilate nourishment as well as "an inward pith or sap and likewise joints and runnings in their grit and sometimes cells which may well supply the place of veins and fibres." He holds however that these "stone plants" were not fossilized vegetables but were produced by the power of "Nature to express the shapes of plants and animals where the vegetative life is wanting. . . . We can only say that here is that seminal root which in the first generation of things made all Plants, and I may

---

[8] *Memoires sur la Structure intérieure de la Terre*, Zurich, 1752, pp. 102–105.

[9] *Memorabilium Saxioniae Subterraneae*, Leipzig, 1709, p. 51.

Gimma gives an excellent review of the whole question in his *Della Storia delle Gemme, delle Pietre e di tutti i Minerali, ovvero della Fisica Sotteranea*, Naples, 1730, Vol. II, pp. 232–251.

[10] *On Trochites, Phil. Trans. Royal Society of London*, Vol. XI (original edition), 1676, p. 732.

say Animals, rise up in their distinct species: God commanding the
earth and water to produce both. . . . It seemed to be a thing of a very
difficult search to find out what this Seminal Root is, which is the efficient cause of these figures."

John Ray (1627–1705) who, in the opinion of Sir Archibald Geikie
was the ablest botanist and zoologist of his day, treats at length of this
subject in one of his books[11] and was drawn first to one side and then to
the other, but could reach no definite conclusion concerning the origin
of these strange objects. He wrote to Lhuyd (1660–1709), who was the
author of a catalogue of the English fossils in the great collection contained in the Ashmolean Museum at Oxford,[12] and asked his opinion on
the subject. Lhuyd replied that he had speculated on the possibility of
vapors rising from the sea bearing up with them some minute spawn of
animal life and then, condensing into rain at the earth's surface, carrying
this spawn down into the earth's crust, there to produce "These marine
bodies which have so much excited our imagination and baffled our
reason" but went on to say:

The frequent observations I have made on such Bodies, have hitherto afforded
little better Satisfaction, than repeated Occasions of Wonder and Amazement; foras-
much as I have often (I may almost say continually) experienc'd, that what one Day's
Observations suggested, was the next day called in Question, if not totally contradicted and overthrown. Nevertheless, so indefatigable is the Curiosity, and indeed
so successful have been the Discoveries of this present Age, that we are daily en-
couraged to hope, this so important a Question will not much longer want its final
Determination, to the great Advancement of that Kind of real Knowledge which
relates to Minerals.

Plot sets forth his views as follows:

This brings me to consider the great question, now so much controverted in the
World, whether the Stones we find in the Forms of Shell-fish, be *Lapides sui generis*,
naturally produced by some extraordinary plastic virtue, latent in the Earth or
Quarries where they are found? Or, whether they rather owe their Form and Figura-
tion to the Shells of the Fishes they represent, brought to the places where they are
now found by a Deluge, Earth-quake, or some other means, and there being filled
with Mud, Clay, and petrifying Juices, have in tract of time been turned into Stones,
as we now find them, still retaining the same Shape in the whole, with the same
Lineations, Sutures, Eminences, Cavities, Orifices, Points, that they had whilst they
were Shells? I must confess I am inclined rather to the opinion of Mr. Lister, that

[11] *Three Physico-Theological Discourses* (Discourse II, *On the General Deluge in the Days
of Noah, Its Causes and Effects*), Chapter IV, p. 123 et seq. 4th Edition, London, 1721.
[12] *Lithophylacii Britannici Ichnographia* (with 23 plates), London, 1699.

they are *Lapides sui generis;* than that they are formed in an Animal Mould.   The latter Opinion appearing at present to be pressed with far more, and more insuperable difficulties than the former.[13]

Plot asks further why fossils should not have been created as "ornaments" for the inner and secret parts of the earth, since tulips, roses, and other flowers (he says) have been created to adorn the surface of the earth, although like fossils, they are of no use.[14]

Martin Lister, however, a distinguished contemporary of these three writers, an active Fellow of the Royal Society of London and the author of several very important works on recent and fossil shells, refused to believe that these fossil shells ever belonged to living creatures.

In one of these works, after describing and figuring the shells of England,[15] he adds an additional chapter entitled *Cochlitarum Angliae,* in which he deals with the English fossil shells.   This is illustrated by a number of figures, many of them taken from Plot's work, to which reference has just been made, but while admitting that these fossil shells bear a remarkable resemblance to the living forms, he reiterates the reasons advanced for regarding them as having an entirely different origin:

We may easily believe (what I have read in Steno's Prodromus) that along the shores of the Mediterranean there may all manner of shell fish be found promiscuously included in Rocks and Earth at good distances too from the sea.   But for our English inland quarries I am apt to think that there is no such matter as petrifying of shells in the business: but that these cockle-like stones were ever as they are at present, *Lapides sui Generis* and never any part of an animal.   It is certain that our English Quarry shells (to continue that abusive word) have no parts of a different texture from the Rock or Quarry where they are taken, that is there is no such thing as Shell in these Resemblances of Shells but that Iron-stone Cockles are all iron stone; Lime or Marble, all limestone or marble; sparre or Christalline-Shells, all sparre, &c., and that they were never any part of an animal.[16]

A deadly blow to the various theories which accounted for fossils as the result of forces at work within the earth and rejecting the view that they were the remains of living things, one that was all the more effective because it was of the nature of a *reductio ad absurdum*, was given by the work of Johann Beringer, a professor in the University of Würzburg, in

---

[13] *The Natural History of Oxford-Shire*, Oxford, 1705, p. 112.

[14] See Ray's *Three Discourses*, p. 124.

[15] *Historiae Animalium Angliae tres Tractatus*, London, 1678.

[16] Dr. M. Lister, *Fossil Shells in several places of England, Philosophical Transactions,* 1671, p. 2282.

1726. Beringer was keenly interested in collecting the fossils which abounded in the Muschelkalk about Würzburg. Some sons of Belial among his students prepared a number of artificial fossils by moulding forms of various living or imaginary things in clay which was then baked hard and scattered in fragments about on the hillsides where Beringer was wont to search for fossils. These the Professor was delighted to find. He studied them with great care and their discovery stimulated him to continue his search with increased diligence, which was rewarded by the finding of other specimens, more and more remarkable in character, fishes were followed by many strange marine forms, insects, bees in their hives or sucking honey from fossil plants, birds in flight or on their nests. To these followed even stranger forms, figures of the sun and moon, then weird letters, some evidently Hebrew characters, while others seemed to be related to Babylonian or other scripts. (Some of these are reproduced in Plate VI.) One of these which he describes and discusses with all due reverence actually presented the Divine name. He wrote a treatise concerning these discoveries illustrated by 21 folio plates.[17]

As certain persons had ventured to assert that these fossils had been made artificially, he devotes a chapter of the treatise to refuting such statements. The distressing climax was reached, however, when later he one day found a fragment bearing his own name upon it. So great was his chagrin and mortification in discovering that he had been made the subject of a cruel and silly hoax, that he endeavored to buy up the whole edition of his work. In doing so he impoverished himself and it is said shortened his days. Unfortunately he did not destroy the copies which he purchased; they were found in his house after his death and bought by a publisher who provided them with a new title-page and issued them in 1767 as a second edition of his work.

While these ancient views concerning the origin of fossils have long since disappeared among educated persons, they still linger on in many places among simple people. The writer calls to remembrance such a case when many years ago he was acting as assistant to the late Dr. Ells, who was making a geological survey of the Millstone Grit region in the central portion of the province of New Brunswick. Here the sandstones of this formation contain in many places beautifully preserved fronds of ferns and fragments of carboniferous trees. A farmer here one day pointed out to Dr. Ells a collection of large slabs of this sand-

---

[17] *Lithographiae Wircebrugengis*, Specimen Primum, Würzburg, 1726.

PLATE VI

SOME "FOSSILS" MADE BY BERINGER'S STUDENTS
(From Beringer: *Lithographiae Wirceburgensis*)

stone which he had brought together and asked his opinion concerning the nature of the very perfect ferns and fragments of trees which they displayed. Dr. Ells replied that they were fossil ferns and wood and gave their respective names. The farmer then said that he had propounded the same question to a "professor" who he had reason to believe was a man of knowledge and ability only the week before and had received from him quite a different answer, namely, that they were produced by the same cause as that which gave rise to the pictures produced by frost on the window panes in winter. "And how did he say these were formed" asked Dr. Ells, to which the farmer replied, "He said they were due to the reactions of the elements of nature." Sports of nature, exactly the answer that Olivi of Cremona would have given 350 years earlier.

Gradually, however, the true explanation of the origin of fossils gained ground and prevailed—they were the actual remains of what were once living things which in some way had become embedded in the rocks of the earth's crust.

Among the very earliest of those who recognized and supported this view, and won the battle for it, were Leonardo da Vinci[18] (writing in 1508) Palissy[19] (in 1580), Fracastoro[20] (in 1538) and Cesalpino[21] (in 1566). Later Hook (1635–1703), Vallisnieri, (1661–1730), Scheuchzer[22] (1672–1733), Guettard (1715–1786), also upheld this view.

Jussieu[23] in 1718 drew attention to the remarkable fact that the remains of fossil plants found in the rocks at Saint-Chaumont (Lionnois), belonged to types found only in far distant hot countries such as India, and must have been carried to their present position by the waters of some great ocean.

The word *fossil* which had hitherto been applied to all bodies which were dug up out of the earth's crust, including minerals and metals, gradually became restricted to organic remains. These were at first termed *extraneous fossils* by many writers, to indicate that they had not originated in the earth's crust where they were found, but later (under

[18] Leicester, M. S., quoted in McCurdy's *Leonardo da Vinci's Note Books*, London, 1906, p. 108.

[19] *Discours Admirables de la Nature des Eaux et Fonteines*, Paris, 1580, pp. 212 et seq.

[20] *Homocentrica*, Venice, 1538.

[21] *De Metallicis*, Book 88, chap. 44 (in the Nuremberg edition of 1602).

[22] *Piscium Querelae et Vindiciae*, Zurich, 1708.

[23] *Histoire de l'Académie Royale des Sciences*, Paris, 1718, p. 287.

Lamarck's influence) the qualifying term was rejected and the word "fossil" was confined to "Still recognizable remains of organic bodies."

The controversy which raged over the question as to whether these fossils were the remains of animals and plants had an element of bitterness infused into it owing to the fact that it was brought into the field of theological controversy. That fossils were of organic origin, it was contended, was contrary to the teaching of Holy Writ, for, according to the account of creation in the book of Genesis, living things were not created until the fabric of the earth itself had been completed. Such being the case, how could the remains of living things be found enclosed within the earth's crust?

When it was definitely established that fossils were the remains of animals or plants, the champions of the theological cause shifted their position and put forward the view that these organic remains had been brought into their present position by the Noachian Deluge. So great was the violence of this flood, they declared, that the ocean was stirred to its lowest depths, while all the soil on the earth's surface was at the same time washed away, and the ocean, after rising to such a height that the summits of the highest mountains were covered, redeposited its burden of sediment with all the living things which were mingled with it, and these sediments, hardening into rock, are now seen as the stratified rocks of the earth's crust with the fossils which they contain.

It is not necessary here to enlarge further upon the controversy. White[24] gives an extended account of it and observes that "It was held by the great majority of theological leaders for nearly three centuries and in the sixteenth century it received such powerful support from a number of men of such eminence that nothing seemed to be able to stand against it."

Before leaving the consideration of this theory, which has now passed into the limbo of exploded hypotheses, it is interesting to recall an echo of it here in America. It is from the pen of no other than Thomas Jefferson, who, in one of his letters writes as follows:

Immense bodies of Schist with impressions of shells near the eastern foot of North Mountain recall statements that shells have been found in the Andes 15,000 ft. above sea level which is considered by many writers both of the learned and unlearned as a proof of a universal deluge.[25]

[24] *A History of the Warfare of Science with Theology*, New York, 1896, Vol. I, p. 226.
[25] Francis W. Hirst, *Life and Letters of Thomas Jefferson*, New York, 1926, p. 186.

To this view Jefferson opposes various scientific objections and after disproving several false theories concluded:

We must be contented to acknowledge that this great phenomenon is as yet unsolved. Ignorance is preferable to error and he is less remote from the truth who believes nothing than he who believes what is wrong.

The fossils in the rocky strata of the earth having at length been recognized as the remains of living things, representing the faunas and floras of the earth in former ages, the science of palaeontology may be said to have come into being, and as one of the first outstanding masters of this science in its early days, the great comparative anatomist, Baron George Cuvier is worthy of special mention.

Cuvier was born in 1769 in the little town of Montbéliard, 40 miles from Besançon. This town, which is now in France, at the time of Cuvier's birth belonged to Würtemberg. Both the French and the Germans have therefore claimed Cuvier as their countryman. The family, which was originally Swiss, coming from a village of the Jura which still bears the name of Cuvier, was driven to Montbéliard because of their profession of the reformed faith. When still a boy his father took him to Caen in Normandy to be educated in a Protestant family. Later on he pursued his studies at the Académie Caroline in Stuttgart. He lived in a time of great political upheaval in Europe. Napoleon Bonaparte, Wellington, Chateaubriand, Canning and Sir Walter Scott were born in the same year as Cuvier, and Cuvier died in 1832, the same year in which Goethe passed away.

His father intended that he should enter one of the administrative branches of the French Government service, and the preliminary training which he received for this walk in life proved to be of much advantage to him later, but his family circumstances turned the youthful Cuvier's path toward the study of science. His earliest scientific labors were in entomology, which led him on to comparative anatomy, and so to vertebrate palaeontology. He was an excellent draftsman and not only prepared some of the illustrations for his earlier works, but even engraved the plates himself.

In 1795 he went to Paris and became associated with Mertrud, Professor of Comparative Anatomy at the Jardin des Plantes. In 1800 he was appointed to a professorship in the Collège de France, in 1803 became Perpetual Secretary to the division of Physical and Mathematical

Sciences of the National Institute and in 1818 was elected to membership in the Académie Francaise.

The commanding position to which Cuvier had risen in the world of science, combined with his great literary gifts and his high administrative abilities, indicated him as a person who could render important services to the state, and he was selected to occupy many high government offices.   In 1814 he was named Counsellor of State by Napoleon, and in the autumn of the same year was appointed to the same position by Louis XVIII.   He was also repeatedly called upon during various periods of his life to act as Commissaire du Roi, the duty of the holder of this office being to defend all bills brought before either House by the ministry.   His success in public life raised up many enemies who asserted that he changed his opinions to conform to those of the changing regimes, and that he held more public offices than any man had a right to monopolize.   He was moreover a Protestant at a time when the Church in France was very aggressive and intolerant.   Cuvier seems however to have been a man of high character and exalted aims.   If he did occupy many positions, he filled them ably and well and, as has been said, "His career can show how the income of the statesman furnished the savant with the means of carrying on his labours; how the counsellor of his sovereign protected the naturalist, and how the new Aristotle became his own Alexander."[26]

As Perpetual Secretary of the National Institute, Cuvier wrote a number of most interesting and valuable papers.   In 1808 he prepared a *Rapport Historique sur les Progrès des Sciences Naturelles depuis 1789*, and each year until the time of his death he presented to the Institute an annual report on the labors of its members and correspondents, as well as of many others who communicated the results of their work to the Institute, which reports form a valuable contribution to the history of the natural sciences and allied departments of knowledge during this period.   As Secretary of the Académie Francaise it was the duty of Cuvier to read at a public meeting an Eloge or eulogy on deceased members of that body.   Three volumes[27] of these were published covering the period from 1800 to 1827, and present most interesting biographies of a large group of the men of science, including De Saussure, Pallas, Werner, Desmarest, Rumford, Priestley, Sir Joseph Banks and Haüy. As Geikie says, "Eloquent and picturesque, full of knowledge and sympathy, these biographical notices form a series of the most instructive

[26] Mrs. R. Lee, *Memoirs of Baron Cuvier*, London, 1833, p. 229.
[27] *Recueil des Eloges Historiques de l'Institut Royal de France*, Paris, 1819–1827.

PLATE VII
PORTRAIT OF BARON GEORGES CUVIER

and delightful essays in the whole range of scientific literature and should be read by all who are interested in the history of geological sciences."

Cuvier's most important contributions to geological science were, however, in the field of vertebrate palaeontology. These were most remarkable, so much so, that he may almost be said to be the founder of this science. His work dealt chiefly with the Chalk, its overlying Tertiary formations, and with the Alluvium of the Paris Basin. In working out the stratagraphical succession of this region he had the assistance of Brongniart, whose training as a mining engineer had given him a knowledge of field geology which Cuvier did not possess.

Cuvier's attainments in comparative anatomy enabled him, often from a few bones only, to reconstruct, in their chief outlines at least, the great vertebrates of the Tertiary and Quaternary times, which not only made an immense contribution to the knowledge of the earth's past history, welcomed by men of science, but awakened a sense of wonder and delight in the popular mind and greatly increased the interest in and respect for the methods of scientific research which could achieve such remarkable results. A further very important contribution to the progress of science made by Cuvier was the establishment of the fact that these bones were not the remains of animals belonging to species which still survived but that they belonged to allied forms which had now become extinct. He also recognized that certain forms were confined to certain beds or formations. This caused him to seek for the reason why certain animals had been abundant on the earth's surface at one period and had then disappeared.

The results of Cuvier's geological work is comprised chiefly in two of his publications. The first of these is his *Essai sur la Géographie Minéralogique des Environs de Paris, avec une Carte géognostique et des Coupes de terrain, 1811*, a second and enlarged edition of which appeared in 1822. This work was due largely to the labors of Brongniart.

The second is his *Recherches sur les Ossemens Fossiles* which appeared in 1812. This great work, illustrated by many plates, ran through several editions, increasing in size until it finally grew to 12 large volumes. The *Discours Préliminaire* in the first volume, was later published separately under the title *Discours sur les Révolutions de la surface du Globe*, went through many editions and was translated into several foreign languages. The English translation was by Professor Jameson of Edinburgh.[28]   In this discourse Cuvier sets forth his belief that the

[28] *Essay on the Theory of the Earth*.   By Baron G. Cuvier with Geological Illustrations by Professor Jameson, 5th Edition.   Edinburgh, 1827.

cause of the disappearance of the faunas which characterize certain formations was due to the fact that the history of the earth has been marked by sudden and widespread catastrophes.

He had found evidence of frequent elevation and subsidence in the Paris Basin with the disappearance of old faunas and the appearance of new ones and to explain these he invoked a succession of sudden and overwhelming catastrophes.   Concerning these he says:

> The changes which resulted in the appearance of dry land in this region were not due to a more or less gradual and wide-spread subsidence of the waters.   There were many sudden uprisings and many successive retreats, which resulted however in a final lowering of the general level.   But it is very important to bear in mind that those repeated advances and retreats were not slow and gradual in character—on the contrary most of the catastrophes to which they gave rise were sudden, and this is especially easy to prove in the case of the last one, which by a double movement, consisting first of an inundation and then of a retreat of the waters, left our present continents essentially as we see them to-day.
>
> It left behind also in northern countries, the carcasses of the great quadrupeds which are found embedded in the ice and preserved down to the present day intact with their hair, hides and flesh.   On the other hand, this perpetual frost did not previously occupy the areas where we now find it, for those animals could not live at so low a temperature.   It was therefore at one and the same instant that these animals perished and that the glacial conditions came into existence.
>
> This change was sudden, instantaneous not gradual, and that which is so clearly the case in this last catastrophe is not less true of those which preceded it.   The dislocation and overturning of the older strata show without any doubt that the causes which brought them into the position which they now occupy, were sudden and violent; and in like manner testimony to the violence of the movements which influenced the waters is seen in the great masses of debris and rounded pebbles which in many localities are found intercalated between beds of solid rock.
>
> Life upon the earth in those times was often overtaken by these frightful occurrences.   Living things without number were swept out of existence by catastrophes. Those inhabiting the dry lands were engulfed by deluges, others whose home was in the waters perished when the sea bottom suddenly became dry land; whole races were extinguished leaving mere traces of their existence, which are now difficult of recognition, even by the naturalist.   The evidences of those great and terrible events are everywhere to be clearly seen by anyone who knows how to read the record of the rocks.

It was to such opinions that Lyell's theory of Uniformitarianism was opposed.   His *Principles of Geology* gave the death blow to the views of the great "Catastrophic school" of geologists.

Cuvier was a warm admirer of Buffon and when a young man studied his works with the greatest assiduity.   He was attracted to this author both by the great zeal which he evinced in the cause of science and by the magic of his style. He, like Buffon, made many friends for the science of geology and drew widespread attention to geological studies by his exalted position, his brilliant style, and his personal charm. It has been said that "Science throughout the world acknowledged in him its chief contemporary ornament."

Surprise has sometimes been expressed that Cuvier could find time to produce such an immense body of valuable scientific work while discharging at the same time the many and exacting duties of his official positions in the government. Lyell visited Cuvier in Paris in 1829 and some light on this question is to be found in a letter which Lyell wrote to his sister from Paris at this time:[29]

I got into Cuvier's sanctum sanctorum yesterday, and it is truly characteristic of the man. In every part it displays that extraordinary power of methodising which is the grand secret of the prodigious feats which he performs annually without appearing to give himself the least trouble. But before I introduce you to this study, I should tell you that there is first the museum of natural history opposite his house, and admirably arranged by himself, then the anatomy museum connected with his dwelling. In the latter is a library disposed in a suite of rooms, each containing works on one subject. There is one where there are all the works on ornithology, in another room all on ichthyology, in another osteology, in another *law* books! &c., &c. When he is engaged in such works as require continual reference to a variety of authors, he has a stove shifted into one of these rooms, in which everything on that subject is systematically arranged, so that in the same work he often takes the rounds of many apartments. But the ordinary studio contains no book-shelves. It is a longish room, comfortably furnished, lighted from above, with eleven desks to stand to, and two low tables, like a public office for so many clerks. But all is for the one man, who multiplies himself as author, and admitting no one into this room, moves as he finds necessary, or as fancy inclines him, from one occupation to another. Each desk is furnished with a complete establishment of inkstand, pens, &c., pins to pin MSS. together, the works immediately in reading and the MS. in hand, and on shelves behind all the MSS. of the same work. There is a separate bell to several desks. The low tables are to sit to when he is tired. The collaborateurs are not numerous, but always chosen well. They save him every mechanical labour, find references, &c., are rarely admitted to the study, receive orders and speak not.

Brongniart, who in imitation of Cuvier has many clerks and collaborateurs, is known to lose more time in organizing this auxiliary force than he gains by their work, but this is never the case with Cuvier. When I went to get Mantell's casts, I found that the man who made moulds, and the *painter* of them, had distinct apartments, so that there was no confusion, and the dispatch with which all was executed was admirable. It cost Cuvier a word only.

---

[29] *Life, Letters and Journals of Sir Charles Lyell, Bart.* Edited by his sister-in-law, Mrs. Lyell, London, 1881, Volume I, pp. 249–250.

Cuvier, as has been mentioned, believed that the several divisions of geological time were separated by sudden cataclysms by which the living things upon the earth were destroyed wholly or for the most part, to be replaced by new floras and faunas in the next succeeding period. One of these took place five or six thousand years ago, and might be considered as coinciding with the Mosaic Deluge. This opinion coming from so eminent an authority was welcomed by the Church, as was also the fact that he was entirely opposed to the idea of the evolution of species, regarding species as permanent although varieties might arise.

In this he was opposed by his great contemporary, the Chevalier de Lamarck (1744–1829), who commenced life as a soldier and having later studied medicine, in his fiftieth year was appointed "Professor of Zoology of Insects, of Worms and of Microscopic Animals" at the Museum in Paris. Lamarck was engaged at the same time as Cuvier in the study of the fossil remains, more particularly of the invertebrate fauna, of the Paris Basin. If Cuvier may be said to be the great founder of *vertebrate palaeontology*, Lamarck may be said with at least equal justice to be the founder of *invertebrate palaeontology*. His work on the fossil shells of the Paris Basin[30] led him to the consideration of the higher problems of biology and thence to the enunciation of opinions in support of organic evolution set forth in his *Philosophie Zoologique*, which has become a classic. Furthermore, in his admirable little volume entitled *Hydrogéologie*, which appeared in 1802, Lamarck considers the action of water in modifying the character of the surface of the earth, the origin of mountains through its erosive action on tablelands, the limitless vista of past time, the evidence afforded by organic remains that the sea for very long periods covered large portions of the surface of the globe and many other geological questions.

This brief sketch of the birth of palaeontology may be fittingly brought to a close by the consideration of the work of William Smith (1769–1839).

He was a contemporary of Cuvier and it is of interest to contrast the personalities of these two great men and the conditions under which they carried out their work. Cuvier lived and worked surrounded by the pomp and circumstance of the most brilliant court in Europe. He was a man who had been trained in the methods of scientific investiga-

---

[30] *Memoire sur les Fossiles des Environs de Paris*, Paris, 1802–06. *Historie Naturelle des Animaux sans Vertèbres*, Paris, 1815–22.

tion, was a renowned comparative anatomist and was throughout his career associated with the great university centers in Paris. William Smith on the other hand received no further instruction than that which he could obtain at a village school. His knowledge of geology was acquired by his own study and observations in the field. He earned his own living from the first, and rose through his outstanding discoveries to the high position which has been universally accorded to him, that of the *Father of English Geology*. Füchsel, Lister, Hooke, Soulaire, Buffon and especially Cuvier had a presentiment that fossils might indicate the age of the strata in which they were found. Arduino[31] indeed in referring to the calcareous strata of the Alps, says that they abound in sea shells of various kinds, "Die sich meistens nach den verschiedenen Lagern unterscheiden," but William Smith first recognized the real significance of fossils as a means of determining the age of the strata in which they occur.

He was born at Churchill in Oxfordshire and, having lost his father while yet a child, came under the protection of his uncle, who was a farmer and observed with but scant approval the interest which his nephew manifested in the collection of the fossils which occurred abundantly in the rocks of the neighborhood. At eighteen he became the assistant to a surveyor who had been engaged to make a survey of the parish in which he lived and he later acquired an extensive practice as a surveyor and engineer on his own account. Projects for canals in various parts of England were being discussed at this time (1793), for the steam locomotive had not yet been invented, and Smith travelled extensively through many parts of the country, being engaged by various companies to ascertain the best course for such canals, and in this connection he carried out surveys and lines of levelling for these and subsequently supervised the construction of the Somerset Coal Canal. As these canals were intended in many cases for the transportation of coal, Smith was employed to report upon coal deposits in various parts of the country. A few of these reports are still in existence, one of them in the library of the Geological Society of London, entitled "The State of the Collieries at and near Nailsea" addressed to the Committee of Management of the Bristol and Taunton Canal, bearing the date 1811. In this he reports that the tonnage of coal present in the area is large and favorably situ-

[31] See *Sammlung einiger Ablandlungen des Herrn Johann Arduino aus dem Italiänischen übersetzt durch A. C. von F*, Dresden, 1778, p. 277 (reproduced from a work which appeared some nineteen years earlier).

ated for extraction, and states that the company may "with the greatest safety" proceed at once to build their canal to these collieries.   While carrying out this work he studied the nature and succession of the geological formations of the various parts of England which he visited and found that he could apply to great advantage his new found knowledge in selecting the best lines for his canals when passing through a district presenting tracts underlain by porous rocks, as well as in choosing good foundations for bridges, and supplies of good stone for purposes of construction.   He observed also the distinctive surface characters of tracts of country underlain by the formations, which often enabled him at a glance to interpret the underlying structure.   He also devoted much attention to the movements of underground waters and to questions of drainage and irrigation.   In 1801 he succeeded in draining the great Prisley Bog and converting it into very valuable agricultural land.[32]   This led to his being called to all parts of England to undertake similar work, travelling sometimes as much as 10,000 miles a year.   He was thus enabled to see new parts of England and to further enlarge his knowledge of the geology of the country.   He always kept full notes of what he saw and thus accumulated an enormous mass of data, which was invaluable when he came to prepare the great geological map of England which was to be the crowning labor of his life.   In 1810 the celebrated hot springs at Bath failed and Smith was sent for to restore the flow of water to the Baths and Pump Room.   Not without much opposition he was allowed to open the hot spring to its bottom and having done so he found that the water had not failed but was escaping through new channels which had been opened up.   These he closed and as a result the baths filled even more quickly than formerly.

Smith's studies on the course of underground waters again met with an important practical application when in the very wet season of 1799 landslides took place in the vicinity of Bath, carrying large areas of cultivated land with houses, trees and lawns down the hill slopes.   By driving tunnels into the hills and draining off the excess of underground waters these devastating movements were in many cases arrested.

While his numerous engagements called Smith to study the strata in many parts of England where the services of a practical geologist were

---

[32] One of his earliest published works deals with this subject.   It bears the title *Observations on the Utility, form and Management of Water Meadows and the draining and irrigation of Peat Bogs with an account of Prisley Bog.*   Norwich, 1806.   On the title page he is described as "Engineer and Mineralogist."

required, he extended his observations at his own expense to other parts of England, Wales and even Southern Scotland, with which it was necessary for him to become familiar in order that he might complete his geological map of England on the construction of which he had set his mind. His life, as Phillips says, was one long wandering. He made no attempt however to keep his discoveries secret until the time came when he could complete and publish the results of his work. He talked enthusiastically and continually about them to everyone who was ready to listen to him and came to be known colloquially as "Strata Smith." Among his most intimate friends were the Rev. Benjamin Richardson and the Rev. Joseph Townsend, referred to in one of Smith's letters as "My two first pupils," both of whom were well versed in natural history and had made extensive collections of fossils, and to them Smith refers in his *Tabular View of the Order of Strata in the vicinity of Bath with their respective organic remains*, prepared and distributed to a number of persons and societies in manuscript form in the year 1799.

Before Smith published any of his conclusions in print these were set forth by Townsend, where one might least expect to find them, namely, in one of his books bearing the title *The Character of Moses established for Veracity as an Historian, recording events from Creation to the Deluge*, which appeared at Bath in 1813. Townsend had himself travelled extensively, visiting every part of England, and had also visited Ireland and many parts of the Continent making geological observations. His book is largely devoted to a description of the succession of the geological strata of England, and gives Smith full credit for having supplied him with the facts which he presents. The object of Townsend in this book is to show that the account of the Mosaic Deluge given in the Book of Genesis is borne out by what is to be seen in the geological deposits of England and the other lands which he had visited and he is convinced that he has shown this to be the case. Townsend and Richardson rendered valuable assistance to Smith, both being well versed in conchology, while Smith himself, working out the succession of the strata and recording the fossils which characterized them, depended on his two friends to name these fossils and to inform him with regard to their true character.

After long continued labor in the field and in the face of great pecuniary difficulties, William Smith completed and published his great map in 1815. It was on a scale of five miles to an inch and consisted of 15 sheets making, when mounted, a map of eight feet nine inches by six feet two inches in size.   It bore the title, *A Map of the Strata of England and Wales with a part of Scotland, exhibiting the Collieries, Mines and Canals, the Marshes and fen lands originally overflowed by the sea and the varieties of soil according to the variations in the substrata.   by W. Smith, Aug. 1, 1815.*   About 400 copies of this map were issued.   Many, but not all the copies, bear a number and are signed by William Smith himself.   This signature, with the number of the map when present, is to be found on the area occupied by the North Sea opposite to the mouth of the Humber and is accompanied by a section entitled *Sketch of the Relative succession of the Strata and their Relative Altitudes.*   The two copies in the writer's possession are numbered 46 and 96 respectively. No fewer than 20 colors are used on the map and an interesting method of coloring is employed, in that the base of each formation is deeply colored and this color shades upward becoming paler toward the outcrop of the next overlying formation.

To accompany and explain this map, Smith in the same year published his *Memoir to the Map and Delineation of the Strata of England and Wales with a part of Scotland*, a thin quarto volume of 51 pages.   On the title page of this he signs himself as "Engineer & Mineral Surveyor."   This was followed in 1816 by another thin quarto volume entitled *Strata identified by Organized Fossils containing Prints on Coloured Paper of the most Characteristic Specimens in each Stratum.*   In the introduction he says that having in his map shown the "courses" of the several formations he may "Now confidently proceed with a general account of those organized fossils, which I found imbedded in each stratum and which first enabled me more particularly to distinguish one Stratum from another."   He uses the word "stratum" here for formation.   The book, of which 250 copies were printed, is illustrated by 19 interesting and curious full page plates.   The frontispiece represents the tooth of a mastodon.   The following 18 plates engraved by Sowerby each present the characteristic fossils of a formation commencing with the "London Clay" and ending with the "Fuller's Earth Rock."   Each plate is printed on a sheet which has the color of the strata of which the formation is composed, the idea apparently being to show the fossils on a background of the same color of that of the rocks in which they are found. In the following year, 1817, a third volume of approximately the same size appeared.   This was entitled *Stratigraphical System of Organized Fossils with reference to the specimens of the Original Collection in the*

British Museum explaining their State of Preservation and their use in identifying the British Strata. It describes a collection of some 700 specimens of fossils from England and Wales which Smith had deposited in the British Museum, "The chief object of the work being to show the utility of organized fossils in identifying Strata." It includes a colored *Geological Table of British Organized Fossils which identify the courses and continuity of the Strata in their order of superposition as originally discovered by W. Smith, Civil Engineer, with reference to his Geological Map of England and Wales.*

These volumes together with several geological sections issued between 1817 and 1819 and some geological maps of individual counties are the chief works of this remarkable man.[33]

W. H. Fitton,[34] one of the outstanding geologists at the time in Britain, in a review of Smith's work, speaks of his map as "The production after the labour of more than 20 years, of a most ingenious man who has been singularly deficient in the art of introducing himself to public notice." Woodward[35] remarks that "Geology in the early years of the nineteenth century had hardly become a profession, unless in the case of mining engineers and surveyors, and it is noteworthy that the three prominent authorities in practical or applied geology at that time, William Smith, John Farey[36] and Robert Bakewell, were not members of the Geological Society of London. It is interesting to bear in mind the lamentation of William Smith, uttered in 1816, that "the theory of geology was in the possession of one class of men, the practise in another." The Geological Society, however, made full amends to Smith for this neglect, by awarding to him in 1831 the first Wollaston medal given by the Society.

[33] The best treatise on Smith's publications is Thomas Sheppard's, *William Smith his Maps and Memoirs*, published at Hull in 1920, and the most complete and authoritative account of his life is to be found in the work of John Phillips, entitled *Memoirs of William Smith, LL.D.*, London, 1844. Phillips was Smith's nephew, who inspired by Smith, devoted his life to the study of geology and as a youth assisted his uncle in some of his geological investigations. Phillips rose to be one of the leading geologists of his time, was the author of many books and treatises on the science and became Professor of Geology, first in King's College, London, then in Dublin and subsequently succeeded Buckland in the chair at Oxford. The results of a critical study of William Smith's map of England by V. A. Eyles of the Geological Survey of Great Britain (Scottish Office) will be found in the *Annales of Science* for April 1938.

[34] *Edinburgh Review* (No. LVIII), Vol. 29 and with some additions in *Notes on the History of English Geology, Philosophical Magazine*, Vol. I, 1832, pp. 147, 443.

[35] *The History of the Geological Society of London.* London, 1908, p. 53.

[36] Author of *The General View of the Agriculture & Minerals of Derbyshire* (2 Vols.), 1811–13

Sedgwick was President of the Society in that year and made a memorable address on the occasion, of which the following is an extract:

His great and original works are known to you all, and I might well refer to them for our justification, and without any further preface place the prize in his hand, offering him my hearty congratulations. But since his arrival in London, within the last few hours, he has has given me a short account of his early discoveries, and has shown me a series of documents of no ordinary interest to this Society and important to the correct history of European geology.

I for one can speak with gratitude of the practical lessons I have received from Mr. Smith; it was by tracking his footsteps, with his maps in my hand, through Wiltshire and the neighboring counties, where he had trodden nearly thirty years before, that I first learnt the subdivisions of our oolitic series, and apprehended the meaning of those arbitrary and somewhat uncouth terms, which we derive from him as our master, which have long become engrafted into the conventional language of English geologists, and, through their influence, have been, in part, also adopted by the naturalists of the Continent.

After such a statement, Gentlemen, I have a right to speak boldly, and to demand your approbation of the Council's award, I could almost dare to wish, that stern lover of truth, to whose bounty we owe the 'Donation Fund'—that dark eye, before the glance of which all false pretensions withered, were once more amongst us. And if it be denied us to hope that a spirit like that of Wollaston should often be embodied on the earth, I would appeal to those intelligent men who form the strength and ornament of the Society, whether there was any place for doubt or hesitation? whether we were not compelled, by every motive which the judgment can approve, and the heart can sanction, to perform this act of filial duty, before we thought of the claims of any other man, and to place our first honour on the brow of the *Father of English Geology*.

If in the pride of our present strength we were disposed to forget our origin, our very speech would betray us; for we use the language which he taught us in the infancy of our science. If we, by our united efforts, are chiselling the ornaments, and slowly raising up the pinnacles of one of the temples of Nature, it was he that gave the plan, and laid the foundations, and erected a portion of the solid walls, by the unassisted labour of his hands ...

I think it a high privilege to fill this Chair, on an occasion when we are met not coldly to deliberate on the balance of conflicting claims, in which, after all, we might go wrong, and give the prize to one man by injustice to another; but to perform a sacred duty where there is no room for doubt or error, and to record an act of public gratitude, in which the judgment and the feelings are united.

Two years later W. H. Fitton[37] referred to Smith in the following words:

Fortunately placed in a country where all our great secondary groups are brought near together, he became acquainted in early life with many of their complex relations: he saw particular species of fossils in particular groups of strata, and in no

[37] *Philosophical Magazine*, Feb. 1833.

others; and giving generalization to phenomena, which men of less original minds
would have regarded as merely local, he proved, so early as 1791, the continuity of
certain groups of strata, by their organic remains alone, where the mineral type was
wanting. . . . Having once succeeded in identifying groups of strata by means of their
fossils, he saw the whole importance of the inference,—gave it its utmost extension,
seized upon it as the master-principle of our science;—by the help of it disentangled
the structure of a considerable part of England. . . . If these be not the advances of
an original mind, I do not know where we are to find them.

William Smith received a further recognition of his merits and his
services to science when in 1835 Trinity College, Dublin, conferred upon
him the degree of LL.D. Incidently an amusing reference to Smith
which occurs in one of Murchison's letters, may here be quoted:

Among my intimates and correspondents of the first year of my geological career
I must not omit to mention George William Featherstonhaugh. He has played a
bustling and useful part through life, has published on a vast variety of subjects, and
was a most lively, agreeable companion. He was the first to introduce our modern
ideas of geology into the United States, which he did with great energy in the year
1831. Afterwards he induced General Jackson, then the President, to appoint him
"State Geologist" in which capacity he made two extensive tours, illustrating them
with long sections. . . . In the French Revolution of 1843, when Louis Philippe fled
from Paris and was hid in a cottage with Queen Amélie on the south bank of the
Seine opposite to Havre, it was Featherstonhaugh, then British consul at Havre,
who managed to get the family of "Mr. Smith" over by night, and popped them into
a British steam-packet. Even in this act the Consul was the Geologist, for he
passed off the ex-King as his uncle William Smith, the Father of English Geology.[38]

While the designation *Father of English Geology* was given to William
Smith by Sedgwick, a still wider appreciation of his labors came to him
many years later from America, when Professor Charles Schuchert[39] of
Yale referred to him as the Father of English Geology *and of Stratigraphy.*

A crowning testimony to William Smith's achievements, however, is
given by one of the greatest living authorities on palaeontology—Professor Alfred von Zittel, of Munich[40] in the following words:

Der englische Ingenieur William Smith zuerst den unumstöszlichen Beweis lieferte
dass die geschichten Gesteine Englands am sichersten nach ihren organischen Ein-
schlussen erkannt und chronologogisch angeordnet werden können, ergänzte er das
vorzugsweise auf mineralogische Merkmale und Lagerungsverhaltnisse begründete
System Werner's in glücklichster Weise und wurde durch Einfuhrung des paläonto-
logischen Elements, *Vater der historischen Geologie.*

[38] Archibald Geikie, *Life of Sir Roderick I. Murchison.* London, 1875, Vol. I, p. 215.

[39] Pirsson & Schuchert, *Text Book of Geology,* New York, 1915, p. 844.

[40] *Geschichte der Geologie und Paläontologie bis ende des 19.* Jahrhunderts, Munich, 1899, p. 77.

It was then William Smith's work that finally won for palaeontology the very high place which it now occupies among the geological sciences.

In the first quarter of the nineteenth century a number of interesting and important contributions to geological knowledge and especially to the geology of North America, based largely on the work of Werner and Hutton, were made by Maclure, Silliman, Cleaveland, Hitchcock, Eaton, Bigsby and others. An excellent account of this early work has been given by Merrill in his *Contributions to the History of American Geology, Washington 1906* which renders unnecessary further reference to it here.

While Hutton, as Zittel says, may be called the *Father of Historical Geology*, Maclure on account of the importance of his work is often and properly referred to as the *Father of American Geology*.

THE ORIGIN OF METALS AND THEIR ORES

As has already been mentioned, the older geologists were accustomed to designate as "fossils" all things which were dug out of the earth's crust. Among these "fossils," the metals and their ores always attracted especial attention, chiefly because of the bright luster which distinguished them from other stones, and because from the ores it was found to be possible through the agency of fire to extract metals.

Men undoubtedly began very early to indulge in speculations concerning the origin of these rare and useful bodies, discovered here and there among the common stones of the mountains. What these speculations were will probably never be known, but the earliest writings which have been preserved, from any European source at least, in which reference is made to this subject are those of Aristotle (B. C. 384–322).

THE VIEWS OF ARISTOTLE

Even Aristotle says but little on the subject of the origin of metals and their ores. He, however, touches upon it briefly in his *Meteorologica* when referring to the origin of stones in general.

In his opinion (already referred to in Chapter IV) the rays emitted by the sun, falling upon the earth's surface, cause the earth to give off two kinds of exhalations—a "Light Exhalation" and a "Dark Exhalation." The first of these consists chiefly of water, and rising into the heavens appears as mist and clouds. The second, or "Dark Exhalation," consists of fumes given off by the rocky surface of the earth itself, and can be seen when the horizon is scanned on hot days or in stormy weather.

[1] This was published essentially in its present form in the *Bulletin of the Geological Society of America*, Vol. 45, 1934.

This also rises into the atmosphere, forming dark clouds. It causes thunder, lightning, and earthquakes, and settling down again upon the surface of the earth becomes consolidated into stones of various kinds. Lucretius[2] refers to these two exhalations in the following words:

> When in the morning the golden light of the beaming sun first blushes over grass jewelled with dew, when the lakes and the ever-flowing streams exhale a mist, aye, even as the very earth seems sometimes to smoke; all these exhalations come together on high above us and now cohering weave a texture of cloud under the sky.

But some of these solar rays penetrate below the surface and pass deep down into the earth's interior. Here they act in the same manner as at the surface, developing these two exhalations.

When these rays from the sun which penetrate into the earth's crust meet with conditions where the earthy element largely preponderates, they produce stones. In other places, where the Aristotelian element of earth is mingled with that of water in suitable proportion, these rays engender metals or metallic ores, because, according to Aristotle, a metal is a combination of the elements of earth and water, the presence of the latter element being shown by the fact that the metal possesses certain properties of water, as, for example, a measure of fluidity, for metals are malleable when hammered, and when heated they will melt and flow. These incipient deposits, being fed and nourished by further exhalations, grow into larger ore bodies in the course of time.

The gems found in the earth's crust have a similar origin, but in their case it is the fixed stars more particularly, whose pure, serene and heavenly rays give birth to these bright and precious stones.

Thus, Aristotle considered the deposits of metals and their ores found in the earth's crust to have a celestial origin.

From the time of Aristotle for the next 1800 years practically nothing was written upon the origin of ore deposits; nothing, at least, is to be found in the few books dealing with natural science which appeared in this long succession of centuries. Avicenna, Theophrastus, Dioscorides, Pliny, Lucretius, Isidor of Seville, and Marbodus, all of whom made some reference to ores, say practically nothing about their origin. Seneca,[3] in referring to the analogy between the earth and the human body, writes:

> So, too, in the earth there are several different kinds of moisture. There are some kinds that grow hard when fully formed. Hence arises all the metalliferous soil from which our avarice seeks gold and silver. There is a kind which turns from liquid into stone.

[2] *De Rerum Natura*, Book V, vv. 460–466.
[3] *Quaestiones Naturales*, Book III, para. XV.

And so, when Albertus Magnus and his pupil, Thomas Aquinas, in their attempt to weld together the teachings of Aristotle and those of medieval theology, founded the Scholastic Philosophy, the old Aristotelian teachings concerning the celestial origin of metals and their ores passed into scholastic literature and continued to appear in successive scholastic treatises down through the ages.

THE FIRE AT THE CENTER OF THE EARTH

It was the alchemists who revived an interest in the subject. With the invention of printing in the latter half of the fifteenth century, their books were multiplied at a rate hitherto unheard of, and their teachings were spread wherever books were read.

Many of the alchemists, and others of their time who thought upon these subjects, were willing to allow that the celestial rays, penetrating deeply into the earth were, or might be, an active cause in the production of metals. It was, however, the general opinion among them that the evidence pointed rather to the existence of a great body of fire at the center of the earth, giving off dense clouds of metal-making vapors, as the primary cause of the development of ore bodies. The existence of this fiery region they believed to be demonstrated by the floods of molten rock, the clouds of ashes, and the great volumes of steam and intensely heated vapors which were ejected from Etna, Vesuvius, and other active volcanoes which they saw in Europe and from others whose existence was reported in various parts of the world. How this fire was lighted and furthermore how it was maintained, however, presented difficult problems and gave rise to a diversity of opinions. Some writers contended that it originated and was maintained through the action of the heavenly bodies. Frascatus[4] is a supporter of this view, his argument being briefly as follows: The earth stands at the center of the universe—he is of course speaking of the Ptolomaic universe—consequently its center is the center of the whole universe. Both are spherical in form. The sun, the moon, the planets, and all the fixed stars shoot out rays of heat which tend to concentrate at the center of the universe—that is at the center of the earth—as the spokes of a wheel meet in its central hub, so that the center of the earth must be the hottest of all places in the universe. Some think that these rays find their final goal at the earth's center, others that when they reach the center they are reflected back along the same course or at an acute angle, but whichever view be true,

[4] Gabrielis Frascati Brixiani, *De Aquis Returbii Ticinensibus.* Pavia, 1575.

these rays, colliding with one another and rubbing against each other in the narrowed space about the earth's center, by their friction generate fire, which is nothing else than an excess of heat. The heat thus derived from celestial sources will be maintained so long as the universe endures.

This view, so widely held in medieval times and so quaint in the light of modern physics, is by Frascatus further developed by the elucidation of the differing properties of terrestrial and ethereal fire and by many other considerations which, although interesting, are no longer relevant.

Thomas Norton, one of the best known of the English followers of what he calls the "Subtill science of holy alkimy" in his *Ordinall of Alchimy*,[5] written in verse about the year 1477, when referring to the metals and metallic ores found in the earth's crust, says:

> For cause sufficient Mettals finde ye shall
> Only to be the vertue Minerall
> Which in everie Erth is not found,
> But in certaine places of eligible ground:
> Into which places the Heavenly Spheare,
> Sendeth his beams directly everie yeare
> And as the matters there disposed be
> Such Mettalls thereof formed shall you see.

The alchemist, Aurelio Augurelli,[6] in his poem entitled *The Golden Fleece and the Art of Making Gold* says:

The place where metals take their origin is the inmost and unmoved centre of the earth, which in shape is like a marble goblet hollowed out of the deep lying rocks, a vaulted chamber into which the rays of the sun penetrate and the heavenly lights shoot into it innumerable rays which ripen and mature the collected vapours which from thence pass out and fill up all spaces in the rocks and all crevices traversing them. If however this vapour or moisture becomes condensed and cannot therefore pass on further through the secret cavities of the rock, it after unnumbered years finally becomes hardened into those unripe metals which are found filling veins in the earth's crust.

And these, he goes on to say, are eventually, through the alchemy of nature, transformed into gold, silver, copper, iron, lead, or tin, as the case may be.

Aurelio Augurelli, like so may of the alchemists, while claiming that

---

[5] See the facsimile reproduction of a copy published in 1652, with an introduction by E. J. Holmyard (Edwin Arnold), London, 1928.

[6] *Vellum Aureum et Chrysopoeia*, Basel, 1518. There is a German translation by Valentinus Weiglius, Hamburg, 1717.

he had discovered the art of making gold, remained a poor man, a circumstance which in his case led to a rather amusing incident. In the work above mentioned he makes the boast, "Were the ocean made of quicksilver, I could turn it into gold." But on presenting a copy of his poem to Pope Leo X, in the hope that he might receive some substantial recognition of the value of his work and at the same time an alleviation of his poverty, the pope sent him in return a package which when opened proved to contain a large bag in which there was nothing but a slip of paper on which were written the words, "He who can make gold requires only a purse in which to keep it."

More than one hundred years later the renowned alchemist, Glauber,[7] wrote:

Not alone I but many others, among them the celebrated Sentivagius, consider that there is an empty space in the earth's centre where nothing is at rest, into which the powers of all the stars are poured forth and where their mutual reactions give rise to an intense heat, from thence these virtues are continually turned back, seeking the circumference, where meeting with moisture and the material of pure earth they give birth to metals.

Others, of whom Georgius Agricola[8] is a representative, contended that this view of the origin of subterranean fire was purely fantastic, and that the heat was derived from the combustion of beds of coal, bodies of bitumen, or in some cases, masses of sulphur which existed beneath volcanic centers, and that these were ignited by intensely heated vapors which, in their turn, derived their heat from friction which was set up within the gaseous mass itself or by its contact with the walls of narrow spaces through which it was forced when in rapid movement within the earth.[9] The cause of these movements, and how the fires could be maintained for centuries at individual volcanic centers, is not explained.

[7] Johan Rud. Glauber: *Operis Mineralis*, Pars. II. *De Ortu et Origine omnium Metallorum & Mineralium*, Amsterdam (1652), p. 10.

[8] *De Ortu et Causis Subterraneorum*, Basel, 1546. Book II, p. 35.

[9] Milton sets forth the same idea:

As when the force
Of subterranean wind transports a hill
Torn from Pelorus, on the shatter'd side
Of thundering Etna, whose combustible
And fuel'd entrails thence conceiving fire,
Sublim'd with mineral fury, aid the winds
And leave a singed bottom, all involv'd
With stench and smoke.

Paradise Lost, Book I, line 230.

Agricola does not seem to have held that the whole central portion of the earth was in a highly heated condition but rather that the subterranean fires were confined to areas beneath the volcanic vents. His explanation of the source of this heat in the combustion of beds of coal became, however, quite generally accepted, and was that promulgated by Werner and many of the Neptunists as late as the opening of the nineteenth century. Agricola's views on the source of the highly heated vapors which, according to him, played such an important rôle in may of the processes which went on in the subterranean world were however vague, for he speaks of them, on the one hand, as having been derived from the vaporization of water and, on the other hand, of water as having been formed by their condensation.

It was the general opinion of the alchemists that the planets, of which the sun was one, rather than the fixed stars, were the particular heavenly bodies from which the creative influence that developed the metals proceeded, and that, furthermore, each planet developed a special metal and to a certain extent impressed upon it its own peculiar characteristics. Thus the sun influenced the production of the bright yellow metal gold; the moon the white shining metal silver; the planet Mercury, deriving its name from the ever moving messenger of the gods, the mobile metal quicksilver or mercury; Venus, the metal copper; Mars, the war god, the metal iron; Jupiter, the metal tin; and Saturn, the dull and heavy metal lead. This opinion had the support of Roger Bacon,[10] "The Admirable Doctor," and also of the great leader of the Schoolmen, "The Angelical Doctor," Thomas Aquinas.[11]

Growing out of the belief that the sun's rays were the active agent in the development of gold in the earth's crust was the idea, which became widespread and is met with everywhere in the literature of the time, that gold was to be found most abundantly in tropical countries where the sun's rays fell upon the earth's surface with the greatest intensity. "Thus we see," says Lehmann,[12] "that gold, which is the most perfect of all metals comes most abundantly in hot climates, silver, copper and

[10] Robert Steele, *History of Science*, vol. 2, p. 49.

[11] *De Esse et Essentia Mineralium*, Cologne, 1592, p. 132.

See also Fabius Colonne, *Les Principes de la Nature ou de la Génération des Choses*, Paris, 1731, p. 315.

[12] Johann Gottlob Lehmann, *L'Art des Mines* in his *Traité de la Physique*, Paris, vol. 1, 1759, also in his *Kurtze Einleitung in einige Theile der Bergwerkswissenschaft*, Berlin, 1751, p. 11.

lead require less heat in their formation and come in cooler lands."
Many authorities went so far as to hold that it was in fact useless to
search for this metal in far northern or far southern latitudes.[13] This
belief had an important practical bearing on the settlement of a number
of international questions. It probably played a part in the decision of
the Spaniards not to press their claim to what is now British Columbia,
after the British expedition under Captain Vancouver met the Spanish
admiral with his fleet off the west coast of Canada:

The treaty signed on 28th October 1790, made an end of the Spanish claim to sole
sovereign rights over American waters north of the Gulf of California, a claim that
had not been disputed for more than a century ... Already the spirit of Mañana
was sapping the vitality of a once great people. The indifference of Spain to this
vast unexplored territory is probably explained by her preoccupation with the quest
for gold and the commonly held theory that the nearer the Equator the better the
chances of finding it.[14]

The astronomical signs of the several planets were by the alchemists
transferred to their corresponding metals, and so deep a hold did the
belief in this connection take upon men's minds, that these seven metals
continue to be designated by these same symbols down to the present day.

In the Middle Ages it was regarded as a fact of peculiar significance,
and one which exemplified the divine plan of creation through which
ran the mystic number seven, that while there were seven planets in
the heavens, there were seven metals in the earth, a harmony which, of
course, ceased to exist when additional metals and new planets were
discovered; but by that time alchemy itself had suffered a transmutation
into chemistry, and these old ideas of mystical coincidence had faded
away.

The alchemists and others who had busied themselves with the question of the origin of ores in these early times, believed furthermore that
every metal first appeared in the form of a soft plastic material which
was called "Gur." Mathesius in his book *Sarepta* referring to "Gur"
says:

The matter of Metals before it be coagulated into a metalline form, is like butter
made of the Cream of Milk, which I also have found in Mines where Nature hath
produced Lead.

[13] Gabriel Plattes, *A Discovery of Subterranean Treasure*, London, 1838.
[14] George Godwin, *Vancouver, A Life, 1757–1798*, 1930, p. 27.

Webster[15] assures us that he has in his possession some pounds weight of this "Metalline Liquor." But it was not lead alone that was formed in this way, for Erastus[16] (quoted by Webster) says that he has in his possession some of this Gur "Of an ironish color, that may, like butter, be wrought with the fingers, from which notwithstanding hard and good iron may be extracted by the fire."

Boyle,[17] quoting Gerhardus, says that "At Anneberge a blue water was found, when Silver was yet in its first being or Ens, which coagulated was reduced into the powder or Calx of fixed and good Silver." With reference to gold and antimony, Paracelsus[18] says these are to be found in their "Ens primum," or first being, in the form of a red liquor or water which is afterwards coagulated and exalted into gold. This "Gur" is also described at length by Trebra.[19]

It is interesting to note that Paracelsus, who was perhaps the greatest of the later alchemists and who believed that the planets played a part in the development of metals in the earth's crust, considered that this influence was effective only during the earliest stages of the metal's development, before it had actually taken upon itself its true metallic nature.

First, we must know that every metal so long as it lieth hid in its first being, or *ens*, hath its peculiar stars. So gold hath the star of the sun, silver hath the star of the moon, &c. But so soon as they come to their perfection and are coagulated into a fixed metallic body their stars recede from every one of them . . . whereby we may perceive that when the ore of lead is from its liquid and soft substance coagulated into an hard metallic body, then the star of Saturn doth leave it and so the rest; but it must be understood that when the star of an imperfect metal hath left it hardened, that yet the star of gold or silver may operate in it to a greater perfection.[20]

The idea set forth in this statement of Paracelsus will be referred to again later.

As time went on, however, with a wider knowledge of the subject, the belief that the metals were developed in the earth's crust through

[15] *Metallographa or an History of Metals*, London, 1671, p. 50.

[16] Op. cit., p. 44.

[17] *Sceptical Chymist*. London (Everyman's Library), p. 192.

[18] *Liber de Renovatione et Restauratione*, pp. 43–45. (See Thomas Sherley: *A philosophical essay declaring the probable causes whence stones are produced in the greater world, etc.*, London, 1672, p. 56.)

[19] *Erfahrungen über das Innere der Gebirge*, Dessau and Leipzig, 1785, pp. 42, 43.

[20] Paracelsus taken from Webster's *Metallographa*, p. 124.

influences emanating from the several planets passed away.   Indeed, Johann Joachim Becher[21] (1635–1682), one of the most celebrated of the alchemists of later times, and a somewhat unmannerly individual, did not hesitate to express in unmeasured terms his opinion of those who still held this, now discredited, theory:

The Planetists, however, who wrongly assign to every metal or every mineral group some planet as its progenitor or generative cause, we dismiss offhand from consideration.   Some of these, although otherwise persons worthy of respect, are so barefaced that they do not hesitate to declare openly that they can see in the planet of each metal the colour of the metal and its chemical symbol.   It is a wonder that they do not see a lion in the sun, the figure of a woman in Venus, or indeed wolves and salamanders which creatures have been assigned to represent certain minerals. I believe that what they really have seen were donkeys, in that they have been looking at their own reflections.

Although Becher denied that the heavenly bodies had any influence on the development of metals, he believed, with Augurelli, that at the center of the earth there was a large vaulted space which was intensely hot on account of the ignition of great masses of sulphurous and bituminous materials which existed about it.   This ignition took place owing to friction set up by rapid movement in these masses, but when ignited they probably did not burn with visible flames.[22]   The intense heat thus developed, acting on great volumes of saline water which, he believed, found access to this fiery center of the earth through fissures in the ocean bed, gave rise to immense volumes of steam and to vapors and exhalations of many kinds, all of which mingled in a raging, swirling, reverberating mass, constituting a fiery chaos.   Certain passages in Holy Writ led Becher to believe that this place of subterranean fire was really "Hell," which, in the opinion of may learned theologians, was situated at the center of the earth, a conclusion which Becher says is supported by the consideration that the damned could not possibly be subjected to greater pain and torment than that they would experience in this dreadful place.

[21] *Chemisches Laboratorium*, Frankfort, 1653, pp. 89, 326.

See also Fabius Colonne, *Les Principes de la Nature ou de la Génération des Choses*.   Paris 1731, p. 287.

[22] See also Milton, who was a contemporary of Becher:

A dungeon horrible on all sides round
As one great furnace flam'd; yet from those flames
No light, but rather darkness visible.

—*Paradise Lost*, Book I, line 61.

Glauber,[23] who discusses the same subject in the chapter on the origin of ore deposits in one of his books, expresses the opinion that if this is not the actual lake of fire prepared for sinners the latter must at least be situated somewhere in the immediate vicinity.

Passing from these somewhat depressing theological considerations to the more cheerful and constructive geological aspects of the question, Becher expresses the opinion that under the influences of the intense pressures thus set up at the center of the earth, steam, vapor, and exhalations of various kinds would be "sweated" out of the chaotic mass and forced through the pores and fissures in the walls of the earth's great central vault, and thence outward through the body of the earth to its external surface.   Some of these vapors—those of purer nature than the rest—would, condensing, appear at the earth's surface as springs of clear, freshwater; others, less pure, as saline springs; still others, partaking rather of the nature of fumes or exhalations, and having present in them a "metallic seed," would condense in fissures and pores of the rocks through which they passed, in the form of metallic ores.

Becher[24] also holds that such being the origin of ores, these are to be found in ever increasing abundance in passing down into the depths of the earth, the exhalations from the central fires being more abundant, richer, and more active as the center is approached.

The Golden Tree

This brings us to the conception of the "Golden Tree," an idea which was present in the minds not only of many of the alchemists but which was believed to be true by many of the miners of old time as well.

It is set forth in concise form by Johann Gottlob Lehmann,[25] writing as late as 1753, as follows:

I hold that the mineral veins which are opened up in mining are nothing but offshoots from an immense trunk which presumably goes down into the very depths of the earth and which, on account of its great distance from the surface, cannot be reached in mining operations.   The great mineral veins are the large boughs of this tree, the smaller ones the slender branches and twigs of these great metal-bearing boughs.   This will not seem incredible to anyone who reflects upon the fact that all considerations point to the belief that it is in the innermost parts of the earth that Nature has located the workshop wherein all metals are produced and that here she

[23] *Operis Mineralis*, Pars. I, Amsterdam, 1651, p. 17.

[24] *Physica Subterranea*, Frankfort, 1681, p. 88.

[25] *Abhandlung von den Metall-Müttern und der Erzeugung der Metalle*, Berlin, 1753, p. 178.

gives them their primitive form and that from thence as mists or vapours, like the sap in a tree, they rise through the minute fissures which correspond to the vascular structures of the wood and are carried up to the earth's surface.

There is a fine passage by Peter Martyr of Angleria[26] (or Angliera) "A learned and grave councillor of Charles the Emperor fifth of that name," who, writing long before Lehmann and shortly after the discovery of the Western World, sets forth the theory of the Golden Tree as exemplified by the mode of occurrence of gold in Hispaniola—that is to say, in Haiti. It runs as follows:

There is also another region in Hispaniola named Cotoby, after the same name: this divideth the bounds of the provinces of Ubabo and Caibo. It hath mountaignes, vales and playnes: and because it is barren it is not muche inhabited. Yet it is richest in golde, for the originall of all the abundance of golde beginneth therein, insomuch that it is not gathered in small graines and sparkes as in other places: but is found whole massie and pure among certain softe stones and in the vaynes of rockes, by breaking the stones whereof they folowe the vaynes of golde. They have founde by experience that the vayne of golde is a lyving tree and that by all wayes that it spreadeth and springeth from the roote by the softe pores and passages of the earth putting foorth branches even unto the uppermost part of the earth and ceaseth not untill it discover itself unto the open ayre: at which tyme it sheweth foorth certaine beautiful coulours in the steede of flowers, rounde stones of golden earth in the steede of fruites and thynne plates in steede of leaves. These are they which are disparcled throughout the whole island by the course of rivers. They say that the roote of the golden tree extendeth to the centre of the earth and that there taketh noorishment of increase. For the deeper they dygge they fynde the trunks thereof to be so much the greater, as far as they may follow it for the abundance of water springing in the mountains. Of the branches of this tree they fynde some as small as a thread and others so bigge as a man's finger according to the largeness or straightness of the ryftes and clyftes. They have sometimes chaunced upon whole caves susteigned and borne up as it were with golden pyllers, and thus in the wayes by which the branches ascend: the which being fylled with the substance of the trunk creeping up from beneath, the branch maketh it selfe waye by which it may pass out. It is oftentymes divided by encountering some kind of stone: yet it is in clyftes noorished by the exhalations of the roote.

It is a pretty conception, that of the leaves and spangles of native gold as parts of the foliage of a golden tree growing up from the center

[26] *The History of Truayle in the West and East Indies and other countreys lying either way toward the fruitful and ryche Molucceas, gathered in parte and done into Englyshe by Richard Eden—Newly set in order, augmented and finished by Richard Willes*, Third Decade, Imprinted at London, 1577.

See also George Meyer: *Bergwercks Geschöpff und Wunderbare Eigneschaft der Metalsfrüchte*, Leipzig, 1599.

of the earth and the nuggets of alluvial gold as its fruit washed down into the streams from its outcropping twigs or branches, the whole tree deriving the material for its growth from auriferous vapors and exhalations emanating from the deep central portion of the earth where the tree is rooted.

Glauber[27] also speaks in similar terms of the arrangement of the metallic veins in the earth resembling a tree with its branches growing up from the center of the earth.

König[28] also refers to a certain mineral deposit which in form was like the branch of a tall slender tree, and after further discussion of the subject says:

Wherefore we reach the conclusion that metals and other minerals are subterranean plants having their birth within the earth.

The frontispiece of Becher's *Natur Kündigung der Metallen* (Frankfort, 1705) presents an interesting allegorical picture of this metallic tree (Plate IX).

The rays from the sun are beating down on the roots of the tree. They bear the inscription "Gigno" (I beget). The trunk of the tree is inscribed with the word "Concipio" (I conceive). In the foliage of the tree are seen the signs of the metals—gold, silver, mercury, copper, tin, and iron. The alchemical signs for sulphur and salt are in the earth at its roots. At the foot of the tree on either side stand two figures—on the right that of a young man resting on his spade, which bears the word "Elaboro" (I obtain by labor); on the left that of an old man scantily clothed, who has lost one leg and is supporting himself on a crutch with one hand, while in the other he holds a pot from which he is watering the ground about the foot of the tree. This pot bears the symbol assigned by the alchemists to the metal lead. This figure represents the god Saturn, the planet bearing whose name was always coupled with the metal lead, in the writings of the alchemists. That this is the true interpretation of the figure is seen from the fact that essentially the same figure appears on page 26 in Becher's *Parnassi Illustrati, Pars III, Mineralogia* published at Ulm in 1662. Here, the old man is leaning on a scythe instead of a crutch, while over his head appears the word "Saturnus." This attribute apparently takes the place of the pruning

[27] *Operis Mineralis.   De Ortu et Origine omnium Metallorum & Mineralium.*   Amsterdam, 1652, Part II.

[28] *Regnum Minerale*, Basel, 1703.

PLATE IX

ALLEGORICAL REPRESENTATION OF THE METALLIC TREE WHICH WAS SUPPOSED TO BE
ROOTED AT THE EARTH'S CENTER AND OF WHICH THE MINERAL VEINS APPEARING AT
THE EARTH'S SURFACE ARE TWIGS AND BRANCHES

(From Becher: *Natur Kundigen der Metallen*, Frankfort, 1705)

knife with which the god is often represented, as being the patron of agriculture; he is also regarded as a deity of the nether world. From an examination of the accompanying text, which is couched in alchemical phraseology, it may be gathered that the figure is intended to represent the idea that water passing downward from the earth's surface and containing the "Besonder Geist" of the "imperfect" metal lead—not itself represented in the branches of the tree—is met by uprising exhalations containing the alchemical elements, salt and sulphur, and from this union, under the stimulus of the heavenly influence the tree grows, and by a subtle alchemy the baser and "imperfect" metal is elaborated or "matured" into the higher and more perfect ones, among which are silver and gold. This conception of the elaboration or development of the base into the noble metals will be referred to again later.

THE "SEED OF METALS"

Among the men of learning who busied themselves with the study of metals and metallic ores there were many who held that these bodies, having once come into existence, in whatsoever manner it might be, were endowed with the power of growth and propagation. The belief that metals grew and propagated themselves within the earth's crust was held as far back as the time of the ancient Greeks. Anaxagoras taught it, and Agatharchides refers to gold as spreading itself through the earth's crust like the roots of trees. Theophrastus and Dionysius refer to metallic veins in the same terms, and not a few others speak of the metals as hidden subterranean plants.[29]

All through the sixteenth, the seventeenth, and the eighteenth centuries, there was keen controversy with reference to the manner in which this growth took place. On one side were those whose position might be stated as follows: There are three kingdoms in nature—the animal the vegetable, and the mineral. The animal kingdom displays terrestrial life in its highest development; the vegetable kingdom manifests life of a lower type; the members of the mineral kingdom are also alive, although their life is on a still lower plane and, indeed, quite rudimentary in character. The members of the two higher kingdoms have the power of propagating themselves, which they do by "seeds" given off from their own bodies. Arguing by analogy, it must be believed that the species

[29] See Blasius Caryophilus, *De Antiquis Auri, Argenti, Stanni, Aeris, Ferri, Plumbique Fodinis.* Vienna, Prague et Trieste, 1757, p. 126.

of the mineral kingdom, metals, ores and other "stones," propagate
themselves by "seeds" also, although these are so small that they cannot
actually be seen.   The several metals, therefore, have their "metalline
seeds" by means of which they reproduce their respective species.

On the other hand, there were those who stoutly denied that the
members of the mineral kingdom had any life or that they possessed
"seeds."   This controversy was carried on with much ingenuity and with
great vigor.   Of the first group the great Cardanus was a member.

Among the earlier writers who supported the view that metals and
metallic minerals propagated themselves by seeds was Bernard Palissy,[30]
the renowned potter, who describes the occurrence of pyrites in the form
of little crystals or concretions in a clay bed.   Their mode of occurrence
was such that it was certain that they had been deposited there from
water, but their seed, he says, was evidently liquid and could not be
seen.

Thomas Sherley,[31] "Physician in Ordinary to his Majesty," wrote a
book on the subject, in which he sets forth and develops at length the
thesis that all stones and metals have seeds from which they grow, as in
the two other kingdoms of nature—the vegetable and the animal.   The
seed, however, is an invisible one, and is often present in vapors which
rise from the center of the earth.   His thesis of the development of this
invisible seed into the visible, or *native* metal, he supports by quotations
from many illustrious authorities.   "The Seeds," he says, "having gotten
themselves Matrices in the Earth and Rocks (according to the appointment of God and Nature) acting on water with which they come into
contact, by a process analogous to that which takes place when the seeds
of plants are moistened, commence to sprout, transmuting this water
into a 'Mineral juice' called Bur or Gur from which by degrees it formeth
Metals."

Webster,[32] a supporter of this same view, in his *Metallographa*, which
is one of the best books which has been written on these ancient opinions,
propounds the question: "If Nature (as is most probable) contains in
her Cabinet the secret seed of Minerals, why may she not, meeting with
fit matter and adjuvant causes, have these small seminary particles

[30] *Discours admirables de la Nature des Eaux et Fonteines tant naturelles qu'artificelles, des
Metaux, des Sels & Salines, &c.* Paris, 1580, pp. 122, 134.
[31] *A Philosophical Essay, declaring the Probable Causes whence Stones are produced in the
Greater World*, London, 1672.
[32] *Metallographa or an History of Metals*, London, 1671, p. 70.
See also Boyle, *The Sceptical Chymist*, London, (Everyman's Library), pp. 194, 202.

stirred up and put into motion, grow up and expand themselves in the manner of plants and by taking on new matter grow and increase?"

Jorden[33] ("Our learned and ingenious countryman, Dr. Jorden" as Webster calls him) waxes indignant at those who deny that metals "vegetate." "What prerogative" he says "have Vegetables above Metals, that God should put seed into them and undeservedly exclude these? Are not Metals of the same dignity with God that Trees are?"

The members of the other group, however, vigorously opposed this view. Among them was Aubertus[34] who, writing as early as 1575, inveighed against Cardanus and all his followers in most unmannerly language, displaying the contempt felt by a man who employs the keen weapon of logic alone, in solving all the problems presented by the natural world, for the man who begrimes himself by resorting to the hard labors of observation and experiment. He calls the "Experimental chemists" (alchemists) of his day "Charcoal burners," "Ash blowers," "Smoke swallowers," and a number of even more opprobrious names. "Cardanus" he says "thinks metals are alive, but it is clear to anyone who has any brains, that this is pure madness, the dream of a mere fanatic. If Cardanus is still alive he ought to have a dose of hellebore."

Quercetanus,[35] who takes up the cudgels on behalf of the chemists, after some gracious talk about the greatness of the science, remarks: "But I return to Aubertus, I do not know that this man can be better compared with anyone than with that famous Phormio to whom Hannibal said, when they were discussing certain military affairs, that he had seen many old men in delirium but had never seen one who could rave better than he did." The whole is quite reminiscent of many of the theological controversies of the time. Long afterwards, Oxford said the same thing to Robert Boyle,[36] that chemistry was no proper avocation for a gentleman; but Boyle thought otherwise, and the "Brother of the Earl of Cork" became the father of scientific chemistry.

The next forward step was the recognition of the fact that whatever might be the genetic processes which resulted in the appearance of metals and metallic ores, there were certain situations or places which were more favorable to their development than others.

These were referred to as the Matrices or Wombs, of the metals.

[33] Edward Jorden, *A Discourse of Natural Bathes and Mineral Waters*, London, 1669, p. 51.

[34] *De Metallorum Ortu et Causis contra Chemistas brevis et dilucida explicatio*, Lyons, 1575.

[35] *Ad Jacobi Auberti de Ortu et Causis Metallorum contra Chymicos Explicationem*, Lyons. 1575 (Introduction).

[36] D'Arcy Thompson in *The Legacy of Greece*, Oxford, Clarendon Press, 1922, p. 143.

Lehmann,[37] who is one of the best known exponents of this view although his book is largely a translation of Hoffman's *De Matricibus Metallorum*, calls them the "Mothers of Metals" and classifies and describes them in his book.   He states that—notwithstanding the opinions of many of the older writers—although metallic particles may exist which are so extremely minute that they may be diffused through the air, he does not believe that these become aggregated into metallic masses in the air, so that this element is not to be regarded as a "Metalmütter."   Neither does he believe that the bodies of animals or of plants act as "Mothers of Metals."   These "Matrices" are to be found, however, in many places and under many different conditions in the earth's crust.   They are also common in the "Sahlbände" of veins; that is in the rocks bordering the veins on either side, into which the vapors, rising through the fissure which when filled constitutes the vein, have penetrated and deposited their metallic content.

He also describes various cases where fossiliferous and other limestones, as well as sandstones, have acted as matrices for the deposition of ores, and illustrates these by numerous figures.

There is an interesting discussion of the question of the growth of metals, metallic ores and other minerals by the great chemist, the Hon. Robert Boyle, in his well-known work, *The Sceptical Chymist*, which appeared in 1661, when alchemy was losing its hold on men's minds, and chemistry was being born.

Boyle was one of the founders of the Royal Society of London and this book was written to uphold the method of observation and experiment as the only means of arriving at a true understanding of nature, as opposed to the dialectic subtleties of the Schoolmen or the "Dark Writings" of the alchemists of his day.   Yet even he was unable at this late date to free himself entirely from the thought that possibly there might be something of truth in Aristotle's teaching concerning the four elements and the transmutation of these into one another.   He describes the experiment carried out by the renowned alchemist Helmont, who took 200 pounds of earth which had been baked dry in an oven, and having placed this in a vessel, planted in it a small willow which weighed five pounds.   This he watered, as need required, with rain or distilled water.

---

[37] Johann Gottlob Lehmann: *Abhandlung von den Metal-Müttern und der Erzeugung der Metalle*, Berlin, 1753.

See also the chapter entitled *Examen de la question: Si les mines se forment ou croissent encore journellement dans la sein de la terre?* in Lehmann's *Traité de Physique*, Paris, tome I, 1759, p. 380.

At the end of five years he took the tree out and found it to weigh 169 pounds, while the earth, after being once more baked, was found to weigh 200 pounds less two ounces. It would seem difficult, therefore, he thought, to avoid reaching the conclusion that the tree had grown by transforming the water into its own vegetable tissue, wood, and bark. Boyle then describes some similar experiments of his own "Concerning the growth of pompious mint and other vegetables," which grew freely and to a large size when placed in vessels containing pure water only.

Boyle then passes to the consideration of the third kingdom of nature and proceeds to give examples of the transmutation of water into various mineral bodies. Instances of this, he states, can be observed in many places. Stones can be seen growing from water dripping from the roofs of many caverns. Furthermore, "That sober relator of his voyages, Van Linschoten, informs us that in the diamond mines in the East Indies when having digged the earth they find the diamonds and take them quite away; yet in a very few years they find in the same place new diamonds produced there since." Fallopius relates how in sulphur mines exactly the same phenomenon is observed.

The continuous growth of iron ore in the mines of the coast of Ilva (i.e., Elba) off the coast of Italy is referred to by Pliny, Strabo, Agricola, Fallopius, Baccius and a host of other writers. Agricola also states that at a town called Saga, in Germany, they dig up iron ore in the fields by sinking ditches two feet deep, which in the space of ten years become filled with iron ore which is again dug out, and the ditches become refilled with iron ore at the end of another ten years.

"This occurrence," says Boyle,[38] "is very notable because from thence we may deduce that earth, by a metallic plastik principle latent in it, may be in processe of time changed into a metal." Indeed, that metal and metallic ores grew through the transmutation of earth and water was a widely accepted belief at the time.

A striking example of this phenomenon, was the growth of pure metallic silver in arborescent forms resembling the twigs and leaves of plants on the walls or timbers in certain old abandoned mines, observed by Dr. Schreter in the Joachim Valley where "Silver in the manner and fashion of grass had grown out of the stones of the mine as from a root, in length of a finger these veins being very pleasant to behold." There was on record also a "Mighty stone or lump, which stood in the middle of a cleft, in shew like an armed man but consisted of pure fine silver,

---

[38] Robert Boyle, *The Sceptical Chymist*, London, (Everyman's Library), p. 191.

having no Vein or Ore by it, but stood there free, which lump held in weight above 1,000 marks which according to the Dutch account makes 500 pounds weight of fine silver."[39]

The belief in an organic growth of ores and metals within the earth's crust finds expression in the works of many learned men toward the close of the Middle Ages and later[40] and in the old German miner's prayer:

"Es grüne die Tanne, es wachse das Erz,
Gott gebe uns allen ein fröhliches Herz."

Barba in his Art of Metals[41] says:

All of us know that in the rich hill at Potosi the stones, which divers years we have left behind us, thinking there was not Plate enough in them to make it worth our Labour, we now bring home and find abundant Plate in them, which can be attributed to nothing but the perpetual generation of Silver.

Boyle[42] also stated that he had been informed by several men having had long experience in the concentration of tin ores in Cornwall, that the ore as it came from the mine was stamped to a fine powder and its cassiterite (which was known as black tin) was separated from it by careful and repeated jigging and washing, and the exhausted tailings were thrown aside. But that, if after the lapse of a number of years these tailings were washed over again, more tin could be extracted from them. Boyle suggested to the miner that probably the ore had not been treated with sufficient care in the first instance, and that some tin had been left in the tailings by the former operators. The miner, however, replied that it was a well known fact that the workmen in former times used the greatest care and skill in the treatment of these ores and effected a much more complete separation of the tin from them than that which was secured by the workers in Boyle's time, but that notwithstanding this, when the old tailings were worked over again, more tin could be, and was, separated from them. Tin ore it seemed had grown anew in the refuse material.

It is interesting to note in this connection that shortly after it was formed, the Royal Society of London decided to have a questionnaire

[39] Quoted from Webster, op. cit., p. 44.
[40] See Stötzel, *Die deutsche Bergmannsage*, Essen, 1936, p. 50.
[41] *The Art of Metals*. Translated by the Earl of Sandwich, London, 1669, p. 49.
[42] Robert Boyle, *Observations about the Growth of Metals in their ore exposed to the air*, London, 1674.

drawn up and sent to the managers of a large number of mines in various parts of Europe, in order to obtain further information than was then available with reference to many questions concerning the nature and mode of occurrence of ores and other minerals, and Boyle was appointed by the Society to draw up the questions to be submitted. Among these questions the following appear, evidently asked in order to ascertain whether this statement of the Cornish miners was supported by the experience obtained in other mining areas:[43]

Whether it was observed that the Ore in tract of time may be brought to afford any Silver or Gold which it doth not afford or more than it would afford if it were not so ripe?

Whether it have been found that the Metalline part of the vein grows so that some part of the mine will afford one or more metal in tract of time, that it did not so before?

And whether to the Maturation of the Mine, the being exposed to the free air be necessary, or whether at least it conduce to the acceleration of it or otherwise.

A still more remarkable story of the spontaneous generation of metals —in this case gold—is one which is met with again and again in the works of the ancient writers who treat of this subject—namely, that of the growth of gold among the vines in the Hungarian vineyards.

The alchemist Libavius says that in the vineyards of Pannonia, near the city of Firmicum, little sprigs of metallic gold grow out of the ground and that these are gathered and coined into money: "Certain and faith-worthy authors have testified that in the nearer Germany between Danubius, there are vines which do sprout forth little branches and for the most part whitish leaves, of pure gold which are given to Kings and Chief Commanders as a rare gift."[44]

Huber in his doctor's thesis entitled *De Auro Vegetabili Pannoniae*, submitted to the University of Halle in 1733, presents a comprehensive study of the literature of this subject, gives a description of the occurrence, and sets forth his conclusions with reference to this remarkable phenomenon:

And just as in these places there are growing plants, members of the Vegetable Kingdom, so not infrequently, by a natural spectacle which is altogether wonderful and delightful, it comes to pass that gold, as if joined with these vegetable growths by a bond of consanguinity, laying aside, as it were, its own metallic character, grows after the fashion of plants out of the same lap of Mother Earth. Between the gold

[43] Roy. Soc. London, *Phil. Trans.*, vol. 1, 1655–66, p. 330.
[44] John Webster, *Metallographa or an History of Metals*, London, 1610, p. 56.

and the vine, indeed, these observers relate that there exists so close an intercourse: that the gold not only embraces the vine externally under the form of threads after the fashion of a climbing plant: but that even the vine sometimes puts forth little shoots and tendrils of pure gold, sometimes little berries of the same metal between its leaves. Gold is found intimately associated not only with the vine but with other vegetable growths: occurring either twisted up in various manners with their roots, or else growing near them in the form of little strings or threads. And this species of gold springing after the manner of vegetable growths, or in the midst of them, we designate by the name of Vegetable Gold.

Toward the close of this quaint thesis, the author, in discussing the probable origin of the gold, gives it as his opinion that the metal was brought into being about the roots of the vine, "By the same living principles as the vine itself," and that it was drawn up, along with the nutritive sap into the stem, and that there finally, through the coming together of the molecules of the gold, it had become aggregated into the forms above described.[45] This idea that metals and their ores grew, in a manner analogous to plants, lingered on into the early years of the nineteenth century, being held not only by Buffon,[46] but by Faujas St. Fond[47] and Fourcroy.

THE "MATURING" OF THE BASER METALS INTO GOLD

Not only was it the general opinion that the metals and their ores grew and increased in the earth's crust by a process analogous to that seen in the vegetable kingdom, that is by virtue of their "metalline seed," but it was the almost unanimous opinion of the alchemists, and one which was generally shared by other writers on mineral deposits, that within the earth's crust the baser metals were continually undergoing a gradual transmutation or change into nobler metals, and that it was the "Aim of nature" ultimately to transform all metals into gold, which aim in due time was attained in all cases where the metallic ores had been developed in suitable "matrices" and when sufficient time had elapsed to allow this change to be completed. Thus, for instance, lead gradually in the course of years changed to silver, and silver into gold. When lead ores, as is often the case, were found to contain more

[45] Many other similar occurrences are referred to in *Neue Sammlung Merckwürdiger Geschichte* by C. E. F. Breslau & Leipzig, 1756, p. 183.

[46] *Histoire Naturelle des Mineraux*, Paris, tome I, 1749, p. 10.

[47] *Essai de Géologie ou Mémoires pour servir a l'Histoire Naturelle du Globe*, Paris, vol. 3, 1809, p. 343.

or less silver, they were regarded as presenting an instance of such a transmutation not yet completed. When assays showed silver ore to contain some gold, these were regarded as occurences in which the former metal was in the act of changing into gold, although it was not as yet fully "matured" or "ripened," as the alchemists said, into that metal, the final product of this universal process of transmutation. This was in agreement with Aristotle's statement[48] that "Nature always strives after the better."

Aristotle believed that it was possible for one metal to change into another, and Avicenna,[49] writing about 1022, seems to have taught this transmutation in some of his works, although he denies its possibilities in others.

Gabriel Frascatus[50] in 1575 set forth his belief that the baser metals gradually "mature" into the nobler ones as follows:

In the earth the heat is not everywhere of equal intensity—and as this heat is the agent which "matures" the metals, different metals are found in different places in the earth according to the intensity and duration of the heat to which they have been submitted. But since Gold is what in the first instance is (really) aimed at by nature—as being the perfection and the ultimate endeavour in the way of metals—and since Silver is next in order to Gold: not only is each always found intermingled with the other, but also both are wont to lie hidden in all the other metals, even if it be only in minute quantity, these being as it were the nursery or seed plot of the wished for offspring. On this account it has been noticed that, not only in the long lapse of years is pure silver transmuted to some slight extent into gold, but also lead is changed into silver.

Ludovicus,[51] writing at the opening of the next century, gives it as his opinion that it is erroneous to say that each metal "Has its own primary principal"; i.e., that each is a separate element in the modern sense of the term. This "Is contrary to nature and to experience." Metals are gradually matured by heat, passing through successive lower forms and eventually becoming gold. There are "Mature metals and immature metals, breathing metals and expiring metals." He goes on to say that Zacherus and Roger Bacon, "Highly skilled philosophers," record that a certain mine of silver, which had been closed and remained

---

[48] *De Generatione et Corruptione*, Book II.
[49] See E. J. Holmyard and D. C. Mandeville: *Avicennae de Congelatione et Conglutinatione Lapidum*, Paris, 1927, pp. 5, 8.
[50] *De Aquis Returbii Ticinensibus*, Pavia, 1575, leaf 16.
[51] *Pyrotechnia Sublimis*, Vienna, 1778.

closed for thirty years, when opened up once again was found to have become a gold-bearing mine.

Later in the same century the renowned alchemist, Glauber,[52] states that there is a continuous transformation of the baser into the nobler metals going forward in nature, which in the course of time will result in the final passage of all these into gold. Hence it is that miners, when they come upon some mineral substance which is still "immature" or "unripe"—bismuth, for example, or cobalt or zinc—and assaying it for silver find that there is in it none of the precious metal, always say that they have opened up the vein too soon and before it had reached maturity. If, however, some years later, they resume work on this ore body which, in the meanwhile, has been exposed to the air, they frequently find that is is now quite rich in silver.

It was indeed the aim of the alchemists, believing firmly that this transmutation of the baser into the nobler was taking place in all parts of the earth's crust where the conditions are favorable, to bring about in their laboratories the same transmutation of the baser metals into gold but to carry out the change more rapidly.

Although this belief in a progressive change within the earth's crust of the baser metals into gold was, as has been said, held by the alchemists as far back at least as the 15th century, it was evidently also held and handed down by successive generations of miners in Europe from early times. It is set forth in the earliest book on ores and ore deposits, printed in Europe, which is probably the earliest treatise dealing exclusively with this subject in any language. It is a most interesting little volume in small octavo size, consisting of 24 unnumbered leaves and written in old German.

The author's name is not stated in the book, but it is now established beyond doubt that it was written by Ulrich Rülein von Kalbe, who in 1497 was Stadtarzt in Freiberg and later became Bürgermeister of this city, just as Agricola was a Stadtarzt in the old mining town of Joachimsthal and later Bürgermeister of Chemnitz. Agricola mentions the book by Kalbus (or Calbus) in his *De Re Metallica*.

The second edition bears the following title: "*Eyn wolgeordnet unt nutzlich buchlin, wie man Bergwerck suchen und finden sol, von allerley Metall, mit seinen figuren nach gelegenheyt dess gebirgs artlich angezeygt. Mit anhangenden Berctnamen den anfahrenden bergleuten vast dinstlich.*"

Like many treatises of this period it is written in the form of a dialogue;

[52] *Operis Mineralis*, part II, 1652.

in this case, one between a Master Miner and his apprentice ("Knabe") in which the former explains to the latter the mode of occurrence of ore deposits, and the way in which they originate.

The first edition of this work by Calbus is undated, but it was published about the year 1505. Of this edition only two copies are known to have survived. One of these is at the Bibliothèque Nationale in Paris and the other in the library at Augsburg. A second edition[53] was published in 1518 at Worms, of which, so far as is known, only four copies have survived, one of which is in the Royal Library at Dresden, a second in the State Library at Munich, a third (imperfect) in the library of the Freiberg School of Mines, and the fourth in the possession of the writer (see figure 56). Other editions appeared in 1527, 1534 and 1539.

At the close of the Middle Ages and during the Renaissance when there was a thirst for knowledge, and education was becoming more widely diffused through the various classes of the community, a large number of little treatises on various technical arts appeared and were apparently in great demand.

This little book on ore deposits (*Bergbuchlein*) was one of them. There was another on assaying (*Probirbuchlein*) which was published in 1518 which went through several editions, and was succeeded by other treatises on the same subject, as well as others on various technical subjects.

Hoover[54] mentions a number of these but the most complete account of them is that given by Darmstaedter.[55] He however has not listed all of these works. The writer has in his collection a copy of a hitherto unrecorded edition of the *Probirbuch* published in Strassburg in 1530 and a copy of the *Bergbuchlein* which is prefaced by the earliest printed text of the mining laws of Freiberg and Iglau, entitled *Der Ursprung Gemeiner Berckrecht*, which was printed by Knobloch in Strassburg about 1519–1520. This latter is a very rare work of which only some five copies are known.

In discussing the origin of ore deposits this little book shows in a most interesting manner how, by the close of the 15th century, among the mining fraternity in Europe the views of Aristotle had become modified by or, rather, interwoven with the later teachings of the astrologers and alchemists.

The influences emanating from the heavenly bodies, according to

[53] Hoover lists an edition stated to have been published in Worms in 1512. He had not seen this nor is it mentioned by Darmstaedter. If it exists *it* would be the *second* edition.
[54] Translation of Agricola's *De Re Metallica* (appendix B), London, 1912.
[55] Ernst Darmstaedter, *Berg, Probir und Kunstbuchlein*, Munich 1921.

Calbus, pass down into the depths of the earth, following more especially
cracks and fissures in the surface which are favorably oriented, with
reference to the courses of these celestial bodies, to receive these emanations.   Each planet develops its own respective metal.

Eyn zvolgeozdent vnd nütz=
lich büchlin / wie man Bergwerck suchen vñ
finden sol / von allerley Metall / mit seinen figuren /
nach gelegenheyt deß gebirgs artlich ange=
zeygt / Mit anhangenden Bercknai
men den anfahenden bergleut=
ten vast dinstlich.

FIG. 56. Title page of the *Bergbüchlein* by Calbus, Worms, 1518.   (Second edition of the
work.)

These influences, in the depths of the earth, give rise to the Aristotelian
"exhalations," and the latter, passing upward toward the earth's surface,
develop and deposit metals or metallic ores in the rocks or rock fissures
through which they take their course.

The metallic bodies so formed, having once come into existence, are not only fed and increased by these exhalations but are by them "matured," that is to say, gradually developed or transmuted into the ores of the nobler metals and eventually into metallic gold. But when gold has been so produced, should these powerful exhalations continue to pass through the deposit thus "Brought to perfection," they will gradually destroy the gold which they have produced, leaving eventually a reddish earthy residue in which perchance, a few flakes of gold survive, representing, in fact, what we now know to be the gossan outcrop of a vein, oxidized by the atmosphere and presenting a "weathered" surface in our present acceptance of the term.

The book is illustrated by a number of quaint woodcuts, and in certain of these, representing the outcrops of mineral veins, curious wreathlike forms are shown in the air above the vein, apparently rising from these outcrops. These are stated in the text to represent "Witterung." Two of these woodcuts are reproduced in figures 57 and 58. Witterung is clearly not used in its modern sense of "weathering" but in another sense which is now obsolete. In this latter sense it is frequently employed in works dealing with mining, written in the sixteenth, the seventeenth and even in the eighteenth centuries, and was evidently a term current among German miners in very early times.

Berwardum[56] defines Witterung as follows:

(1) A name given by miners to the natural heat which disintegrates and wastes away metals and especially ores when they have reached their "perfection."

(2) The vapour or exhalation which at times rises out of the earth from rich veins.

Hubner[57] gives the following definition:

(1) The vapour which sometimes, and especially after rain, rises into the atmosphere from rich veins and presents a fire-like appearance.

(2) The natural or subterranean heat, which brings ores to their "perfection," and when they have reached this condition wastes them away again.

It is also referred to by Glauber,[58] who says that "Witterung" or "Coruscatio" is a bluish lambent flame often associated with sulphurous

[56] *Interpres Phraseologiae Metullurgicae*, Frankfort, 1684.

[57] *Natur-Kunst-Berg-Gewerb und Handlungs Lexicon*, Leipzig, 1792.

[58] Glauber, op. cit., p. 28.

See also Theobaldus, *Arcana Naturae*, Nuremberg 1627, and Ludovicus, *Pyrotechnia Sublimis*, Vienna, 1778. The latter author speaks of veins of mercury ore as often overshadowed by a dense cloud at their outcrop.

FIG. 57. "Witterung" (w) rising from the outcrop of a rich vein (g). The poorer vein adjacent to it shows no "witterung." (From the *Bergbüchlein* of Calbus.)

FIG. 58. "Witterung" (w) rising through the morning mist (n) from the outcrops of two rich veins (g). (From the *Bergbüchlein* of Calbus.)

vapors, which may be taken as indisputable evidence of the presence of mineral veins and which, when it comes to the surface, causes the grass in the vicinity to become scanty in growth and the trees to assume stunted forms and to put forth fewer leaves and these of a paler color than others in the district.[59]

According to Melzern[60] the fact that when spring is coming the snow disappears from the land about the St. Georgen Mine in Schneeberg much sooner than it does elsewhere in the vicinity, is undoubtedly due to the "witterung" rising from its ore bodies.

FIG. 59. Man killed by poisonous exhalations from rich mineral veins, given off during a thunderstorm.   (From the *Historia* of Olaus Magnus.)

Olaus Magnus[61] speaks of the terrible thunder and lightning which are experienced in mountains which contain mineral veins, caused by sulphurous and other exhalations which issue from caves in these mountains.   The exhalations are extremely poisonous in character, and in his book there is a woodcut showing such a storm on one of these mountains and a man who has been killed by the exhalations issuing from rich mineral veins.   (Figure 59.)

The "witterung" then, represented in these illustrations from the treatise by Calbus and shown as issuing from certain mineral veins, conveys in a striking manner the idea held by the miners of those early times that the ores which they found in the veins worked by them,

[59] Agricola (in his *De Re Metallica*, Book II), refers to "a warm and dry exhalation" emitted from the outcrops of veins.

[60] *Beschreibung der Stadt Schneebergt*, Schneebergt, 1684.

[61] *Historia Olai Magni Gothi Archiepiscopi Upsalensis de Gentium Septentrionalium variis conditionibus &c.*   Basel, 1567, Book 6, chap. 11.

originated in exhalations of some kind, rising from the depths of the earth through the cracks and fissures.   The foul air which often accumulated in these old mines, was considered by the miners as due to such exhalations escaping from the ore bodies on which they were at work.

It is interesting to note that the belief in this phenomenon of "witterung" or "auswitterung" gradually passed away toward the close of the 18th century.   Thus, in the textbook prepared for the classes in mining at the Mining Academy at Freiberg and published in 1772,[62] the following passage appears:

A certain more or less doubtful indication of the presence of mineral veins is afforded by Auswitterung.   This is seen usually at dawn in hot weather, as a vapour rising from the outcrop of certain veins.   It is also noted that where this appears the hoar frost is lighter, the snow melts earlier and the grass comes to maturity more quickly, than over the country rock which underlies the vein.

Trebra,[63] writing in 1785, refers to:

The phenomenon which the miners call Auswitterung, the appearance of flames, sometimes small and at other times large, which according to their description resemble the lambent flame of burning alcohol.   Although I have never seen it myself, this appearance of flames according to the testimony of trustworthy people, can be seen at times on sultry mornings or evenings, on the surface, or in the mines playing about the exposed portions of veins which hold rich ores.   Such an 'Auswitterung' was seen in August 1776 in the dim light of early morning by a mine foreman called Schreiber at a certain place on a mountain slope at Marienberg, where a few years later a deposit of very rich silver ore was discovered.

Werner,[64] however, a few years later says:

After thirty years of careful observation and research I find that there is no reliance whatsoever to be placed on the signs indicating the existence of ore bodies which were believed in by the miners in old times, such as "Witterungen" or appearance of lambent flames accompanied by the melting of the snows and the stunted growth of trees at places where this was to be observed.   The old generation of miners, which has now for the most part passed away, had many remarkable stories to tell about these things but now these phenomena are never seen and people have almost ceased to mention them.   They have taken their place with the Gnomes and Kobalts who have vanished away to make place for realities.

The present writer when at Goslar in 1931 had a conversation with an intelligent old miner, who, having worked in the Rammelsberg mines

[62] J. G. Kern, *Bericht von Bergbau*, Freiberg, 1772, p. 29.

[63] *Erfahrungen von Innern der Gebirge*, Dessau and Leipzig, 1785, p. 41

[64] *Neue Theorie von der Entstehung der Gänge*, Freiberg, 1791, p. 185.

ever since he was a boy, had been appointed by the authorities to conduct visitors through that portion of the mines now open to the public. When shown the illustration in Kalbe's book, he said that he had never heard of the term "Witterung," nor did he know what the cuts were intended to represent. It would seem, therefore, that among the miners in the Rammelsberg, where the ore body has been worked for the past 1,000 years and where at one time "Witterung" was believed in by every miner, the very word itself has passed out of knowledge.

THE VULGAR OPINION

As far back as the 16th century,[65] and probably much earlier, there were some people who held the opinion that the world was created just as we now see it, and that neither it nor any of its constituent parts had undergone any change or development since it was brought into being by the word of the Creator. These were the Fundamentalists of their time. They based their beliefs, if such they can be called, upon what they held to be the simple and direct teaching of certain passages in Scripture, notably the opening words of Genesis "In the beginning God created the Heaven and the Earth," combined in most cases with a complete ignorance of the "Lessons" to be drawn from the "Sermons in stones" which are preached on every mountain side.

Agricola, the "Father of Mineralogy," states that there was in his time a wide diversity of opinion concerning the origin of ore deposits. The philosophers, the alchemists, and the astrologers, offered different explanations, while the "Common people" had still other views. The opinions of these latter, he says, are so amusing and fanciful that one can scarcely listen to them with patience. They run directly contrary to all experience and observation. They are briefly as follows: Not only the country rock but also the mineral veins which cut it, with their ores and accompanying gangue minerals, were created by God just as we see them at present. No similar deposits are now being formed in the earth's crust, and the Creator has endowed Nature with no power to continue their production. To see that this view is entirely untenable (Agricola goes on to say), it is only necessary to visit a few mines, for there one can see that the growth of rock is actually taking place at the present time, as shown by the development of new material on the walls

[65] See Bernard Palissy, *Discours Admirables de la Nature des Eaux et Fonteines &c.* Paris, 1580, p. 195.

or roof. Here, he probably refers to movements now known to be due to pressure. Iron ores can also be actually seen to be in the course of deposition in the island of Elba and elsewhere.

Franciscus Rueus,[66] in his interesting book written some twenty years later, endeavors by an ingenious explanation of the sacred text to reconcile the theories of instantaneous creation and of development, giving it as his opinion, as well as that of other learned writers, that the statement with which the account of the creation in Genesis opens, is not to be interpreted as meaning that "In the beginning" the earth and all that therein is was brought into existence in its completed form in an instant of time, but rather that when the divine decree went forth that they should come into being, this edict constituted "Creation." Some things at once appeared in their final and perfect form; others in their principles and beginnings, these to reach their completed growth as time went on through the action of secondary causes, which were also put into operation by the creative act.

In his opposition to the views of the common people, Agricola is also supported by the Italian writer, Marco Antonio della Frata et Montalbano[67] who says, in one of the most important of the early books on mining:

At the command "Let the Earth bring forth her increase," not only did all plants instantly appear but there was also given to gold, silver and the other metals the "vegetative" power, whereby they also could reproduce themselves. For since it was impossible for God to make anything that was not perfect he gave to all created things, with their being, the power of multiplication.

Albaro Alonso Barba,[68] the Spaniard, also, after referring to certain differences of opinion between authors concerning the generation of the various metals, says:

Many to avoid disputes of this nature, do hold with the Vulgar: that at the creation of the world God Almighty made the veins of metals in the same condition that we find them at this day: herein doing Nature great affront: by denying her, without reason, a productive Virtue in the matter, which allowed to her in all other sublunary things; moreover that experience in divers places hath manifested the contrary.

[66] De Gemmis, Zurich, 1566, p. 2.
[67] Pratica Minerale Trattato, Bologna, 1678, p. 2.
[68] The Art of Metals, Madrid, 1640. (Translated by the Earl of Sandwich, London 1669.) See also Ferdinandus Ludovicus, Pyrotechnia Sublimis, Vienna, 1778.

He then goes on to give as examples the growth of iron ore in Elba and the deposition of other ore bodies in various localities at the present time.

Löhneyss[69] takes a similar position, and writes:

And so we see that the fissures and the veins now have their seminal faculties, which God by His word has created in the earth and by which He causes the ores continuously to grow within the earth. So God did not create all the ores and metals at one time and all together in the beginning, any more than He caused all corn and fruits to grow once and for all, but He causes the earth each successive year to bring forth these at their proper season that man and beast may have them for their food . . . in short God by His word has sowed the seed of metals in the bosom of the earth, that in the regular course of nature all the various ores should grow perpetually through the action of the sun, moon, stars and elemental powers.

This "Vulgar opinion," which by its re-appearance in one form or another was to give rise to such prolonged and bitter controversies in other fields of geological research as time went on, was thus for the time being duly met and answered by Agricola, Reuss, and their supporters.

THE TRANSITION TO MODERN VIEWS

So long as men continued merely to speculate on the subject, no further progress towards a correct understanding of the origin of ore deposits could be made. Logic and the dialectic method, which were believed by the Schoolmen to be capable of solving any problem if properly employed, were useless when applied to this—one might "philosophize" forever and get no further. It was the introduction of a new method—the "Novum Organum," as Francis Bacon called it later— the basis of which consisted of a close and detailed study of the things themselves, that was required in order to ascertain how these had actually originated. This coming to grips with Nature by observation rather than by seeking to elucidate these recondite subjects by the application to them of the principles of logic was, as already mentioned, regarded with contempt by medieval scholars, and considered as undignified and as unworthy of a man of learning and culture.

The new school, so far as the science of geology is concerned, may be said to have originated with a few men who lived and worked about the middle of the 16th century, of whom Agricola and Palissy were the leaders. The members of this school who made by far the most important new contributions to a knowledge of the subject were a group of men

[69] *Bericht von Bergwerk*, 1617, p. 20.

who succeeded one another in quick succession in the latter half of the eighteenth century, all of whom held important positions in the mining districts of northern Germany, for the most part in and about the Saxon Erzgebirge, and whose knowledge was obtained from a close personal study of the mineral deposits, largely mineral veins, from which the ores in these important mineral districts were derived. It is of interest therefore, to consider briefly the work of these men.

This school, which perhaps may be called the Freiberg School, has Agricola as its leader, and culminated in the great Neptunist, Werner.

A brief account of Agricola's life has already been given in chapter VI. As is there shown, he was a man of many-sided knowledge, which in his case was combined with indefatigable industry and the attitude of mind, inculcated by the "New Learning," which based all conclusions on close and accurate observation alone. Although he was inclined to pay respectful attention to the opinions of the older writers, he adopted these opinions only when he believed them to agree with observed facts. He enriched and extended this older knowledge by a wealth of new observations, the results of his own personal investigations and study in the wide field of Mineralogy, Mining, Metallurgy, and Assaying.

Whenever Agricola writes on any subject, the reader recognizes at once that he was in advance of his age, for although he could not always free himself completely from the opinions of his time, his writings come as a refreshing breeze, which sweeps away many of the clouds of mist and fancy which formerly enveloped the subjects of which he treats.

His best known, which is also his last, work is the *De Re Metallica*, which appeared in 1556. This book, dealing with the actual practice of mining, and being excellently illustrated, appeals to a wide circle of readers, and although written in Latin, has been translated into English, German, and Italian, ten editions of it in all having appeared.

In another of his books, the *De Ortu et Causis Subterraneorum*, the first edition of which appeared in Basel in 1546, he treats, among other things, of the origin of ore deposits. Presenting a critical review of the opinions of the earlier writers upon the subject, he rejects them. "Aristotle," he says, "maintained that the metals are formed from watery vapors. But why does he not submit some basis of proof for this assertion? And likewise he asserts that the infusible rocks are developed out of dry exhalations. Why does he assert this without any evidence? The very mountains themselves refute such a statement. The alchemists tell us in all seriousness that all metals are composed of Mercury and

Sulphur. Experience which is always the best guide both for the teacher and the learner, shows how far removed this statement is from the teaching of reason and experience. All honor to the Manes and ashes of Albertus Magnus but his confused argumentation with its appeal to Avicenna, Hermes and the like ancient authorities is worthy rather of an alchemical disputation than the reasoned system of a student of nature."

Agricola also casts aside as idle dreams the teaching of the astrologers, that the development of the several metals within the earth's crust was brought about through the influence of the stars or the movements of the planets. The fissures in which metallic deposits often occur, he teaches, were formed first, and the ores with their accompanying gangue minerals were then deposited in them. These fissures he recognizes as having been developed by movements which took place during the formation of the mountains in which they occur, although in some cases he thinks they had originated in the disruptive force of subterranean waters when breaking open channels for themselves in the deeper parts of the earth's crust.

Within the earth's crust, he says, there are two kinds of water; the first formed by the condensation of steam or vapor within the earth, and having therefore a subterranean origin, and the second, rain water which has percolated from the surface deep down into the earth's crust. These waters—whether they be of the first or the second class—moving through the earth's crust, take up earthy matter from the rocks through which they pass. When thus impregnated with mineral matter, the water becomes a "Nobilis succus," a "Lapidifying juice" or solution, which deposits among other things, metallic minerals and their gangue stones in cracks and fissures through which they pass, and thus give rise to mineral veins.

In this conception of the origin of ore deposits, put forward by Agricola as far back as the middle of the sixteenth century and based on his close observation of nature, there are the germs of two great theories concerning the origin of ore deposits, which took definite form some three or four hundred years later; namely, the theory of ascension and the theory of lateral secretion.

Passing now to the German Mining Geologists of the eighteenth century, it will be of interest to consider briefly the work of a few of the more important of them, and the conclusions which they reached. They will be referred to in the order in which their most important papers appeared. There were two leading questions which presented themselves for consideration in connection with the study of these German deposits which were, for the most part, mineral veins; first, the manner in which the fissures had been developed, and, second, how these had been filled to produce the veins.

Rösler[70] of Freiberg, who died in 1673, of whom mention may first be made, believed as the results of his studies that the veins were due to the filling of open fissures, and in this connection directed attention to the significance of the druses in the vein.

Henkel[71] is said by Werner to be the "Father of Mineral Chemistry," but chemistry in the modern use of the term was just coming to birth when Henkel wrote. Following his argument closely, it is found to be as follows: All substances give off "Exhalations." He cites the aura which surrounds a magnet as a good example of such an exhalation. In the case of some substances these exhalations will quickly vaporize and pass into the atmosphere. Water and spirits of wine or alcohol are examples of this class. Others, such as rocks or minerals, vaporize very slowly. He cites marl, limestone and quartz as examples. Marl is a soft rock which forms a good soil in which plants grow readily. The water in this soil forms with the marl, when the latter is in a state of fine subdivision, a species of gelatinous material which is carried by the water to the roots of the plants. It is not, however, as yet in a sufficiently fine state of subdivision to enter the tissue of the plant but effects this passage by passing into a state of "vapor" which, mixing with the water, can now enter the plant and mingle with its juices. Possibly, he goes on to say, the marl does not actually vaporize, but at any rate it becomes "attenuated" and is carried off in intimate association with the water. That is, as chemists would now say, marl is to a certain extent soluble in water. Quartz is also similarly soluble, for it is often deposited from water in the form of beautiful crystals.

The emanations which give rise to mineral deposits are derived from three sources: First, from "vapors" given off by various bodies disseminated through the earth's crust which, meeting and mingling with one another, set up certain reactions which result in the production of metallic minerals; second, from the vapors and gases present in saline and

[70] Balt. Rösler, *Speculum Metallurgiae politissimum oder helpolirter Bergbauspiegel*, Dresden, 1700.

[71] Joh. Fr. Henkel, *Pyritologia oder Kieshistorie*, Leipzig, 1725, and *Mediorum Chymicorum non ultimum conjunctionis primum appropriatio, &c.* Dresden and Leipzig, 1727.

sulphurous waters which rise from the depths of the earth's interior and which originate, in part at least, through the waters of the ocean finding access to the central fires and being by them vaporized and set in motion toward the surface; third, from the "witterung" or exhalations of the minerals already in existence within the earth, which in some cases, as for instance alum shale and pyrite, when broken up and allowed to remain exposed to the atmosphere, can be seen to become hot and to give off fumes.

Zimmerman, who was a student of Henkel, explained the veins as due to a complete alteration of the country rock along certain cracks and fissures, through which some mineralizing solutions or vapors had forced their way.   The original rock was thus completely changed in structure and composition, the primary minerals being replaced by ores and their gangues along well defined strips which now constitute the veins.

Von Oppel[72] followed and made a very important advance in clearly recognizing and pointing out for the first time the fundamental difference between "veins" and bedded ore deposits ("flötze"), the former being open fissures subsequently filled with minerals differing entirely in character from those of the country rock which they traverse, and the latter being layers conformably interbedded in a series of stratified rocks, although differing in character from them.   He also states that the fissures, which the veins now fill, were developed either by movements within the earth's crust, which shattered it, or by a thorough drying out of the materials of which the crust is composed.   Also, that the "veins" cut across the structure of the country rock, and that the chief veins in a district usually follow a direction conforming to that of the valleys, which, it is now known, often mark predominant lines of dislocation in a region.   He also notes the fact that parallel to the main vein there is often a series of smaller ones, and that all of them thin out and eventually disappear on the line of the strike.

About twenty years after the publication of Von Oppel's *Anleitungen* there appeared a little book by Delius,[73] which was followed three years later by a more comprehensive treatise[74] by the same author.   His treatment of the subject is rather amusing in his impatient and con-

[72] *Anleitungen zur Markscheidekunst nach ihren Anfangsgründen und Ausübung kürzlich entworfen*, Dresden, 1749.

[73] *Abhandlung von dem Ursprunge der Gebirge und der darinne befindlichen Erzadern*, Leipzig, 1770.

[74] *Anleitungen zur der Bergbaukunst nach ihrer Theorie und Ausübung*, Vienna, 1773.

temptuous condemnation of the older writers on ore deposits and his substitution for their's, explanations of his own, which, although distinctly more modern as a whole, in many cases are as quaint as those which he rejects. Delius commences by stating that the views on this subject expressed by the leading alchemists are so ridiculous as to leave one in doubt whether to be amused or angry. He wonders why all these writers were not chained up as lunatics ("an Ketten gelegt worden"). Their chatter about the action of the Creator ("Archeo"), the influence of the stars, the "Seeds of metals," and the "Witterung" due to vapors arising from the depths of the earth first depositing then burning out ore deposits, is to be entirely disregarded. He inveighs against Lehmann especially, and his theory of "Metalmuttern," saying that so many of these are mentioned that it would be difficult to find corresponding fathers for them, adding that this writer's statements that metals are still being formed in the earth's crust is false, no ores are now being formed, all the fissures are filled, and the workshop is closed up. Occasionally, it is true, percolating rain water may dissolve a little ore here and redeposit it there, but all development of primary ore has long since come to an end.

His own views are as follows: Most of the mountains on the surface of the earth owe their origin to the Mosaic Deluge. They were originally great heaps of more or less soft material. After the retreat of the waters, these were dried out by the heat of the sun, and as the process of desiccation went forward, cracks were formed. As time passed rain water falling on the surface, percolated into the earth and dissolved out of the materials, now hard and rocky, which composed it, metallic substances which were sparsely disseminated through these rocks, in an extremely minute state of subdivision. Minute specks of various ores, he says, can still be found in all mining districts, in what is termed barren rock. These, accompanied by some earthy matter, are by the waters carried into the fissures, filling them up, and thus giving rise to mineral veins.

He rejects the old idea of the Central Fire and the Golden Tree; if it were true, veins would get wider and richer in depth, but, he says, "All veins and fissures in the world grow smaller in depth and eventually pinch out completely. Furthermore they get poorer and poorer as they go down and finally cease to contain any values. The gold veins in all the Siebenbürgen seldom retain their values below 40 to 60 Klastern (240 to 360 feet). This impoverishment is due to the fact that the sun's heat is necessary to the development of ores in veins and this cannot penetrate into the earth to greater depths than those above mentioned. Gold occurs abundantly only in warm countries and for the same reason." He is, in fact, so obsessed with the idea that the sun's heat (together with the concomitant action of air and water) is the agent which is especially active in producing ores, that he persuades himself that in all mines the temperature gets progressively lower as greater depths are reached, except when the rock contains pyrites and allied materials which produce heat.

However, Delius sets forth the facts which underlie one of the greatest discoveries concerning ore bodies which has been made in modern times, although he gives the wrong explanation of the causes to which they are due. This is *the surface alteration of ore deposits*, with the development of secondary minerals, and an *underlying zone of secondary enrichment*. He says that in veins of the Siebenbürgen, malachite and azurite are found extending from the grass roots to a depth of about 120 feet, but not lower. This, he observes, is evidently due to the fact that these minerals require much more of the sun's heat for their production. Gold occurs immediately below them, because it also needs much heat to develop it. Below it comes a zone rich in silver, as this noble metal requires less heat, and still lower down, lead is present abundantly and, being a "cold" body, continues to considerable depths. In fact, he says, most silver veins pass into lead veins in depth.

He describes the gossen cap seen in many ore deposits and gives a reason for its development. Veins which contain sulphur and arsenic minerals, with iron compounds, he says, often present a scorched or "burnt" appearance at their outcrop, owing to the action of the sun.

Charpentier, who was a professor in the Mining School at Freiberg, followed with two excellent books. The first of these appeared in 1778[75] and the second,[76] which sets forth the results of a long period of additional observation and study, some years later.

The *Mineralogische Geographie* gives an excellent description of the veins and other mineral occurrences of Saxony and some of the adjacent areas, and in the last few pages of the book the question of the probable origin of the ores discussed. It is an admirable presentation of the facts gathered during his long years of mining experience. He considers in

[75] Joh. Fried. Wilh. Charpentier, *Mineralogische Geographie der Chursächischen Lande,* Leipzig, 1778.
[76] *Idem*: *Beobachtung über die Lagerstätte der Erze, hauptsächlich aus den sächsischen Gebirgen,* Leipzig, 1799.

succession those facts which have a definite bearing on the question of the genesis of the ore deposits, and, based on them, he offers an explanation of the origin of these deposits, which, he says, cannot be considered as more than a conjecture but which he believes represents the closest approximation to the truth attainable at that time.

He begins by stating that it is inconceivable to him that a multitude of great fissures and cracks could have been developed in the rocks of the earth's crust by the process of drying out, to which cause a number of writers have attributed them. If then the idea be set aside that veins were developed by the materials which fill them flowing into fissures formed in this way, only two possible explanations remain: first, that of the "Ordinary man"; namely, that these mineral deposits were formed as we now see them, by the direct command of the Creator without the intervention of any secondary causes, which explanation, he says, may be at once dismissed; second, that these ore bodies, including the veins, have been produced by some alteration brought about in the original country rock.

He then goes on to state the reasons why he holds the second explanation to be the correct one. Although some veins are sharply defined against the country rock, others pass imperceptibly into it. In still other cases the country rock for some distance on either side of the veins shows a distinct change in character, and the values pass into this altered strip on either side. In tin ores, grains of cassiterite are often distributed through great masses of country rock, constituting a stockwork. These facts taken together seem to indicate that the veins and the ores were developed by changes which took place in the preëxisting country rock. It would rather seem that one kind of material had been altered into another. In looking about in nature for any analogy to this remarkable phenomenon he thinks of the silicification of wood. This he believes cannot have taken place by the removal of the woody tissue and its replacement by silica, because, had this taken place, the change must have proceeded from the surface of the wood inward, and had this been the case the outer part, when silicified, would have prevented the access of the solvent to the interior. There has apparently been in some unexplained way a change of wood into silica, which is, he thinks, more remarkable than the change of gneiss into a stockwork or a vein.

He believes, however, that in the crust of the earth there are solutions or vapors at work which could bring about such changes, and that these can find a passage even through solid rock along the minute cracks and little fissures which are everywhere present in rocks. In this way the agents, be they solutions or vapors, might bring about changes in the solid rock, and in it develop new minerals and even change strips of the country rock into what are known as veins, although no wide fissure was originally present.

Gerhard,[77] who wrote in 1781, held that veins were open fissures which originated through great movements in the earth's crust as well as from other causes, and that they were filled by the action of waters dissolving material out of the surrounding country rock and depositing it in these fissures.

There appeared next in chronological order an important and widely read work by Trebra.[78] This handsome and well illustrated folio is chiefly descriptive in character, but the author, who was a high mining official in the Brunswick Lunenberg Electorate, with long experience in mining operations, devotes a few paragraphs to the presentation of his opinions concerning the genesis of ore bodies. Like Charpentier he says that his view is open to many objections and is to be regarded as an hypothesis rather than as a well substantiated theory. He thinks that ore deposits owe their origin to processes which at that time were known as "Gährung" and "Faulniss," which may be translated literally as "Fermentation" and "Rotting," or "Decomposition." In modern times the former term, at least as applied to processes taking place in the inorganic material of the earth's crust, has little or no meaning. It is of interest, therefore, to ascertain just what was in the mind of Trebra and others of these early geologists when they employed these terms.

Influencing their conception was probably some survival of the ancient idea that the earth bore a certain analogy to the body of an animal, and as the body of the animal was nourished by the blood and other fluids passing through the arteries and the vascular system, so the earth was nourished by fluids circulating in its body. The central fire in the earth supplied the motive power which started and maintained these movements, as the heart causes the blood to circulate in the body of an animal.

When certain animal or vegetable materials are mixed with water and a little yeast or other ferment added, the mass, if kept warm, is seen to undergo certain curious changes, movements are observed to take place, new substances are developed in it, and the whole takes on a new charac-

[77] Carl Abraham Gerhard, *Versuch einer Geschichte des Mineral-Reichs*, Berlin, 1781.

[78] Fried. Wilhelm Heinrich von Trebra, *Erfahrungen vom Inneren der Gebirge*, Dessau and Leipzig, 1785.

ter—through the action of fermentation.   Again, if animal or vegetable substances are merely exposed to the atmosphere, they often decompose or rot away—frequently with the development of gases and evil-smelling vapors.

These early writers believed that analogous processes are at work in the inanimate materials of the earth's crust, acting slowly and quietly, under the influence of heat and moisture.   Owing to such, for instance, granite is changed to gneiss, graywacke is altered into shale, other rocks assume a coarsely crystalline character, and pyritous shale becomes an alum-bearing rock.   These slow but wonderful changes were brought about, they thought, by processes similar to a slow fermentation, which in the eighteenth century were called "Gährung" and "Faulniss," and they acted apart from and independently of, the more violent phenomena of nature represented by volcanoes and earthquakes.

The modern terms, Metamorphism and Weathering, embrace most, if not all, of these operations in nature.

Trebra believed that it was through processes of this nature that ore bodies of all kinds, including mineral veins, are developed, although he makes no conjecture as to the ultimate source from which the materials of the ore body are derived.

Werner,[79] who is the most renowned of all the geologists of this group although his contributions to this particular portion of the science are not of greater value than those of several of its other members, toward the close of his life wrote a small book setting forth what he terms a "New Theory of the Origin of Veins."   He was a professor in the Mining School at Freiberg, and the book presents his conclusions drawn from a study of the subject extending over a period of thirty years, as well as a statement of the grounds on which these conclusions are based.

Although he adds something to the previously existing knowledge of the internal structure of the veins, and the differences between the internal structure of veins and beds, Werner's theory is a new one chiefly in that it offers a novel explanation of the way in which preëxisting fissures were filled to form veins.

Werner, of course, as the leader of the Neptunists, taught that the crust of the earth is composed of rocks deposited from the waters of a primeval ocean.   Some of these rocks were of the nature of mechanical sediments, which the water had held in suspension; others were chemical

---

[79] Werner, Abraham Gottlob, *Neue Theorie von der Entstehung der Gänge, mit Anwendung auf den Berghau besonders den freibergischen*, Freiberg, 1791.

precipitates from the water; to the formation of still others, both processes contributed their share. Flows and sheets of basalt, found in
association with the regular sedimentary succession, were also supposed
to have been deposited from the waters of this same ocean.

As these sediments were settling down and becoming compacted into
rocks, cracks developed in the mass, owing to contraction due to the
expulsion of the interstitial water or through the slipping of masses
down the slopes of valleys which had been formed by movements of the
subsiding waters or by the rending action of earthquakes. These usually
took the form of fissures open to the surface, and if these were developed
in portions of the earth's surface which were still under the ocean, the
water filling them would be at rest and not affected by the ceaseless
movements of the great ocean itself. In such fissures the materials
forming the veins were deposited from the oceanic waters, the druses
representing unfilled portions of the fissures.

It is unnecessary here to follow Werner further in his exposition of the
subject. His theory has been entirely disproved; in fact, the Bergrath
Constantin[80] of Werner's own city, Freiberg, less than fifty years after
the publication of Werner's "Neue Theorie," states that not only this
theory, but Werner's whole system of geognosy, had completely broken
down under the accumulation of a great body of newly discovered facts.
This decline and fall, is the subject of chapter VII.

In treating of ore deposits, as of other subjects, Werner assumes a dog-
matic tone quite different from that taken, for example, by Charpentier.
In the seventh chapter of his book, for instance, he lays down in succession a series of things that are definitely and certainly known, the statement concerning each commencing with "Wir wissen gewis." This list
is instructive because a number of these, which he supposed to be irrefut-
able facts based on indisputable knowledge, have proved in the light
of further discoveries to be false and untenable. In the progress of
science, as in other lines of human experience, it is found that many
"Sure things" are by no means certain.

Werner states that veins did not derive their metallic content from the
country rock through the action of circulating waters (i.e., the theory of
Lateral Secretion) because the country rock in districts where metal-
bearing veins occur, contains no traces of the metals which are found in
the veins; also, that this theory is ruled out of consideration because the

[80] Friedrich Constantin, Freiherr von Benst, *Kritische Beleuchtung der Werner' schen Gangtheorie*, Freiberg, 1840.

first layer secreted from the wall of the fissure would make this imperme-able to the passage of all further solutions. He also denies absolutely that the veins have been filled by materials deposited by vapors rising from the deeper parts of the earth, on the basis that such metal-bearing vapors have never been found in any mine workings, and if this explana-tion be accepted it would carry with it the conclusion that such vapors had also given rise to the ore bodies which occur as beds, a contention which, he says, is absolutely untenable.

In the same manner he marshals "facts" which he believes absolutely disprove all other genetic theories which were put forward by previous writers. Finally, it is interesting, and not a little amusing, to note that in inveighing against Lehmann's "Metalmütter," a theory which modern research has shown to contain a large element of truth, Werner says that its advocates had put forward a lot of so-called evidence to support it, which not only failed to do so but which absolutely disproved the theory in question.

He then goes on to remark, apropos of these writers, that "This un-critical attitude of mind, one which is in the highest degree inimical to the progress of science, is only too common in our day. I could cite many examples of it but will mention only one, the well-known theory of the volcanic origin of Basalt." This was an unfortunate example for Werner to select, because in this very theory he himself presents a strik-ing illustration of the uncritical and perverse attitude of mind which he condemns in others, for throughout the greater part of his life he held, and supported with every conceivable argument, the contention that basalt was an aqueous deposit laid down from the waters of the primitive ocean. Toward the latter part of his career, he was forced to admit that molten lava was poured out of certain volcanic vents, but even then he would not allow that this was any proof of the existence of heat in the earth's interior, but contended that this lava consisted of certain beds of "wacké" which had been melted by the combustion of a bed of coal beneath the volcano, insisting that beds of coal were to be found in the immediate vicinity of all volcanoes.

Werner's personal charm, however, must have been altogether remark-able. As Geikie says, no teacher of geological science either before or since has approached him in the extent of his personal influence or in the breadth of his contemporary fame.

It is not the intention in this brief outline, which deals more particu-larly with the early history of the views which were held concerning the genesis of ore bodies, to follow out the development of these opinions in more modern times; this has already been done by others. It will be sufficient here to touch upon the contributions of six other men whose work was in a measure the outcome of that of some of the writers to whom reference has already been made, and especially of those in the mining regions of northern Germany.

Forchammer, of Copenhagen, in a paper which appeared in 1835, detailed the results of chemical analyses which showed that, contrary to the statement of Werner, minute quantities of the heavy metals were present in almost all rocks. He claimed that circulating ground waters made their way laterally into fissures carrying materials in solution and filled the fissures, thus giving rise to mineral veins.

It was, however, Sandberger,[81] of Würtzburg, who gave the *Theory of Lateral Secretion* its standing in scientific literature. He continued and extended Forchammer's work, devoting some nine years (1873–1882) to detailed chemical investigations, for the purpose of ascertaining how these metallic constituents occurred in the country rock of ore-bearing districts. He found that minute quantities of almost all the metals and rarer elements which were known in mineral veins were present in almost all the country rocks cut by these veins, even such rocks as clay slate being found to contain minute amounts of copper, zinc, lead, arsenic, tin, and cobalt. He then isolated the constituents of the various rocks which he investigated—chiefly igneous and metamorphic rocks—taken from many widely separated mining districts, and assayed the heavier and the lighter constituents separately. He found that the metals were contained chiefly in the dark basic silicates such as augite, mica, and hornblende, and that the feldspars were to be considered chiefly as furnishing the constituents of the gangue minerals. This investigation gave direct support to the theory of lateral secretion, showing that the country rock could supply the materials which constituted the veins.

Stelzner,[82] however, in a subsequent paper pointed out that the precautions which Sandberger had taken to remove minute grains of metallic sulphides, which might have been introduced into the country rock by later solutions passing through it, had been insufficient. Such waters might even be the same as those which, rising from deep-seated sources,

[81] *Untersuchungen über Erzgänge*, Wiesbaden, 1882.
[82] A. W. Stelzner: *Beiträge zur Entstehung der freiberger Bleierz und der erzgebirgischen Zinnerz-Gange*. Zeit. für pract. Geol., vol. 4, 1896, p. 377–412.

had deposited the filling of the vein itself. He showed, when the rock before analysis was treated with bromine, which would remove all sulphides present in it, that lead and silver—the characteristic metals of the Freiberg veins—were found to be absent from the country rock and that the theory of lateral secretion would not account for the origin of the veins of the Freiberg district, at least. The theory, however, in one form or another has met with, and still receives, much support on the part of many authorities.

While these German geologists, living in a part of Europe where there was relatively little vulcanism, naturally showed a more or less marked tendency to adopt the views of the Neptunists, certain other geologists, whose studies had been carried out in countries where volcanic phenomena were general and wide spread, formed other opinions concerning the origin of ore deposits.

Thus, the Italian geologist Breislak, who for many years had studied Vesuvius and Etna as well as other volcanic centers in his native land and who had, in consequence, ranged himself on the side of the Vulcanists, expresses his views on the origin of metalliferous deposits as follows:[83]

Let us then pass to ore deposits, directing our attention first to those which are found in the Primitive rocks. Let us consider that the material of this rock while still in a fused condition contained certain metallic substances. These latter owing to their mutual affinities would have a tendency to unite with one another and to separate themselves from the rest of the mass from which they differed not only in character but also in specific gravity and in all their physical and chemical properties. This separation would in many cases be imperfect, in which cases these metallic portions would remain intermingled with the rest of the rock. But when the rock became consolidated on cooling the metallic substances are found to be separated from the rest of the mass although enveloped by and enclosed in it. If some portions of the rock cooled more slowly than others, the metallic portions would remain longer in a fluid state and would thus occupy spaces between those portions of the rock already cooled although not as yet completely solidified. In this way veins rich in metallic minerals would be seen in the rock after its solidification, which veins might have been formed at successive stages during the cooling of the mass, and thus cut across one another. If we will set aside all bias and try to apply this idea to the phenomena displayed by the occurrence of metallic substances in the primitive rocks, I am inclined to think that an explanation of these will be found which is much less strained and which avoids many of the difficulties which are met with when other explanations are attempted.

[83] Scipion Breislak, *Institutions Géologiques traduites du manuscrit italien en français par P. J. L. Campmas.* Milan, tome II, 1818, p. 282.

Breislak believed that the stratified rocks of the Paleozoic and later times, had separated out in the form of mud or slime from a primeval sea of warm or hot water. While in this primitive muddy deposit, still in a semi-fluid condition, the several constituent materials tended, for the same causes which were operative in the primitive rocks when they were in a fused condition, to segregate themselves into separate masses, the calcareous material forming bodies of limestone, argillaceous material coming together as shales, and so on, having a bedded or laminated structure impressed on them by the movements set up in the mass as it consolidated owing to the drying up of the waters. These movements caused the various masses, at first irregular in shape, to adopt a form conformable to the curved surface of the earth's exterior. During the process the ores of the various metals in solution, or in the form of minute particles held in suspension in the ocean waters, came together and concentrated themselves within those rocks with which they had a special affinity. "Thus," Breislak says, "it need not be a matter of surprise that when the great deposit of galena in Derbyshire was forming the ore concentrated itself in the limestone beds rather than in the inter-stratified beds in which clay or silica preponderated, these substances having less affinity with lead."

Brunner,[84] who was a Bavarian mining official, held somewhat similar views concerning the origin of ore bodies, believing them, however, to have segregated, for the same reasons as those assigned by Breislak, from the primitive chaotic mass constituted of the three primitive "Ur-stoffe," namely "Lichtstoff," "Luftstoff," and "Erdstoff"—which constituted the primitive chaos and out of which the world was finally developed.

The hypothesis of Breislak and Brunner was an early and crude presentation of the *theory of Magmatic Segregation*, which has attracted so much attention in recent years.

Finally, Hutton the leader of the Plutonists, held that all mineral veins had been injected into the fissures which they occupy, while in a fused condition:

All these appearances conspire to prove that the materials which fill the mineral veins were melted by heat and forcibly injected in that state into the clefts and fissures of the strata. These fissures we must conceive to have arisen, not merely

[84] Joseph Brunner, *Neue Hypothese von Entstehung der Gänge*, Leipzig, 1801.

from the shrinking of the strata while they acquired hardness and solidity, but from the violence done to them when they were heaved up and elevated in the manner which has already been explained.[85]

Ask the miner from whence has come the metal in his veins?   Not from the earth or air above, nor from the strata which the vein traverses: These do not contain one atom of the minerals now considered.   There is but one place from whence these minerals may have come; this is the bowels of the earth, the place of power and expansion, the place from whence must have proceeded that intense heat by which loose materials have been consolidated into rocks, as well as that enormous force by which the regular strata have been broken and displaced.[86]

In How Far has Modern Science Advanced Toward a True Understanding of the Subject?

In the opening paper contributed to a symposium on the "Origin of Ore Deposits," in the Section of Geology at the Centenary Meeting of the British Association for the Advancement of Science, held at London in 1931, Professor Cullis remarked that notwithstanding a detailed study by the most able geologists, no consensus of opinion has been reached concerning the origin of most of the great ore deposits of the world, such, for instance, as those of the Rand, Huelva, Sudbury, and Kirunawara, a fact which indicates that the subject of ore deposits is one in which less progress has been made since the time of the Middle Ages than is the case in most other branches of Geology, and that although much has been learned, much still remains to be discovered.

Having, then, considered briefly the views of these older writers concerning the origin of ore deposits, it may be of interest to inquire in how far these opinions have proved to be correct; also, in how far actual advances have been made in our knowledge of this subject by later investigators and especially through the intensive studies of ore deposits by many geologists of distinction in all parts of the world in recent years.

It will be noted, in the first place, that the earliest of the theories put forward to account for the origin of ore bodies—namely that of the influences of the sun and other stars in the production of metals and ores within the earth's crust—has now been definitely relegated to the limbo of exploded hypotheses, together with so many other rejects of ancient learning.

A second theory which obtained general acceptance at one time, that

---

[85] John Playfair, *Illustrations of the Huttonian Theory*, Edinburgh, 1822, p. 76.

[86] James Hutton, *Theory of the Earth*, Edinburgh, vol. 1, 1795, p. 130.

of the growth of metals and their ores from "Metallic seeds," has also been abandoned. The development of iron ores in certain lakes and other bodies of water and the discovery of beautiful arborescent growths of metallic copper or silver in rock fissures or even in old timbers in abandoned mines, were facts which to the ancients afforded conclusive proof of this latter theory. Research in recent times, however, has definitely disproved the contention that these metals were deposited from some waters because these contained "metallic seeds" and were not deposited from others because in these the specific "seeds" were absent. It is, however, interesting to note in this connection that the deposition of ferric hydrate in bogs and from the waters of lakes is not always a simple chemical process, but is now known, in some cases at least, to be effected through the agency of certain bacteria. It is more than likely that the advocates of the "metallic seeds" theory, had they been able to observe the action of these lowly forms of life, would have hailed the discovery with delight and concluded that in them at last they had actually discovered the "seed" of iron, in whose existence they had firmly believed but which they had never been able to see or isolate.

A third theory, that which held metals and ores to have been brought into existence by the transformation of water or some other "element" into these metallic bodies, has also been proved untenable. It has been established that in many cases at least the ore deposits which were supposed to come into existence in this manner, owe their origin to the fact that water under certain circumstances can and does take these metals or rather their compounds into solution and, often after carrying them long distances, under changed conditions redeposits them.

The apparent regeneration of cassiterite in the exhausted tailings of the Cornish tin mines, described by Boyle, which seemed to him to be inexplicable except on the supposition that the earthy matter of the tailings had in the course of years been changed into tin ore, may, perhaps, if indeed the facts as recorded are correct, now find another explanation. Potter's clay, ground and mixed with water so as to form a plastic mass of impalpable grain ready to work upon the potter's wheel, if set aside for a certain time, will be found to have little lumps of siliceous matter developed in the mass owing, it is believed, to the solution of extremely minute particles of silica scattered through the clay and their redeposition upon some larger grain of silica to form these larger concretions. So *possibly*, in a body of finely ground tailings from which the cassiterite had been removed as completely as was possible by the methods of concentration then available, some ore might remain in an extremely finely subdivided state in the tailings, and in the course of time might, in a similar way, aggregate itself together into grains of sufficient size to enable it to be recovered when the tailings were again subjected to the same concentrating process.

A fourth theory, the "Vulgar opinion" that all ore deposits were called into existence together with the whole fabric of the earth's crust at "The beginning," once and for all by the creative word, has also been set aside, it being now definitely established that in many parts of the world, ores of many kinds are at the present time in course of deposition.

A fifth theory, generally accepted for centuries, which taught that within the earth's crust the baser metals are gradually and progressively changing into higher forms and that they eventually "mature" into gold, has also been discarded.

Fifty years ago the idea that one "element" spontaneously changed into another would have been regarded as incredible and absurd. Recently, however, it has been found that the elements uranium and thorium, wherever they have been discovered in the earth's crust, are in the act of disintegrating with the production of metallic lead. The transformation of other "elements" has also been brought about in the laboratory. In fact, the "elements" themselves, as such, it would seem, are about to disappear, each being apparently a special pattern or arrangement of the ultimate entities of that unknown something of which all the "elements" are composed, so that—strange as it may seem—the dreams of the alchemists may eventually come true, and gold may be produced by the transmutation of some other "element."

Had the zonal arrangement of ores, which has been discovered to exist around many intrusive stocks and volcanic centers, presenting many instances where in a single shaft, iron sulphides and carbonates pass upward into lead ores rich in silver, been known to ancient writers, they would undoubtedly have considered this as definite evidence, that the "imperfect" metal iron, under the influence of the sun's heat was in such cases "maturing" successively, as it approached the surface, into the "higher" metals, zinc, lead, and silver.

Again, the fact that in the veins at Cobalt, the native silver is later than the Cobaltite and is seen to replace it, would also have been regarded as a brilliant and a conclusive proof of the "maturing" of the less "perfect" metal cobalt into the precious metal silver.

Certain facts, however, which the alchemists regarded as definite evidence of *transmutation* have now received a very different explanation, as, for instance, that when the water pumped out of certain mines is allowed to flow over scrap iron, the iron gradually disappears and copper is found abundantly in the sludge in the bottom of the conduit.

Modern investigation has shown that these five theories have no basis of observed fact to support them.

On the other hand, three theories which were put forward, or at least adumbrated, by these older writers, have received a large measure of confirmation or support from recent investigations. These may be called respectively—using the terms in their widest sense—the theories of Ascension, of Lateral Secretion, and of Magmatic Differentiation. Around these, controversy centers at the present time, and concerning them there is still a wide diversity of opinion.

The theory of Ascension is a development of the old "Witterung" theory, which has come down to us from the Middle Ages, that the sulphides and similar ore bodies owe their origin directly to highly heated vapors or solutions rising from the deeper parts of the earth's crust. This theory was strongly advocated by Élie de Beaumont in 1847, and in recent years has met with widespread acceptance, especially in view of the fact that many ore deposits are directly associated with volcanic centers or igneous intrusions, in connection with which great volumes of steam, hot water, and highly heated vapors are given off. It seems to be definitely established that in many of these cases there is a direct genetic connection between the intrusion with its accompanying vapors and the associated ore deposits.

The waters and vapors accompanying volcanic activity or associated with igneous intrusion are believed by most investigators to have formed a constituent part of the rock magma itself, and to have been given off from the latter as it cooled on approaching the surface. These are the waters of "Juvenile" origin; that is to say, waters which have come to the surface of the earth for the first time, as Suess believed to be true of the waters of the hot springs of Karlsbad and which have been shown to form a constituent part of the waters of the Yellowstone National Park in the admirable memoir of Allen and Day.[87] It is interesting to note that the recognition of this class of Juvenile waters; although not necessarily given off from a cooling magma, goes back as far as the time of Agricola, who speaks of two kinds of water found within the earth,

[87] E. T. Allen and Arthur L. Day: *Hot Springs of the Yellowstone National Park*, Washington, 1935.

the first being what is now called meteoric water—rain water which has percolated into the earth from the surface—and the second, water which has its origin in deep-seated sources within the earth itself.

This distinction is insisted upon by Pernumia[88] writing in 1570. He discusses at length the distinction between these two classes of water and puts forward the same criteria for recognizing those of the second class as have been suggested for distinguishing Juvenile waters by those who have written upon this subject within the last few years.

One great difficulty in connection with the study of this class of deposits is the almost complete ignorance which still prevails concerning the character of the interior of the earth. From what depth within the earth's crust do these igneous intrusions and effusions arise? Was the late J. W. Gregory correct in the statement which he made in his presidential address before the Section of Geology of the British Association for the Advancement of Science recently, when he said that the study of ore deposits confirms the evidence obtained from earthquakes, that the core of the earth is surrounded by concentric shells, differing in composition, and that the metallic ores arise as gases or solutions from the shells below those composed of plutonic rocks? Is it known that these metal-bearing vapors or solutions come from the metallic barysphere and that they are not derived from the same source as the igneous magmas with which they are so intimately associated? Is this metallic barysphere composed exclusively of an alloy of nickel and iron or is this succeeded, on approaching the center of the earth, by deeper envelopes of the still heavier metals, such as gold and platinum, as held by H. S. Washington and others? If the ascending vapors, giving rise to great gold deposits in the earth's crust, arose from such a golden source there is evidently some element of truth in the ancient conception of the Golden Tree.

The theory of Lateral Secretion, which regards the ore deposits as having been formed by the concentration through solution of minute particles of ore sparsely disseminated through a great body of country rock by the agency of percolating waters belonging either to the "Vadose" or "Deep circulation," also has its origin in the writings of Agricola but took definite form through the researches of Forchammer in Denmark and of Delius and of Sandberger in northern Germany. The evidence in support of it was not conclusive, but the theory has been adopted

[88] Johannis Pernumia, *Philosophia Naturalis*, Padua, 1570, Leaf 70.

by competent investigators in more recent years as the true explanation of the origin of many important ore deposits.

The theory of Magmatic Segregation, which holds that certain ore bodies had their origin in the segregation of the metallic constituents from a fused magma, was, as stated above, put forward by Breislak and Brunner as constituting the most rational explanation of the origin of ore deposits. This theory has been widely accepted in recent times as the true interpretation of the genesis of many important ore bodies. It has taken on many varietal forms and has probably been the subject of more controversy and of a greater diversity of opinion in recent times than any other theory in connection with the origin of ore deposits. In how far it is true must be left for the future to decide.

In addition to these three great theories, which may be said to divide the field between them, there is the theory of the origin of mineral veins through the "injection" of the vein filling, in a fused or fluid condition, as held by Hutton. This does not now meet with very wide acceptance, but would seem to be the true explanation of the origin of those gold-bearing quartz veins which are demonstrably part of intrusions of granite pegmatite. This theory in its wider implications probably has its most important recent development in Spurr's[89] views on the origin of "Vein Dikes."

The ore deposits which are now classed as due to contact metamorphism or to replacement, do not here require special reference, because these, when viewed from the standpoint of the primary source of their metallic content, come under the heading of one or the other of the classes of ore bodies already mentioned.

It is interesting to note further that a distinct advance has been made in modern times in the recognition of the true origin of two great classes of ore deposits whose genesis was misunderstood by the older writers. These are the Residual and the Alluvial deposits. That type of residual deposits, resulting from the surface alteration of primary ore bodies or by the concentration of ores sparsely disseminated through bodies of rock, which have been removed by solution or weathering, is not mentioned by ancient writers.

The origin of alluvial deposits was misunderstood by ancient writers. Gold-bearing sands are mentioned by several writers as having originated in the earth's interior and having been carried out of the mountains by

[89] J. E. Spurr, *The Ore Magmas*, New York, 1923.

streams of water issuing from caverns. Calbus says that the most favorable river sands in which to look for gold are those which contain little grains of precious stones, such as amethysts and rubies, because these owe their origin to the purest and finest exhalations from the earth's interior which, when combined with moisture, produce gold also. Other explanations were also given, but it was only in comparatively modern times that alluvial deposits were recognized as originating in the atmospheric disintegration of great bodies of rock and the concentration, through the action of moving water, of the metallic constituents sparsely disseminated through them.

If then, casting our thoughts back through the long succession of past centuries, we ask why the ancients entertained such quaint and curious views concerning the origin of the metals and their ores, we find that these views were quaint because they were based almost entirely on philosophical speculation and not on observation. It was not until the time of the Renaissance, when men like Agricola arose, whose views were based on practical knowledge gained by a long continued study of the structure of the earth's crust in the great mining districts of Europe, that the commencement was made of a true understanding of the subject.

The study of the earth's crust, thus begun, has since that time been prosecuted, on an even wider scale and with an ever-increasing enthusiasm, by a multitude of men well qualified to undertake the task, having at their command all the resources of modern knowledge. Great advances in our knowledge have been made, but it is evident that much still remains to be discovered.

Looking back upon what has been accomplished and looking forward to what is still to be achieved, we may well sum up the situation in the words of a writer of the eighteenth century, who had a wide knowledge of nature but a somewhat checkered career, and say:

Even those of the present and last age who have been able to discover the Mistakes of these (the writers of early times) and have the Advantage of yet greater and farther Improvements in Science, if they speak frankly and ingenuously must own, that though they have discovered the Errors of their Predecessors, and are certain they are nearer the real Knowledge of the Mysteries of Nature than those of any other Age have been, they yet are sensible, that they are only making farther and farther Advances toward what, perhaps, it is not human Nature ever perfectly to complete.[90]

[90] Sir John Hill—in his notes to his translation of Theophrastus, *On the History of Stones*, London, 1774, p. 5.

CHAPTER X

THE ORIGIN OF MOUNTAINS

INTRODUCTION

The great mountain ranges of the world and its isolated hills and mountains rising from the plains are among the most striking of the physical features of the globe and must from the earliest time have inspired mankind with a sense of wonder and of awe and have called forth a spirit of inquiry concerning the manner in which they had originated. All wild tribes inhabiting mountain regions have stories, traditions and legends concerning the manner in which the mountains came into existence. And as mankind progressed and developed, these questionings would be ever more insistent, in fact they have continued to present themselves to the mind of man down to the present day and will do so for many a long year before the question of their origin is clearly understood and finally settled.

De Saussure in the admirable *Discours Preliminaire* to his great work,[1] which will be referred to later, speaks in glowing terms of the charm of mountain scenery and the inspiration to be derived from an intimate association with these giants of nature.   It is above all, he says, the study of mountains that will most rapidly advance our knowledge of the theory of the earth.   The plains present great level stretches in which we cannot see the underlying beds unless they happen to be exposed in valleys cut by running streams or in excavations made by the labor of man, but these are generally quite inadequate, for they are not deep enough nor are they sufficiently numerous.   The high mountains on the other hand often display enormous transverse sections through the rocky crust, in which are revealed at a glance the order, succession, thickness and direction of beds which constitute them and the relation of these to one another.

In the present chapter it is proposed to review the opinions which have been held concerning the Origin of Mountains—so far as these have survived in written records—from the earliest times down to the beginning of the nineteenth century, and then to mention theories which have been put forward since that time and which are under discussion at the present day.   Having done so an attempt will be made to consider what progress has been made towards a true understanding of the subject and what still remains to be discovered through further study.

REFERENCES TO MOUNTAIN BUILDING IN CLASSICAL TIMES

While mountains are, of course, mentioned very frequently by writers in classical times, it seems strange that scarcely any reference is made by these writers to the manner in which mountains originated.   Several writers among the ancient Greeks refer to the fact that in certain places the rocks of a district contain the remains of fish and other marine animals showing that they must have been raised up out of the ocean, but these were areas of general continental uplift rather than mountain chains.   Apparently the only reference to the actual birth of a mountain made by an ancient Greek writer is that of Pythagoras (580 B.C.) found in the outline of his views on the processes of nature which is given by Ovid in the fifteenth book of his *Metamorphoses*.   It refers to a hill near the city of Troezen in the Peloponnesus.

[1] *Voyages dans les Alpes*, Vol. I, p. vi.

Near Troezen, ruled by Pittheus, there is a hill, high and treeless, which once was a perfectly level plain, but now a hill; for (horrible to relate) the wild forces of the winds, shut up in dark regions underground, seeking an outlet for their flowing and striving vainly to obtain a freer space, since there was no chink in all their prison through which their breath could pass, puffed out and stretched the ground, just as when one inflates a bladder with his breath, or the skin of a horned goat. That swelling in the ground remained, has still the appearance of a high hill and has hardened as the years went by.

This passage, says Humboldt,[2] justifies us in believing that Ovid in it describes, "That great natural phenomenon which occurred 282 years before our era, between Troezen and Epidaurus on the same spot where Russegger has found veins of trachyte."

Aristotle[3] mentions the earth at Heraclea in Pontus and again at Hiera, one of the Aeolian Islands, as swelling up into a mound which finally burst asunder, by "winds" escaping from the earth's interior, and Strabo[4] refers to a mountain near Methone on the Hermionic Gulf, cast up during a fiery eruption.

Lucretius and Seneca make no reference to the subject, and Pliny while saying nothing concerning their origin, in a quaint sentence which appears at the beginning of the 36th book of his *Natural History* tells us why they were made:

All things else which we have handled heretofore even to this book may seem in some sort to have been made for man, but as for mountains, Nature framed them for her own self; partly to strengthen (as it were) certain joints within the veins and bowels of the earth; partly to tame the violence of great rivers and to break the force of surging waves and inundations of the sea and in one word by that substance and matter whereof they stand, which of all others is most hard, to restrain and keep within bounds that unruly element of water.

MEDIEVAL SPECULATION AND DISCOVERIES

Passing on to medieval times it is to be noted that all medieval studies and all medieval opinion concerning the course of nature are dominated by two influences—the Bible and Ancient Authority.[5] Thus the Noachian Deluge comes to occupy a prominent place in the speculations of medieval writers, because with the subsidence of the waters of this deluge the sur-

[2] *Cosmos*, Vol. 1, p. 239 (Bohn's Edition).
[3] *Meteoroligica*, Book II, 8.
[4] *Geography*, Book I.
[5] See Konrad Kretschmer, *Die Physische Erdkunde im Christlichen Mittelalter*, Penk's Geographische Abhandlung, Band IV, Heft I, 1889.

face of the dry land, as we now know it, made its appearance. The question therefore at once presented itself, as to whether the surface so exposed was smooth and even or whether there were hills and mountains upon it. On this question opinion was divided, but it was generally believed that when the waters retired some mountains at least were present. Among these was of course Mt. Arrarat on which the ark came to rest. These elevations were however generally regarded merely as somewhat higher parts of the original surface as then revealed.

The appearance of the dry land at once raised the question as to where the water went when the flood abated. Severian of Gabala thought that the valleys were depressions in the earth's crust made by the Creator to accommodate these waters in order that the dry land might appear. This was the opinion of the Venerable Bede and of Peter Lombard, and also of Milton:[6]

> God said,
> Be gather'd now, ye waters under heaven
> Into one place, and let dry land appear.
> Immediately the mountains huge appear.
> Emergent, and their broad bare backs upheave
> Into the clouds, their tops ascend the sky.
> So high as heav'd the tumid hills so low
> Down sunk a hollow bottom broad and deep,
> Capacious bed of waters: thither they
> Hasted with glad precipitance.

But the process of erosion, observed by a number of these medieval writers, made a marked impression on their minds. Ambrose (A.D. 340–395) thought that in some way the waves of the ocean might have beaten down the land along its coasts and casting aside the debris, have extended its bed to make place for an increased volume of water.

John of Philoponus,[7] who was the librarian of the great library at Alexandria and wrote a treatise on the Creation of the World, while agreeing with Ambrose, was of the opinion that in order to provide sufficient space to accommodate the receding waters, it was necessary to assume that portions of the ocean bottom had in some manner been hollowed out to accommodate the waters in question.

During the next half dozen centuries no one in Europe returned to the consideration of this perplexing question.

---

[6] *Paradise Lost*, Book VII, 282–291.
[7] See Konrad Kretschmer, op. cit., p. 121.

Leaving then, for the moment, the history of the development of opinion in Europe concerning the origin of mountains, let us pass to Persia and glance at a very interesting and important contribution to this subject which was formerly believed to have been from the pen of Aristotle, since in Medieval Latin versions of Aristotle's *Meteorologica* it is often appended to the fourth book of that work, in the form of an additional chapter. It consists of three paragraphs entitled respectively *De Congelatione Lapidum*, *De Causa Montium* and *De Quatuor Speciebus Corporum Mineralium*. There has been much discussion with reference to the authorship of this fragment, but it has now been established, through the researches of Holmyard and Mandeville,[8] that it was not written by Aristotle but by Avicenna, being in part a direct translation and in part a résumé of sections of a work by this latter author entitled *Kitab al-Shifa* or, the *Book of the Remedy*, which was written in Arabic by Avicenna in Persia between A.D. 1021 and 1023.

Avicenna's views concerning the origin of mountains which, as will be noted, have a remarkably modern tone, may best be presented in a translation of his own words by Holmyard and Mandeville:

We shall begin by establishing the condition of the formation of mountains and the opinions that must be known upon this subject. The first (topic) is the condition of the formation of stone, the second is the condition of the formation of stones great in bulk or in number, and the third is the condition of the formation of cliffs and heights.

We say that, for the most part, pure earth does not petrify, because the predominance of dryness over (i.e. in) the earth, endows it not with coherence but rather with crumbliness. In general, stone is formed in two ways only (a) through the hardening of clay, and (b) by the congelation of waters. . . .

Stone has been formed from flowing water in two ways (a) by the congelation of water as it falls drop by drop or as a whole during its flow, and (b) by the deposition from it, in its course, of something which adheres to the surface of its bed and (then) petrifies. Running waters have been observed, part of which, dripping upon a certain spot, solidifies into stone or pebbles of various colours, and dripping water has been seen which, though not congealing normally, yet immediately petrifies when it falls upon stony ground near its channel. We know therefore that in that ground there must be a congealing petrifying virtue which converts the liquid to the solid. . . . Or it may be that the virtue is yet another, unknown to us. . . .

Stones are formed, then, either by the hardening of agglutinative clay in the sun, or by the coagulation of aquosity by a dessicative earthy quality,.or by reason of a dessication through heat. If what is said concerning the petrifaction of animals

---

[8] *Avicennae de Congelatione et Conglutinatione Lapidum*, being sections of the *Kitab al-Shifa*. *Librairie Orientaliste*, Paul Geuthner, Paris, 1927, p. 18.

and plants is true, the cause of this (phenomenon) is a powerful *mineralizing* and petrifying virtue which arises in certain stony spots, or emanates suddenly from the earth during earthquakes and subsidences, and petrifies whatever comes into contact with it.   As a matter of fact, the petrifaction of the bodies of plants and animals is not more extraordinary than the transformation of waters . . .

The formation of heights is brought about by (a) an essential cause and (b) an accidental cause.   The essential cause (is concerned) when, as in many violent earth-quakes, the wind which produces the earthquake raises a part of the ground and a height is suddenly formed.   In the case of the accidental cause, certain parts of the ground become hollowed out while others do not, by the erosive action of winds and floods which carry away one part of the earth but not another.   That part which suffers the action of the current becomes hollowed out, while that upon which the current does not flow is left as a height.   The current continues to penetrate the first-formed hollow until at length it forms a deep valley, while the area from which it has turned aside is left as an eminence.   This may be taken as what is definitely known about mountains and the hollows and passes between them. . . .

Mountains have been formed by one (or other) of the causes of the formation of stone, most probably from agglutinative clay which slowly dried and petrified during ages of which we have no record.   It seems likely that this habitable world was in former days uninhabited and, indeed, submerged beneath the ocean.   Then, be-coming exposed little by little, it petrified in the course of ages the limits of which history has not preserved; or it may have petrified beneath the waters by reason of the intense heat confined under the sea.   The more probable (of these two possi-bilities) is that petrifaction occurred after the earth had been exposed, and that the condition of the clay, which would then be agglutinative, assisted the petrifaction.

It is for this reason (i.e. that the earth was once covered by the sea) that in many stones, when they are broken, are found parts of aquatic animals, such as shells, etc.

It is not impossible that the *mineralizing virtue* was generated there (i.e. in the petrifying clay) and aided the process, while the waters also may have petrified. Most probably, mountains were formed by all these causes.

The abundance of stone in them is due to the abundance, in the sea, of clay which was afterwards exposed.   Their elevation is due to the excavating action of floods and winds on the matter which lies between them, for if you examine the majority of mountains, you will see that the hollows between them have been caused by floods.   This action, however, took place and was completed only in the course of many ages, so that the trace of each individual flood has not been left; only that of the most recent of them can be seen.

At the present time, most mountains are in the stage of decay and disintegration, for they grew and were formed only during their gradual exposure by the waters. Now, however, they are in the grip of disintegration, except those of them which God wills should increase through the petrifaction of waters upon them, or through floods which bring them a large quantity of clay that petrifies on them. . . .

It is also possible that the sea may have happened to flow little by little over the land consisting of both plain and mountain and then have ebbed away from it. . . . It is possible that each time the land was exposed by the ebbing of the sea a layer was left, since we see that some mountains appear to have been piled up layer by layer, and it is therefore likely that the clay from which they were formed was itself at one time arranged in layers. One layer was formed first, then, at a different period, a further layer was formed and piled (upon the first, and so on). Over each layer there spread a substance of different material, which formed a partition between it and the next layer; but when petrifaction took place something occurred to the partition which caused it to break up and disintegrate from between the layers.

As to the beginning of the sea, its clay is either sedimentary or primeval, the latter not being sedimentary. It is probable that the sedimentary clay was formed by the disintegration of the strata of mountains.

Such is the formation of mountains.

Albertus Magnus (1205–1280) is the next writer to whom reference should be made. In his opinion mountains were produced in one or other of two ways. The first of these was through the action of sub-terranean winds which, when they were prevented from escaping to the surface because the earth's crust was too thick or because the way to the surface was blocked by the oceanic waters, blew up the surface into bubble-like elevations, i.e., mountains. It was for the last of the two above mentioned reasons that the highest mountains are found along the sea coast of the world. Volcanoes are formed by these same subterra-nean winds setting fire to local accumulations of inflammable material at certain places within the earth. The other cause which gave rise to mountains was the erosive action of the sea cutting into its bed and heaping up the sand so produced, into mounds along its shore—thus valleys would be formed in some places and hills, or, with a rising sea coast, even mountains in others.

The fullest and most interesting account of the origin of mountains which has been left by any writer up to the close of the thirteenth cen-tury is that given by Ristoro d'Arezzo[9] in his work entitled *La Composi-zione del Mondo*. All that is known of Ristoro are the following few facts which may be gathered from certain passages in his book: He was born at Arezzo, in Italy, he was "Religioso" and he wrote his work in that city, "nel convento nostro." Furthermore he worked in gold and silver, practised drawing and painting and made a study of the science of the stars. This is his only surviving work, if indeed he ever wrote any other, which is doubtful. In commenting upon this work the Abbé Fontani[10] says that Ristoro planned to "Treat of all that nature presents for the consideration of Mankind" and, as a matter of fact, his work does

---

[9] *La Composizione del Mondo di Ristoro d'Arezzo, testo Italiano del 1282 publicato da Enrico Narducci*, Roma, 1859.

[10] See page LVII of the introduction to the above work.

contain all the cosmological doctrine of the time. H. D. Austin, in a dissertation entitled *Accredited Citations in Ristoro d'Arezzo's Composizione del Mondo, a Study of Sources* submitted for the Ph.D. degree at Johns Hopkins University in 1911, discusses the sources from which Ristoro drew his material. Austin examined all the citations which Ristoro makes from earlier authors and found that he made use of the following works:

(1) Alfragan—*Liber de aggregationibus Scientiae Stellarum.* (He shows an intimate and extensive knowledge of this book.)
(2) Zahel—*Inductorium de Principiis Judiciorum.*

He probably made use of the following also:

(3) Isidore of Seville—*Origines sive Etymologiae.*
(4) Artephius—(*Liber?*) *Rerum Praeteritarum* (*Praesentium*) *et Futurarum.*
(5) Albumasar—*Introductorium Maius.*

It is possible that he used the following:

(6) Aristotle—*De Coelo et Mundo.*
(7) Averrhoes' Commentary on Aristotle's *De Meteoris.*
(8) Avicenna—*Liber Canonis.*

There is no cogent reason for believing that Albertus Magnus or Thomas Aquinas were known to Ristoro. In all cases studied, excepting possibly Isidore and Aristotle, Ristoro's information was derived from Latin translations of Arabic works.

The earliest manuscript of Ristoro's work bears the date A.D. 1282. From a close study of the old Italian text of this manuscript, Fontani is of the opinion that the book was written about the middle of the thirteenth century. Ristoro was thus a contemporary of Albertus Magnus. This excellent work has already been referred to in Chapter III. It is necessary here to consider only Ristoro's views concerning the origin of mountains.

Ristoro finds the primary and chief cause of mountain-making to reside in the stars. The heavens cannot, he tells us, exert their generating power upon the land so long as this is covered by water—the dry land must first appear. He therefore first inquires whether the virtue of the heavens has lifted the earth above the sea or whether it has caused the waters to withdraw themselves from that portion of the earth surface which is now dry land.

The virtue of the heavens shed down upon the earth passes first through the Zone of Fire, thence enters and passes through the Zone of Air, when it reaches the Zone of Water, and then the Earth.

The virtue from the heavens first causes the water to withdraw from an area which he says is stated by learned men to comprise about one quarter of the earth's surface and which is known as the habitable earth.

The dry land thus brought into existence lies chiefly in the northern hemisphere, owing to the fact that the northern heavens are more full of virtue, since they contain a greater number of stars and constellations than the southern heavens, so that the earth lying beneath the latter continues to remain for the most part covered by water.

The heavenly virtue of the stars lays bare the land by "drawing away" the water from it.   This action he compares to that of a magnet which draws up iron filings from a table and continues to hold them in suspension, thus laying bare the surface on which they rested.   But the water thus removed from the north "Was gathered together into the larger sea which surrounds the whole earth and which is called by some the Ocean."

The dry land having thus been brought into being, the heavenly virtues proceed to exercise their action upon it.

If on a dark night we direct our gaze upward into the starry heavens, we look through all the lower transpicuous spheres up to the eighth Ptolomaic sphere, that of the fixed stars.   Some of these stars which we see are "low Stars," that is, they are relatively near the earth, others while still in the same sphere are set higher up, that is further back and at a varying but greater distance from the earth.   These are the "higher stars."

If we imagine ourselves to be on one of these lower stars and leaving it take our way to a higher star, we must, as it were, ascend a hill.   If from this we pass to another star situated at an equal distance from the earth, we pass over a plain, if from this we descend again to a lower star we, as it were, go down a hill into a valley.   The heavens thus may be said to have a "montuoso e valloso" character, that is to say, to be full of mountains and valleys.   Having this character, the heavens, through the virtues which they possess, develop in the dry land beneath them an exact reproduction of their own form.   Wherever there is a "mountain" in the heavens, a mountain of corresponding size will be developed in the earth directly beneath it; where there is a "valley" in the heavens a similar one will appear beneath it on the earth's surface, and so the earthly topography is a reflection of the heavenly.   To the modern mind accustomed to think of the attraction which one body has for another, increasing as the distance between the two bodies diminishes, one would expect that the higher mountains would be drawn up by the nearer stars,

but such is not the case, these heavenly "influences" have laws of their own, a distant star draws up a high mountain, a nearer one merely develops a lowly hill, and so the effect is exactly that which is produced on a level surface of wax when a seal (which has a "montuoso e valloso" surface) is pressed upon it.

While this is the manner in which the chief features of the earth's surface were developed, Ristoro recognizes other lesser influences at work which also make mountains. He mentions first the action of running water, which cuts into the land and leaves erosion remnants in the form of hills or mountains, secondly, the action of the waves along a sea coast throwing up hills of sand or gravel, thirdly, the Noachian Deluge which deposited great mountain masses of sedimentary material on certain parts of the earth's surface. Then again earthquakes at times upheave the earth or blow it into bubble-like forms and thus make hills or mountains. A fifth cause is the deposition of calcareous deposits from certain waters which by their accumulation sometimes form hills, and lastly, there are hills heaped up by the labor of man in certain places.

By these various processes mountains come into being, but he points out that whenever in the realm of material things there is "Generatio," there must also be "Corruptio," that is, a passing out of existence—and so the mountains are being destroyed by various forces, among which the chief are the action of running water and the shattering action of earthquake shocks.

But Ristoro[11] must be allowed to describe these interesting orographic processes in his own medieval manner:

And now let us consider the generation and corruption of mountains, how it is that they may be made and how destroyed. We see water washing away the earth and running down from the mountains carrying the rocks along with it, filling up the valleys and raising the level of the plains. On the other hand, we see water cutting into the land and making valleys, and the valleys being thus made, high land remains on either side. And we see water taking earth from one place and depositing it in another place. We see it raising earth from a lower to a higher plane, and on the other hand washing it down from a higher to a lower level, and so it seems to have the power of making mountains and valleys. This is seen in the case of rivers in flood which, after their waters have abated, leave their beds exposed, showing the sand which they carried, deposited in the form of hills and valleys. And this is also to be seen along the sea shore where the water throws out the sand and retreating leaves a succession of little mountains and valleys, which look as if they had been made artificially. And we see water sometimes excavating sand from its bed and raising it to a higher plane which, in comparison to the depression, is a mountain.

And also mountains have been made by the waters of the deluge which covering the earth and retained upon it by high winds or some other cause, dug up the earth in one place and carried it to another, for it is the nature of moving water to churn up the earth and make hills and valleys, and thus leave the earth's surface "montuoso e valloso."

And indeed we dug out of a place near the top of a very high mountain the hard parts of fishes of many kinds which we call "nicchi" but which other people call "chiocciole" similar to those which artists use to hold their colours. And in a similar position on another mountain we found sands of many colours mingled with large stones, and many small round ones, which looked as if they had come from rivers—and this shows that these mountains were made by the deluge. We have found many mountains of this kind.

We have climbed one high mountain the summit of which was formed by a great platform of very hard rock of a ferruginous colour, which looks as if it had been placed there with much care by the hand of man. In this there was carved out a great castle resembling a citadel. This platform rested upon earth laid down in water, as is shown by the fact that when we dug into this mountain top beneath the platform in question, we found in some places sand mixed with earth, and in other places tufa with rounded stones like those found in a river, and in still other places associated with these were bones of fishes of many kinds and other things, which shows that this mountain, like others already referred to, on whose summits sand and the bones of fishes were found, owed its origin to the deluge.

Furthermore the deluge could make mountains without sand and fish bones out of whatever material was at hand.

[11] Ristoro d'Arezzo, *La Composizione del Mondo*, Book 6, chapter 8, *Concerning the Cause and Manner of the Generation of Mountains and their Destruction*. (*"Della cagione e del modo della generazione delli Monti e della loro corruzione."*)

In that district where these mountains occur in which sand and fish bones are found, the evidence is clear that the district in question was once under the sea, or at any rate, under the water of some place like the sea, because in places such as small rivers, great bodies of sand with fish bones would not exist in sufficient amount to build up mountain masses.

Earthquakes also may be the agency by which mountains are made and destroyed. When the force which causes the earthquake is powerful and has its origin beneath the earth, it may throw up the earth's surface and make a mountain; it may also blow up the earth (into bubble-like form) and thus give rise to a mountain, and there would remain beneath it only a hollow space, dependent on the nature of the country. We once ascended such a mountain and as we climbed we purposely stamped on the ground with our feet, and from beneath there came a sound as if the ground were hollow. And such mountains, hollow within, could have been made just as easily by an earthquake as by the deluge, or in a third way, for if the earthquake were a very strong one, it could throw up the hard earth and stones which, remaining joined together, might form a mountain which was hollow, within. In addition to great earthquakes originating in the earth's interior, there are also less violent tremors which can scarcely be felt.

Some of these earthquakes, as we are informed by learned men, were so violent that they shook the whole Province of Italy. These, while they could not throw up earth and make mountains, were able to shatter mountains already in existence, and also bring about other disasters.

And there are also mountains which are all white like snow, these also owe their origin to water which is making stone. A proof of this is that the water welling out from the summit of these mountains and spreading itself over the slopes of the mountains becomes dissipated leaving stone behind, and thus these mountains are growing continually.

At the top of one of these mountains there was a pool of hot water, in which we bathed and our hair when immersed in this water, had stony matter deposited on it, like wax on the wick when candles are made.

And we find, according to the Romans, mountains that are made by men, as they were by the Romans. These people caused earth to be sent from every part and corner of the world as tribute in recognition of their lordship, and this they put together in one place, where it built up a mountain which was called by the Romans, "The mountain made of the earth from all the world." And the citizens of the city of Arezzo, desiring to make a reservoir between two mountains to hold water, made another mountain. And so we see that there are certain mountains made by the hand of man.

Thus we have set forth the origin of mountains and the cause of their origin.

Ristoro d'Arezzo, while offering many quaint explanations of the phenomena of nature, was a man who was far ahead of his time, in that he had evidently personally studied and observed the geological structures displayed in the mountains of Tuscany, and had drawn correct conclusions from what he saw. Thus in the geological section in which he describes a series of hard, compact ferruginous strata, eroded into fantastic castellated forms, so often seen in the Apennines, overlaying a series of softer sandstones, shales and conglomerates, containing rounded concretions and the bones of fishes, he recognizes that this was an aqueous deposit. And those other deposits which he found on the top of a higher mountain from which he dug out many rounded concretions with the bones of "fishes," under which term they then included the shells of various molluscs, he recognizes to be of marine origin. But for three centuries after his time many if not most of the writers on the subject held that these "fossils" mentioned by Ristoro were not the remains of animals at all but had been developed in the rocks by other forces, having no connection with living things. He was evidently on a geological excursion when on the top of one mountain he found the pool of hot water in which while bathing his hair was "petrified," and finally, he clearly recognized the conflict in nature between the forces which make and destroy mountains.

In 1320 no less a person than Dante Alighieri[12] addressed himself to the question of the source of this power by which the land of the habitable globe was raised above the waters of the primeval ocean. This was Dante's only work dealing with natural science and was written about a year before his death. It is in the form of an address entitled *De Aqua et Terra* which was delivered with great pomp and circumstance in the "Church of the glorious Helena in the renowed city of Verona." The whole argument is set forth according to the forms of medieval logic and in Dante's well-known style. "It is a good example" says his translator "of the intellectual exercises or tournaments in which men of intelligence and learning delighted at the time."

Dante begins by stating that while he was at Mantua "A certain question arose which, though often dilated upon for show rather than for the sake of the truth, remained still undetermined. Wherefore I, who have from my childhood continually nurtured truth, could not bear to leave the question unexamined, but it pleased me to show the truth about this matter." Many had held that the level of the ocean was higher than the level of the land surface of the globe. Dante refutes this. When God said "Let the dry land appear" what was the *efficient cause* through

[12] *De Aqua et Terra*—translated into English by C. H. Bromby under the title—*A Question of the Water and of the Land*, London, 1897.

which this command was carried out? He considers the action of each of the elements, fire, air, earth and water in turn, and shows that none of them had the power of producing this result. The cause thus lay outside the earth itself. He then considers the powers resident in the Ptolemaic spheres which in their revolutions around the fixed central body of the earth carried with them their respective planets. He sets forth the powers exercised by each planet and shows that the power sought for was not to be found in any of them. There remained therefore only the Primum Mobile and the Heaven of the Fixed Stars. He shows why it could not be the former and consequently by a process of exclusion reaches the conclusion that it was the eighth heaven, that is to say, the Heaven of the Fixed Stars, from which the power emanated and that this was an elevating power which caused a portion of the earth's surface to rise up out of the ocean waters and thus produce the dry land, on which alone mankind could live. And this was accomplished, "Either by way of attraction as a magnet draws the iron, or by way of compulsion by generating vapors, as in some mountain parts," and he adds "He that does not recognize these things, let him know that he is outside the limits of philosophy."

In this address Dante is dealing with the question of the elevation of the continents as a whole and makes no mention of the origin of mountains except the reference in the passage just quoted. He really adopts an explanation which had already been put forward some years earlier by Ristoro d'Arezzo.

Leonardo da Vinci (1452–1519) did not recognize any force originating in the earth's interior in the genesis of mountains, nor would he allow that they had been produced through any process of elevation. In his opinion they were exclusively the product of the action of the aqueous forces upon the surface of the earth. He saw in them gigantic erosion remnants carved out of the primitive crust of the earth by streams and rivers, whose destructive power had deeply impressed him when as an engineer he had repeatedly studied the devastating action of the Alpine streams.

Agricola briefly considers the question of the origin of mountains in one of his works[13] and touches upon it in another.[14] He mentions five ways in which mountains come into being: (1) Through the eroding action of water. (2) Through the heaping up of sand by winds, espe-

---

[13] *De Ortu et Causis Subterraneorum*, Book III, 1546.
[14] *De Natura eorum quae effluent ex terra*, (last chapter), 1546.

cially along sea coasts and in desert countries. (3) Through subterranean winds. (4) Through earthquakes. (5) Through volcanic fires.

He believes, however, that most mountains have had their origin in the
eroding action of water, and he describes this process as follows:

It is evident that many and indeed most mountains owe their origin to the action
of water. The little brooks first wash away the surface soil and then cut into the
solid rock and carrying it away grain by grain finally cut even a mountain range in
two, removing also great blocks of rock in the process. In a few years they thus dig a
deep depression or river bed across a level or gently sloping plain. This action can
be seen to be taking place even by an inexperienced observer, especially in a mountainous region. In the course of years these stream beds reach an astonishing depth,
while their banks rise up majestically on either side. From these steep banks small
fragments of rock are continually detached by rain or frost, or large blocks, owing to
the transverse cracks and fissures which are seen in the rocks of all mountains, fall
owing to their great weight and are precipitated into the stream below. In this way
the steep cliffs gradually recede and become converted into gentle slopes, and so
the original plain becomes converted into a series of elevations and depressions. The
elevations are called mountains and the depressions valleys. This action is intensified
where larger streams or rivers replace smaller brooks and where they have a steep
gradient and shoot forward with the rapidity of an arrow. When the mountain
cliffs on either side have become progressively lowered in height, wide valleys are
formed and in them fertile fields appear bordering the stream. At this stage the
mountain lies back from the stream on either margin of the valley. It may rise
gently, gradually or even steeply. These mountains were once more or less level
land. The chains or rows of mountains which follow the course of the stream on
either side, are now far distant from it and widely separated from one another. This
is because the stream wears away its banks, sometimes only one of them which is
made of softer materials and at other times both banks. The material which it re-
moves from its banks it deposits either in its bed or carries it off down stream to lay
it down somewhere else. It continues to do this so long as it meets with no solid
rock which offers sufficient resistance to prevent the further action of the stream.
When it does it will eat into the softer portions of the bank and be deflected away
from the harder. And so we see that the stream will soon adopt a sinuous course,
and now swinging from side to side will incise a new bed for itself, and abandon its
former one. This process results in making the valley still wider. In this way
whole mountains are destroyed by the action of water and their debris scattered
far and wide.

All these varied and wonderful processes by which water destroying builds and
building destroys, mightily altering the appearance of the earth's surface, have
been in operation since the most remote antiquity, so far back in the dim distance
of the past beyond the memory of man that none can tell when they had their beginning. Nevertheless, Nature even now is day by day continuing her work in this
same manner, but the sharpest eye is not sensitive enough to follow the slow action
of nature in her creative work.

This description of the earlier stages of the Cycle of Erosion might have been written in the twentieth century.   If it be contrasted with the account of the origin of mountains given by Valerius Faventies written but fifteen years later, it will at once be seen what a keen observer Agricola was and how far he was in advance of his times.

Falloppius[15] writing in 1557 sets forth his views concerning the origin of mountains thus:

And thus the rocky material of which mountains are built up is formed, Aristotle tells us, by the condensation of the dry exhalation clotted together by moisture and unless this was the case, you see, my dear young students, that mountains would not be enduring, for they are always being worn down by the sun and rain, but that they remain in place is due to the fact that they are formed by this exhalation from the earth's interior as shown by the pyramidal form of the mountain, this being the form assumed by all light bodies as they rise.   But if anyone will say: "I do not see how it is that the mountains grow from such an exhalation—I do not see why there should be an exhalation making stones and mountains," I say that it behooves him to call to mind that passage of Aristotle in Chapter 15 of the first book of the *Meteorologica* in which he says:—"In this we have the cause set forth why we cannot actually see the mountains being renewed by the dry exhalation.   The cause is because our lives are so short, if we could live long enough we would certainly see that this is actually taking place, as Strabo testifies is the case in certain excavations from which iron ore has been taken out and which in a hundred years are found to be completely filled up again."   The same process is at work renewing the rocks of the mountains through the action of the dry exhalation.   And so they err who see in the deluge the origin of mountains, a view dismissed by Aristotle in his *Meteorologica*.   Agricola is also in error when he holds that mountains originate through the action of running water cutting channels in level land and leaving the mountains as outstanding remnants of the original plain; they are also wrong who say that mountains are portions of the earth's crust blown up and dilated by winds, and so are those who declare that they have been formed by the heaping up of sands, because as a matter of fact mountains are made of rock, and rocks have their origin in the dry exhalation mixed with moisture in sufficient abundance to make it hard and solid.

Before passing on to the consideration of the opinions of more recent times, attention should be directed to one more work which seems to have been the *first treatise written in Europe dealing exclusively with the origin of mountains*, and one which gives a conspectus of the whole question as this presented itself to the minds of educated men in the earliest years of the Renaissance and affords at the same time a most interesting picture of the manner in which a geological problem was approached and its solution sought during the later years of the sixteenth century.   This book is entitled *De Montium Origine*, its author is Valerius Faven-

---

[15] *De Medicatis aquis atque de Fossilibus*, Venice, 1564, leaf 102.

ties. It is a small quarto of 16 numbered leaves and was printed by the Aldine Press in Venice in the year 1561. It is one of the publications of the Accademia Veneziana which society passed out of existence a few years after its foundation, having conceived an ambitious plan of issuing a long series of books and papers. Of these only a few actually appeared,

D E
MONTIVM
O R I G I N E,
VALERII FAVENTIES,
ORDINIS PRAEDICATORVM,
DIALOGVS.

IN ACADEMIA VENETA,
M D L X I.

s

FIG. 60. Title page of the *De Montium Origine* by Valerius Faventies (Valerio Faenzi).

and this work of Faventies was the last. In the place of the Aldine anchor there is on the title page of the publications of this Academy a small figure of Fame with a ribbon bearing the words "Io volo al ciel, per riposarmi in Dio," on which account the society was often called the Accademia della Fama.

The book is an extremely rare one and seems to have escaped the notice of students of geology.[16]   It is written in good Latin but does not present the beautiful appearance of the earlier Aldines.   Little is known of the author.   He belonged to the Dominican Order, was probably a member of the "Domus" (Monastery) of St. John and St. Paul in the city of Venice and was recognized as an accomplished scholar.

Renouard,[17] and Quetif and Echard[18] state that so far as is known this is his only work.   The first of these writers, however, mentions (page 273) a work entitled *I dieci circoli dell' Imperio* among the works published by the Accademia Veneziana in 1558, as written by Valerio Faenzi (the Italian town of Faenza or in Latin Faventia) who in all probability is the same person.

The book is dedicated "From the lovely Ascanian Hills of Montegallium" to Philipo Maria Campegio, Bishop of Feltri, in sonorous and highly complimentary phrases: "Men of illustrious birth and of surpassing virtues are justly compared to lofty mountains.   Besides we see that you are numbered among those mountains upon which the holy city of God is founded" (that is, the chain of Bishops from Peter the rock onwards) "whereof the saintly David has sung: "Her foundations are upon the sacred mountains" (all these are figurative or spiritual hills).   "And I must not pass over in silence that you are often wont to ascend to the sublimest ridge of divine contemplation wherein according to the prophecy of the seer it hath pleased the most high God to dwell.   Thither you mount up, that when you have quaffed the sweetest waters of holy doctrines from that spring of nectar, lakes and rivers may flow down out of your mouth as from the lofty mountains to water the wilderness of men's minds."

Campegio, he goes on to say, is the very man he needs as a patron.   His universal fame is commensurate with the universal interest in the subject.   Besides his zeal in cherishing the writer's Academy and particularly his liberality to the individual learned professors on whom its glory rests (a graceful hint!) are irresistible magnets of attraction to his person.   A treatise on mountains can be dedicated to no one else than to this peak and paragon of towering excellence and munificence.   They belong together by a certain pre-established harmony.

[16] There is no copy of it in the library of the British Museum and the Auskunftsburo der deutschen Bibliotheken after a search for it in the libraries of Germany report it as being found only in the Stadtsbibliothek of Berlin and in the libraries of the Universities of Breslau and Kiel.   There are copies of it in the Bibliothèque Nationale in Paris and in the Bodlean Library.

[17] Renouard, *Annales de L'Imprimerie des Aldes*, Paris, 1834.

[18] Quetif and Echard, *Scriptores Ordinis Praedicatorum recensiti*, Paris, 1721.

These charming old books are usually dedicated to some patron and in terms of such unblushing adulation and with such a display of literary pyrotechnics that one can but cherish the hope that this worthy gentleman was possessed of that saving sense of humor which would lead him, while fully appreciating the fascinating play of words, not to search too diligently for the needle of truth in the haystack of verbiage.

The treatise is cast in the form of a dialogue between Rudolphus and Camillus, two members of a party of friends who met together from time to time at the invitation of Benedictus, of the ancient and illustrious family of the Rudolphi, at his beautiful estate at Garda on Lake Benacus (now called the Lake of Garda) in the hunting or fishing season. The preface states that on one occasion having enjoyed a morning of most successful fishing in Lake Benacus, the party sat down by the shore in a place commanding a fine view of the lake. They were struck by the wonderful beauty of the great stretch of water spread out before them, bounded in part by gentle slopes of verdure rising gradually from the waters but elsewhere by the frightful and forbidding mountains which ascend abruptly and whose summits tower above the lake. So seated, their minds turned to the question as to how these mountains had come into being. The matter was discussed by Rudolphus and Camillus, while the rest of the happy party sat quietly on the sward and listened. The "mise en scène" resembles in a way that of Boccaccio's *Decameron*, but the date is some two hundred years later.

Rudolphus is the chief speaker, and in the subsequent dialogue puts forward in succession *ten theories* comprising all the more important explanations of the origin of mountains which had been suggested up to that time. The discussion of the subject is characteristic of the times, the basis for the acceptance of the theories is always either a supposed logical demonstration or the authority of various ancient writers, chief among whom is Aristotle, or else some statement in Holy Writ. It will be noted that neither disputant gives any indication that he had ever climbed a mountain in his life or that he had any desire to do so.

The reasons assigned by Valerius Faventies for the origin of mountains are the following:

(1) Earthquakes.
(2) The swelling up of portions of the Earth which had been moistened with water.
(3) The uplifting power of air enclosed in the Earth.
(4) Fire.
(5) The Souls of the Mountains.

(6) The Stars.
(7) Erosion.
(8) Wind.
(9) Moisture in the Earth being drawn upwards by the Sun.
(10) The work of man.

These in his book are treated in the order set down above. That the discussion may be clearly presented, while at the same time the distinctive medieval flavor of the work is preserved, a series of extracts have been translated from the Latin. These follow one another in the regular order of the text and represent about one half of the book, the intervening portions of the original which have been omitted, consisting chiefly of philosophical discussions or discursive matter which contributes but little to the elucidation of the line of argument.[19] Camillus is the first speaker.

Camillus: Most learned Rudolphus, seeing that you have so deep a knowledge of all the noble arts, pray explain to us in a few words the opinion which we should hold concerning the origin of mountains. I know of no one who has discussed this matter fully. Whence did mountains take their rise, and in what manner and from what source have they come into being?

Rudolphus: The learned Boccaccio says that mountains are elevations consisting of earthy matter; these are of different heights, some including all the higher ones are composed of rock, others and especially the low ones consist of earth. A very important agency in the development of mountains is that of earthquakes. Owing to these many islands have been raised out of the sea. This was the way in which Pliny states Delos and Rhodes came into being, and the same is true of the islands of the Grecian Archipelago. Since therefore it has been proved, that by the agency of earthquakes, certain islands have emerged into light, that certain others have retreated from the mainland while some have been united with it, and since owing to that agency countless cities and mountains have collapsed, Aristotle[20] reached the conviction that in some cases mountains are raised up and formed by earthquakes.... Besides there is another fashion in which mountains arise from the same cause. Whenever that shuddering of the earth occurs without being accompanied by any fracture, there is the most favourable condition for the elevation of mountains. And it is a thoroughly established fact that fissures are not always produced by an earthquake. The island of Delos once experienced a ghastly quake without fracture, which Callisthenes thought astonishing. Cyprus, Paphos and Zacinthos are constantly quaking without ever suffering such fractures. The reason

[19] For the translation the author is indebted to Dr. John MacNaughton, Hiram Mills Professor of Classics at McGill University (now retired).

[20] What Aristotle says is that wind enclosed within the earth and giving rise to earthquakes, sometimes blows up a portion of the earth's crust into a mound-like elevation—which may burst and thus allow the wind to escape. *Meteorologica*, Book II, 8.

for this is that the solid portions of earth (the rocks) do not split readily; therefore it is that islands at times emerge into light, or mountains are formed.... If however the less solid portions give way, a monstrous gap appears and a tumbling down of islands, cities and all things resting on that superstructure. Therefore, Seneca says, expressing this view—"The earthquake stirs the earth to its innermost roots, it opens out veins of boiling waters, sometimes chills these down to the freezing point and sets a thousand marvels moving, raises up plains, bulges valleys on high, lifts new islands out of the deep, and in short nothing in all nature is more puissant, nothing of more mordant efficacy than an earthquake. It can scatter great tracts of earth, lift up unheard of mountains from the depths and set down islands never seen before, in the midst of the sea.

Camillus: I agree that mountains sometimes arise through the agency of earthquakes. But the idea that the loftier ones should be assembled in such fashion arouses the greatest astonishment in my mind. Atlas on account of his height is said to hold up the platform of heaven, Olympus is so sublime that he towers above the clouds, whereof saith Lucan "He passeth beyond the clouds," Lebanon and the mountains of Armenia I put to one side, I come to our own peaks. There are the Apennines, there are the Pyrenees, which it seems hard to think were piled up by movements of the earth, particularly as they sometimes resemble waves of the sea, one crest following upon another in height, in arrangement and in position, just as if they had been set in due place by some marvellous artistry.... I cannot believe that earthquakes have produced such systems of mountains. Earthquakes take place chiefly in places where the underlying rocks are cavernous, where they are exposed to tremendous blows from the sea waves lashed to fury by violent storms.... The violent force of the sea storms seems necessary to produce earthquakes, in that the rocks smashed by the impact must be soft. But many mountains supposed to have been thrown up by earthquakes are found far from the sea and are often made of hard rock. Proximity to the sea and its violent storms may be necessary, but mountains are often far from the sea.

Rudolphus: This argument seems at first sight to have some weight, but when examined more closely it is seen to be fallacious: the artistic lay-out and the just proportion of size shown by mountains when the earth has settled down into its permanent solidity may be accounted for by a uniform motion in its various parts, just as the waves of the sea move in admirable order under a steady breeze. Nor is it only along the sea coasts (where the rocks are soft) that earthquakes are experienced, they quite often take place in many other places, though it is quite true, as you say, that they occur more frequently where the earth is smitten by the waves of the sea. Such disturbances too can occur more readily in caverns though they are sometimes found in well filled solid places.

Camillus: To tell the plain truth, the more I hear you discussing this subject, the more I become involved in difficulties. However, now that by common consent of all we have decided upon launching forth upon this ocean of perplexities, I beseech you to declare what other causes may be assigned.

Rudolphus: There are many others, but before I mention them I think it worth while, by way of preamble, to lay down certain points. In our sphere (the sublunary, that is) the elements never occur in a pure or simple form. All four elements are present in all creatures endowed with sensory and vegetative powers and all things devoid of life or soul are put together out of the four elements. It must be admitted that no element can be found in our sphere in its purity or simplicity, but that all the elements are always mixed together there. And the Divine author of all things has so constituted the order of things that no animal can exist in any one of the elements if this be pure and free from admixture of others. And so no animal can exist for any length of time on Mt. Olympus on account of the purity of the air, and anyone desiring to reach the summit can only do so by placing a sponge soaked with water before his mouth, so that the air of the mountain top will be mixed with *water* before he breathes it. And no animal can be produced in the center of the earth nor live there. The clear water of the ocean has an admixture of earthy matter through the presence of salt in it, the mere fact that it precipitates salt makes it quite clear that sea water is mixed with earth particles.

Camillus: Quite true, but I should like to know what bearing this has on the subject in question.

Rudolphus: Because it follows from this and other considerations that earth which is by nature dry, when mixed with water and other elements may expand and rise up in the form of certain protuberances which, after they have hardened, take the shape of mountains, but this may be a matter of many years owing to the amount of earth which is affected. The earth may thus rise in some places and sink in others, where the virtue has gone out of it, while elsewhere plains may be developed, while mountains sink as we shall see later.

Camillus: What you say may be quite true, but I cannot see that anything can be deduced from it, because everywhere in the earth the elements are commingled, why therefore do not mountains rise up in all places?

Rudolphus: Minerals are made up of the four elements. But it seems that a certain "mineral virtue" also plays an essential part in the composition of mountains. Fire, air, earth and water must furthermore be present in those proportions required to form minerals.

Camillus: When I think of the beginnings of things it comes to my mind to question whether air, water, earth and fire each contribute something to the building up of mountains.

Rudolphus: Earth cannot be the "efficient cause" of mountains. It is no doubt the "material cause" because it undoubtedly forms in large part the material of which the mountain is composed. But we are not inquiring into the material, formal or final causes which, indeed, lie on the surface. The question we are dealing with is what is the force which throws up great mountain masses.

Camillus: What part do you think air plays?

Rudolphus: According to some philosophers, it is air that keeps the earth floating in the waters of the sea. Without the air which fills certain concavities over its center it would sink in the much larger element which surrounds it. But just as a ship is only partly submerged by its weight and partly kept floating by its concavity, so the earth is kept afloat by the air at its center.

Camillus: But what has that to do with us?

Rudolphus: Listen: The air, through this disposition of nature, having rushed in and formed the cavities, the earth at once bulged out, and so mountains were made.

Camillus: I am not at all concerned about the problem of the earth's floating, but I have good reasons to support me in quite refusing to believe that air has the power of thrusting out masses of earth. It is an element which readily yields to resistance and can easily be driven this way or that by a heavenly body or by any other impelling force. Since then its nature is such, who shall say that it is air which piles up mountain masses, consisting as they do of the heaviest material, though it is true that Diogenes of Apollonia did not hesitate to maintain that this same air partook of divinity? However, if it does produce some effect, it cannot be of itself alone but only in so far as it is reinforced by other elements. For when it is hot and moist it may steal into the bodies and dilate them. But let us pass from air. I should be glad to hear what effect water has.

Rudolphus: We will reserve our treatment of the power of water until later in order to avoid unnecessary repetition.

Camillus: Well, then, what about fire?

Rudolphus: Those who think that it is the element of fire in them, which confers their lasting quality upon stones must necessarily believe at the same time that the formation of mountains is due to the same cause. For nearly all mountains are crowded with stones. Nay, if one looks more closely, most of them are made of heaps of the same material as stones. However, these fire devotees, to prevent any suspicion that they spoke without reason, have established their view as follows: Fire, they say, as we read in Pliny and Isidore is hidden in all things. It lurks in stones and iron. Pieces of wood if rubbed together for a long time give out flame. Concave mirrors if placed in face of the sun's rays set fire to tow and like materials exposed to them. Fire issues out of a stone on Mt. Nimphaeus. In the lands of Aricia if a coal falls on the ground the earth instantly takes fire. In the Sabini and Sidicini territory there is a stone which gives out fire if it be anointed. In the town of Selentium there is a certain rock to which if a piece of wood be attached there follows an instant kindling of fire. Even lakes and human bodies have been known to have broken into flame. We are told that the whole of Lake Trasimine, for instance, took fire. Lucius Martius in Spain while haranguing the soldiers after the fall of the Scipios in battle and spurring them on to vengeance, went ablaze before their eyes, as related by Valerius Antias. Sailors assure us too that after storms at sea the masts of vessels are all ablaze (St. Elmo's fire). Since then fire is found in all things, it follows if we choose to believe certain philosophers that it is owing to the virtue of fire that mountains are raised. Everybody knows of Aetna, Chimaera in Lycia, the Apennines not above a mile from Florence, Vesuvius in Campahla, a marvel which I would that our countryman Pliny had never coveted the sight of. He got too near in his curiosity and met his death. ... But there have been some who give a different account of the process. For according to the testimony of these the fire once shut up within the moist earth makes it boil up (in mountain-bubbles) just as water boils, especially if its kettle be well-caulked when a fire is set under it.

Camillus: A pretty analogy, but I do not yet see how mountains can be made that way.

Rudolphus: I will explain. If by effect of some external agent the portions of the boiling earth should harden they stay there as mountains. But when it happens that the fire gives out before they have settled down in due rigidity they tumble down instantaneously and come to view only as mountain precipices and screes with fire often issuing out of them.

Camillus: Does the potency of fire pile up mountains without any distinction as regards time?

Rudolphus: Far from it. It happens especially in the period when fire deluges are stirred to life. For certain[21] philosophers mark a date for universal conflagration, and maintain that all things on earth will be consumed with fire on the day when all the constellations meet in the sign of the Crab.

Camillus: Those who assert this are, in my opinion, quite wrong, for all mountains are not composed of rock, some consist of softer materials. Furthermore all rocks are not hardened by fire or by heat—on the contrary, as pointed out by Aristotle, Pliny and Albertus, rock crystal is water which has been congealed on great mountain heights by the complete abstraction of heat from it. Mountains upon which the snow lies all the year round—and many rocks are generated in intensely cold waters. And it is not easy to explain why mountains are raised up in one part of the earth and not in other parts. It is only the smaller number of them that seem to cherish fire in their bosoms, the loftier ones send forth mighty rivers and retain their snow in all seasons almost without a break.

Rudolphus: I think there is good sense in your arguments. But I would not deny that some small mounds may occasionally be raised by the force of fire. Posidonius says that as a certain island was rising out of the Aegean, the sea foamed for a while and smoke swept up to the sky with fire issuing from its rifts. Then rocks rolled up to view and crags appeared, some of them eaten hollow by fire and crushed, and last of all the peak of a burning mountain came into sight, which grew and grew until it reached the size of an island. I will say nothing of the similar work of nature which came to light near Naples.

Camillus: I have heard the story and believe it, and so I think that some small hills are formed in that way, but so much for fire.

Rudolphus: Aristotle and Hippias tell us that Thales of Miletus believed that all things had souls (anima) as he inferred from the magnet and amber. Orpheus agreed with them for he suspected that the earth could not possibly fail to possess a spirit of life, seeing that the plants and animals which are born from it are endowed with it. It could not communicate to other things what it did not first have itself. Now this spirit or soul of earth when it streams along within, without let or hindrance, gives peace to its element. But if it should encounter opposition it is violently agitated and strives with all its force to push the obstruction on, and so at length it lifts up some portions of earth, just as in the bodies of animals if the spirit be untowardly affected, we see eruptions breaking out. . . . But in spite of all that the hypothesis is a mere delusion. There is

[21] All things began with fire and will end in it.—*The Stoics.*

no sign of soul in mountains.   Certainly not of the intellectual soul, of sympathy, imagination, thought or memory, nor of the sensitive soul, they do not hear, see, taste or smell; and finally, not even the vegetative or reproductive soul.   The mud that may swell their bulk is not really assimilated but only mechanically laid upon them.   Neither is it generated by them, though it is true that it does adapt itself to their texture, being strong or soft when they are so.   That the earth does not beget mountains, that they are no more children than they are parents, is plain from the fact that they do not resemble their supposed parent.

Camillus: Pliny says that the Babylonians considered that earthquakes, rifts in the earth and other similar phenomena, were produced by the stars, so that the elevation of mountains might originate from the same source.

Rudolphus: The celestial bodies dominate inferior things and set them in motion, except those things which depend on the human will.   Philosophers have discovered by many observations that all the stars without exception exercise great power.   I think it very much to the point to show in particular how great is the power of the moon, for this star is nearer us than the others.   The moon, as we learn from Pliny, saturates the earth with her rays.   Her approach plumps things up, her departure shrinks them.   As she waxes shell fish grow.   Her influence is most felt by bloodless creatures.   But there are certain men, too, and birds and quadrupeds, such as wolves and dogs, whose brain is sometimes affected by her with manifold whimsies.   Since she rules over moist things it is only reasonable that she should preside over the element of water and set it in motion.   (He then goes on to show the moon's influence on fishes and on the raising and lowering of the level of the sea.)

Camillus: But what has all this to do with us?

Rudolphus: The action of the heavenly bodies and of the moon above all, causes the disappearance of sea and rivers in many places and floods in others.   Egypt in the days of Philip of Macedon was completely buried under the water of the Nile for a long time, as Aristotle tells.   On the other hand we are credibly informed that the Caspian Sea once went dry so that navigation ceased upon it.   The famous migration of the Cimbri and Tentines was forced upon them by devastating encroachments of the ocean on their lands.   One of the writers who maintain that the sea shifts its bed, was once digging a well.   He came to the clay, went on digging and found a ship's rudder embedded in it.   And in Venice a workman digging a cistern came upon the parts of a skiff along with the small chain by which, as was usual, it was attached.   The Senators of Venice, to ensure a steady flow of the waters of the Adriatic, which are the very life of their splendid city, spare no expense or labour in constantly dredging away the silt. . . . Then there are the great historic floods, Noah's for one, when on account of the sins of men, the most high God swept away all things and made an expanse of sea and sky, the proof of which is seen in the rocks in which we find shell-fish and oysters in remote mountains.

Camillus: What bearing has all this on our subject?

Rudolphus: I wish you to realize the truth of the doctrine bequeathed to us by Aristotle that the world is subject to vast changes, and that in consequence the day shall come when there will be water where we now see earth; when

the moist shall be drained dry and the dry become moist. Such changes may
come about through the familiar agency of the moon; but more readily perhaps
in times of the mightiest transformations, through the conjunction of the moist
planets in the rainy signs of the Zodiac; or, as some will have it, in consequence
of the revolution of the sun's orbit and above all of the conjunction of all the
planets, and the revolution of the fixed stars. Others have thought that the
assemblage of a swarm of constellations in the station of Capricorn is what
brings about such inundations. . . . Parts of the earth are soft, others are hard,
the soft are carried away by the wash of the water and by the winds, the firmer
parts stay in their place and hardening in the course of time arise in the shape
of mountains. . . . But to return to what I was saying about the excavations
effected by force of water, there are in the bowels of the earth tremendous caves
extending back to an immense distance, abysmal clefts overhung by mountains
on every side, with tarns too and wide spaces shrouded in midnight darkness
about them, precipitous, unplumbed chasms where whole cities have been
engulfed, plunging down in thunderous ruin. It is quite certain that all this
is mainly the work of water. That the dizzy depth of valleys seen between
mountains is due to the same cause, is plain from the waters that race down
through them. Anyone who takes the journey from Verona to Trent will
conclude that the great plain has been fashioned by the impact of racing waters.
I have often speculated upon this splendid work of nature in the district of
Planurae by the banks of the Tellio. This is a tiny streamlet that purls on its
cunning way hugging the roots of the hills in the bottom of a deep gorge, the
last small relic of a mighty flood of waters which once filled the valley plain, as
if artfully hinting to the spectator by its significant course: "My waters were
the hand that raised these hills!"

Camillus: Do recall, please, what you originally decided to expound. All this seems
to me to be a digression to the element of water from our stars.

Rudolphus: Well, to keep within the bounds and not roam too widely from our
subject, efficient causes are a genus including two distinct species, first, remoter
or general causes, second, the proximate and particular. The celestial bodies
rank among the former; their influence is mediated by the waters set in motion
by them, which are the proximate and particular cause of the rise of mountains.

Camillus: For my part I agree. But there are two difficulties that trouble me.
First, the earth occupies a more elevated position than the sea. Secondly, I
hear that the earth which is in its own proper place, may change into some
other place, which if true, seems to present difficulties. The ocean never entirely changes its position, but the Mediterranean Sea and the springs, rivers
and marshes in its neighborhood will be transferred to other positions unless
time fails. I infer from Aristotle and your own words that the operation of the
winds has some effect.

Rudolphus: Wind has marvellous power indeed. Its breath can set in motion and
shatter all things that stand fast on the earth's surface. It has an effect on fire,
sometimes blowing it out, sometimes fanning it to fiercer flame, sometimes it is
kindled into a blaze by its own agitation, as happens in the thunder cloud,
sometimes it enters the pores of the earth and kindles fire by friction in certain
materials, such as sulphur; on a level surface like the sea it lifts the water on

high and with its warring blasts makes billows mountains high, causing tidal waves that drown the land. It sucks down ships, tears trees from their roots, overthrows houses and makes hay of all that meets its path. So mighty is the force of the wind, says Isidore, that it not only tears up stones and trees, but convulses earth and sky and stirs the seas to their depths. . . . It may readily be believed, then, that winds sometimes raise up new mountains and sometimes lay old ones flat with the plains.

Camillus: Cambyses of Persia once despatched an army against the tyrant Amon, which was surrounded and buried by sand set in motion by the winds. It would be no wonder then if some mounds or hillocks should be raised by their breath. We cannot however think it could happen in the case of loftier mountains.

Rudolphus: If you have the patience to listen, I will expound to you the views of a very great philosopher upon the sun and vapours.

Camillus: Don't think you can possibly bore me. Go ahead.

Rudolphus: The sun by its light and motion attracts vapours to itself from earth and water. When water is heated it smokes, so the earth is said to evaporate when its particles rise into the air. But if considerable moisture forms beneath the earth and is powerfully drawn upwards by the sun's attraction, then being forced to find its way out, it separates the particles of earth unless they can hold their own against it, and so hollow apertures are formed and rents, as constantly happens in marshy places at the time of the highest temperatures. But if the vapour shut in be very great in quantity, and if it cannot break the resistance of the earth, then it lifts it up and forms a mountain. Therefore, since there is an abundance of this imprisoned vapour in the South, there are more mountains to be seen there than in the North. But that moisture and vapours are enclosed within the bowels of the earth is quite plain, even from the view of Democritus, who used to say that the earth was hollow and contained inside of it the sources of all waters and great stretches of seas and rivers. It is no objection to this ascent of the earth to say that it is, of course, a heavy body which by its own nature, sinks downwards: for just as water which is also heavy, mounts as high as the peaks of mountains because the hot vapours shut up in it are drawn upwards by the rays of the sun thrust on it, in the same way the earth, forced up by vapours, mounts in spite of its nature, and so mountains are formed.

Camillus: This is a plausible view, as easy to refute as it is to support. For I don't see why, as vapours can be raised by the sun's power every day, mountains should not rise every day as well. But to leave nothing at all out of our discussion, I think it worth inquiry, my learned friend, since so far all you have told us has come from the workshop of the philosophers, whether or not mountains were created by the Supreme Architect at the the time when He created the Heavens and the elements.

Rudolphus: I congratulate you, my scientific friend, for naming God as the Supreme Architect! For I had a suspicion that you shared the error of certain philosophers who deny the divine origination of the world, and spurn the name of creation.[22]

[22] N.B.: A very interesting sop to Cerberus, the inquisition here.

Camillus: I confess that I have some philosophers' books[23] in which is expressed an insane belittling of the idea of creation.   But I prefer to follow the lead of Plato, Trismegistus, Orpheus and Hesiod.   Nay, even if none of these proud pundits made any mention of creation, I bow to the divine view which we have before us in the Sacred Scriptures.   It would be impiety to oppose the truths made known to us by our reading of those divine communications.   But that has nothing to do with our subject.   We are discussing not creation but the origin of mountains.   I maintain that all things were made by the Most High; the one thing I desire to know is whether the mountains and the earth are of the same age or not.

Rudolphus: Some think that when the Supreme Architect created the world, it left His hands a smoothly rounded sphere, without mountains or valleys.   These appeared in consequence of several distinct inundations at different times, so that before the first flood there were no mountains.   My own view differs only in one point from this.   I agree that the earth was round in the beginning.   Nay, I believe it is round now.   But the idea that no mountains had come before the deluge is, as I will show later, repugnant both to reason and to inspiration.

Camillus: Right learnedly and eloquently, my friend, have you discussed the origin of mountains.   But to make a frank confession, I cannot distinguish which of these views you have expounded has the best claim upon my final acceptance.

Rudolphus: I will tell you what I think.   Some mountains arise directly at the command of God.   Others through earthquakes, deluges, the stars (especially the sun), by the force of winds, all of these, with the Mineral Virtue as a contributory factor.   Some are heaped up by human hands.   The Testaceum at Rome has been formed by the piling up of fragments of earthenware which have been brought together here.   Some of them were mortuary urns, others contained the tribute brought from various cities and provinces.   As to the mountains erected by the sacrilegious hands of the Giants, I do not mention them because all that is mere fable and quite unworthy of being mixed up with philosophic problems.

Camillus: There is one point which is still unexplained and that is, why has no mortal ever seen a mountain rise?

Rudolphus: How could any man see the birth of these which God himself made when mankind had not yet appeared upon the earth?   Some mountains, however, as a matter of fact, have been seen to rise up and by many people, as for instance, that on the island of Corinth, to which I have already referred.   Others have undoubtedly risen which have not been seen by man, human life is short, as is also for that matter, the existence of nations.   Peoples and nations furthermore migrate often from place to place and leave vacant great stretches of country without any inhabitants to see such things.

Camillus: I must confess, however, that I do not understand why some mountains are high and others are low.   Why, furthermore, one solitary mountain is found in one place while in other places there are many mountains.

[23] Even that needs apology!

Rudolphus: The same is true of plants and animals, they differ in size owing to the abundance or deficiency of the food by which they grow or for certain other reasons.

Camillus: It would complete your excellent and charming exposition if you would tell us how plains and valleys have been made.

Rudolphus: Valleys are merely hollow spaces between mountains or hills; higher cliffs are formed when parts of the earth on either side of the valleys are elevated or when softer portions of the rock are washed away by waters. The most beautiful plains are made where by the descending waters, the surface of the earth is dried out equally and all the higher parts are carried away.

Camillus: This is sufficient.

Rudolphus: I have I think taken up all those points which there is time to touch upon.

Camillus: You have treated the whole subject with incredible and quite laconic brevity. Short discourses, if they contain substance are by far more satisfactory than long ones and, if they do not, they are less tiresome.

THE PERIOD OF CLOSE STUDY AND OBSERVATION IN THE FIELD BETWEEN 1600 AND 1830

Having reviewed at some length the speculative opinions put forward to account for the origin of mountains, down to the close of the sixteenth century, we now pass to the consideration of the actual study of mountains in the field, undertaken in the subsequent centuries by a succession of able and diligent observers.

As we have seen, the record of past centuries was not one of speculation only, there were some few writers such as Avicenna, Ristoro d'Arezzo, Leonardo da Vinci, and especially Agricola, whose opinions were based in part at least on observation in the field. These, however, were outstanding exceptions, lights in a barren desert of speculation in which the tree of knowledge was not to be found. Their very appearance is so unexpected that it takes us by surprise. But in the period which we are now to enter, speculation is for the most part set aside, and it is realized that by observation only can a true understanding of this difficult subject be attained.

Steno—the origin of mountains by faulting and erosion

The first outstanding figure in this new era is Nicolaus Steno, who, leaving the domain of speculation and conjecture, takes a long step forward toward a true interpretation of the origin of mountains based on a study of the hills and mountains of Tuscany, and who sets forth in his

Prodromus[24] an outline of these investigations and of the conclusions to be drawn from them.

Steno,[25] the son of a goldsmith Steen Pedersen, was born in Copenhagen on January 10th, 1638. He entered the University of Copenhagen in 1656 and took up the study of medicine. Four years later he went to Amsterdam to continue his studies in anatomy under Blaes. While there he discovered the parotid duct still known as the *Ductus Stenonianus*. After staying in Amsterdam a few months he went on to Leyden where he remained until 1664. Here the brilliant young Dane, who seems to have had a genius for friendship, was thrown into the company of a group of men of great distinction in the world of learning, among whom was Spinoza. "But," Winter observes, "great as was the influence of these men, it was perhaps less telling for his spiritual development than the religious tolerance which Holland alone of European countries then afforded."

After a short stay in Paris, Steno in 1665 went to Florence, where he received an appointment at the Court as physician to the Grand Duke Ferdinand II. He was also appointed to a position in the Hospital of Santa Maria Novella. This was the happiest and most productive period of Steno's life. He was enrolled as a member of the celebrated Accademia del Cimento in Florence, a learned society which embraced among its members some of the most distinguished scholars of the time. One of these, Redi the poet and naturalist, in a letter to Athanasius Kircher written at this time says, "I have the honor to serve at a court where distinguished men gather from all parts of the world. In their wanderings they bring and seek in exchange the fruits of high endeavor, and so warm is their welcome that they fancy themselves transported to the mythical gardens of the Odyssey." It was at this time (1668) that Steno carried out his geological work and wrote the *Prodromus*.

[24] *Nicolai Stenonis de Solido intra Solidum naturaliter contento dissertationis prodromus*, Florence, 1669.

A facsimile edition of the original was published by Edw. W. Junk of Berlin in 1904.

An English translation with an introduction and explanatory notes by John Garrett Winter and a foreword by William H. Hobbs appeared in 1916 as one of the University of Michigan Studies (The Macmillan Co., New York).

The facts concerning Steno's life have been taken from Winter's admirable introduction to this translation, in which references to the original documents are given.

[25] Niels Steensen, the Danish form of the name, in accordance with the learned custom of his day, was Latinized by the bearer as Nicolaus Stenonis. The spelling in French is Stenon and in Italian Stenone. Following the usual custom, Steno took his surname from his father's Christian name.

Through the influence of Maria Flavio del Nero, a nun in charge of the pharmacy connected with Santa Maria Novella and certain other persons, Steno was now induced to join the Roman Catholic Church. From this time his interest in science rapidly waned. Later he took Orders, was appointed a bishop *in partibus*, wrote a few short papers on anatomy and a number of theological treatises, followed an extremely ascetic life, and lived in poverty to which he succumbed on November 26th, 1686 in Germany. His remains were brought to Florence and laid to rest in the Church of San Lorenzo.

Steno's life is a mirror of the age in which he lived; the time of the Renaissance in which literature had come to a new birth and when science, turning anew from the paths of barren speculation to the royal roads of observation and experiment, was entering upon that career through which it eventually was to change the whole aspect of society. But it was a time also of intense religious controversy, intolerance and persecution, one in which the student of nature, should his conclusions not conform to the teachings of the church, pursued his work with his life in his hand. It is evident that Steno in certain of his conclusions was influenced by ecclesiastical authority. This is especially the case in dealing with considerations of geological time. He considers it to be certain that the sea shells found in the rocks of Tuscany were brought into their present position by the Noachian deluge "Four thousand years, more or less, before our time," and by the same consideration of time was led to the still more fantastic conclusion that the remains of elephants and other great bones dug up from the fields of Arezzo, and which he says "Do not belong to animals of this climate," were the remains of elephants and pack animals of Hannibal's army.

Steno's writings which have come down to us, consist of 24 papers on anatomy, 14 on theological questions and a single treatise on Geology (the *Prodromus*) which, together with two letters written to Cosmo III in which he describes certain grottos that he had visited, embodies all his geological work.

The *Prodromus* is addressed to the "Most Serene Grand Duke" Ferdinand II of Tuscany and Steno commences by telling him—what every student of nature knows so well—that one who undertakes a scientific investigation is like a traveller into an unknown land following a path which promises to lead him speedily to the desired goal but that, owing to the fact that unseen obstacles are continuously encountered, this goal seems ever to be receding and more and more difficult of attainment.

That he will "Never be able to form in advance a due estimate of the time which will be necessary for loosing the knotted chain of difficulties which by coming forth one by one from concealment, delay them that are hastening toward the end." In fact, to use the simile of Democritus, such a research was like undertaking to bail out a well, since no one could tell how long or how difficult this would be as no one could determine in advance the number and volume of the hidden springs which must be encountered during the work. But having made certain progress in his research and not knowing when, if ever, he might be able to complete it, he would do like the debtors, "When they have not the means to pay in full, they pay what they have." In this *Prodromus* then he presents an outline of the research which he had undertaken and a summary of the results which he had already obtained. He speaks of the extended work which he had hoped to present to the Duke as the *Dissertation*. This was never completed.

The title of the work, *Concerning a solid body enclosed by the process of nature within a solid*, seems at first sight singularly inappropriate for a geological treatise, but when the contents of the book are examined its true significance is seen. The *Dissertation* was to be divided into four parts. Steno's attention apparently was first directed to the subject by the presence of shells and other "sea objects" in the rocky strata in the hills about Florence. The consideration of these was to be taken up in part one of the *Dissertation*. These, it will be noted, are solids enclosed in the solid crust of the earth. In the second part, the "Universal Problem" was to be considered, "Given a substance possessed of a certain figure and produced according to the laws of nature, to find in the substance itself evidences disclosing the manner of its production." The third part was devoted to the investigation of different solids contained within a solid, in view of the laws discovered and set forth in part two. The fourth division was to treat of the subject of the Universal Deluge.

It will thus be seen that the *Prodromus* deals with a variety of geological questions. Further reference will here be made only to Steno's views concerning the origin of mountains. Under the category of solids contained within solids, Steno includes not only agates, all crystals of minerals enclosed in rocks and all fossils, but also the strata which make up the earth's crust in the district about Florence, for they lie on some lower strata and are overlain by others which are higher. From a study of these he draws his conclusions concerning the origin of mountains.

The unevenness of the surface of the earth manifested in mountains, valleys, high level plains and lower plains, as well as elevated bodies of water, are due to movements causing a change in the position of strata. He remarks:

That alteration in the position of strata is the chief cause of mountain formation is clear from the fact that in any given range of mountains there may be seen:
1. Large level spaces on the summits of some mountains.
2. Many strata parallel to the horizon.
3. Various strata on the sides of mountains inclined at different angles to the horizon.
4. Broken strata on the opposite sides of hills, showing absolute agreement in form and material.
5. Exposed edges of strata.
6. Fragments of broken strata at the foot of the same range, partly piled into hills, and partly scattered over the adjoining country.
7. Either in the rock of the mountains themselves or in their neighborhood, very clear traces of subterranean fire.
8. Mountains can also be formed in other ways, as by the eruption of fires which belch forth ashes and stones together with sulphur and bitumen; also by the violence of rains and torrents, whereby the stony strata, which have already become rent apart by the alternations of heat and cold, are tumbled head-long, while the earthy strata, forming cracks under great blasts of heat, are broken up into various parts.

Hence, he continues, it could easily be shown:

1. That all present mountains did not exist from the beginning of things.
2. That there is no growing of mountains.
3. That the rocks or mountains have nothing in common with the bones of animals except a certain resemblance in hardness, since they agree in neither matter nor manner of production, nor in composition, nor in function, if one may be permitted to affirm aught about a subject otherwise so little known as are the functions of things.
4. That the extension of crests of mountains, or chains as some prefer to call them, along the lines of certain definite zones of the earth, accords with neither reason nor experience.

The strata forming the earth's crust may undergo a change of position giving rise to mountains in two ways. First, a violent upthrusting due to the sudden escape of burning gases from the earth's interior, or of air forced out from the earth's interior by the collapse of air-filled cavities within the earth, or, secondly, by the spontaneous slipping or downfall of the upper strata after fissures have developed in them in consequence of the withdrawal of the underlying substance, or foundation. Hence by reason of the diversity of the cavities and cracks the broken strata assume different positions; while some remain parallel to the horizon, others become perpendicular to it, many form oblique angles to it, and not a few are twisted into curves because their substance is "tenacious." This change can take place either in all the strata overlying a cavity, or in certain strata only, the upper strata for instance being left unbroken. This second process is illustrated by a plate showing six diagrammatic sections (nos. 20 to 25) drawn from occurrences in the mountainous region of Tuscany. These latter are reproduced in figure 61. They illustrate Steno's views concerning the manner in which an uneven and accentuated surface may be developed by the forces of nature acting upon a plain underlain by a series of horizontal strata, while at the same

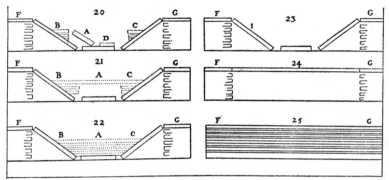

FIG. 61. From Steno: *Prodromus*, representing six successive stages in the geologica history of Tuscany.

time representing six successive stages in the geological history of Tuscany itself. These, while diagrammatic in character, probably represent the earliest geological sections ever prepared.

Steno's explanation of these sections is as follows:

The six figures, while they show in what way we infer the six distinct aspects of Tuscany from its present appearance, at the same time serve for the readier comprehension of what we have said about the earth's strata. The dotted lines represent the sandy strata of the earth, so called from the predominant element, although various strata of clay and rock are mixed with them; the rest of the lines represent strata of rock, likewise named from the predominant element, although other strata of a softer substance are sometimes found among them. In the Dissertation itself I have explained the letters on the figures in the order in which the figures follow one another; here I shall briefly review the order of change.

Figure 25 shows a vertical section of Tuscany when the rocky strata were still whole and parallel to the horizon.

Figure 24 shows the huge cavities eaten out by the force of fires or waters while the upper strata remained unbroken.

Figure 23 shows the mountains and valleys caused by the breaking of the upper strata.

Figure 22 shows new strata, made by the sea, deposited in the valleys.

Figure 21 shows a portion of the lower strata of these new beds destroyed, while the upper strata remain unbroken.

Figure 20 shows the hills and valleys produced by the breaking down of the upper sandy strata.

When Steno says that "There is no growing of mountains" he is referring to a belief prevalent in his time and long after, that isolated rocks, hills and even mountains grew up out of the surface of the earth as trees do.

Marie-Pompée Colonne,[26] a distinguished and erudite Italian gentlemen, writing only 65 years after the appearance of Steno's work, states that mountains grow like plants—"Je suis porté à croire que les montagnes végétent et qu'une partie de celles que nous voyons à présent ont végéte sur la terre comme les arbres." This may seem remarkable, he goes on to say, but we know that coral grows in the sea. The mountain grows up like a tree but when the growth ceases in any part, the surface of the mountain breaks up into little fragments and the mountain thus suffers "corruption" and eventually disappears. This growth takes place very slowly: he thinks that a hill might not grow more than "cent pas" in a thousand years, one cannot in fact notice that it is growing when observing it day by day any more, to use his own expression, than you can see your hair grow. The heat which contributes to this growth is that of the interior of the earth.

While it is now known that mountains do not "vegetate" as Colonne supposed, that mountain ranges do grow was one of the outstanding facts to be established by later investigation.

When Steno says that mountain chains do not run in certain definite zones across the earth's surface, he had in mind the statement made by Kircher, four years previously, in his *Mundus Subterraneus* to the effect that there are a series of mountain chains running from east to west across the earth's surface crossed by one very high chain which runs from north to south.

It will thus be seen that Steno had recognized three types of moun-

[26] Monsieur Colonne, Gentilhomme Romaiin, *Histoire Naturelle de l'Univers*, Paris, 1734, Vol. I, Pt. II, p. 204.

tains: (1) Block or Fault Mountains, (2) Volcanic Mountains, (3) Mountains of Erosion.   The two latter types had already been mentioned by Agricola and other of the older writers, but Steno's field studies had shown him clearly that many of the Tuscan mountains belonged to a new type which owed their origin to faulting, and his observations that the downward sloping strata on the mountain sides were in many cases "Twisted into curves because their substance is tenacious" led him to the brink of the discovery that many of these mountains owed their origin

Fig. 62. Illustrating Kircher's opinion that a series of mountain chains runs east and west across the earth's surface, crossed by a high chain which runs north and south. (From his *Mundus Subterraneus.*)

to the folding of the strata of which they were composed, and constituted still another class, namely, Folded Mountains.

Steno's defection was a distinct loss to science.   Had he continued his researches, his accuracy of observation and scientific insight would probably have led him to make further very important contributions to geology.   As it was, his *Prodromus* remained almost unnoticed at the time and it was not, Zittel remarks, until Élie de Beaumont and Alexander von Humboldt drew attention to it that the importance of his work received due recognition.

Lazzaro Moro. Mountains are due to the fire at the center of the earth

Passing over the wild hypotheses and conjectures of the "Cosmogonists," the next writer whose contribution to the question of the Origin of Mountains should be considered is the Abbé Anton Lazzaro Moro of Venice[27] (1687–1740).

As has been seen, from very early times the uprush of fiery gases from the earth's interior had been regarded by most writers as one of the causes of mountain making. The ultimate source of these was generally believed to be the "Fire at the Center of the Earth." It was in volcanic regions that the belief in the existence of such a fire was deeply stamped into the minds of men. This fire Moro regarded as the primary cause to which all mountains owed their origin. He may be considered as the leading Plutonist of his time. Although in the dedication to Giovanni Emo Procurator of San Marco, Moro says that his book contains not a little that is new, he as a matter of fact brings forward little or nothing in the way of new fact, but has built up his discussion of the subject on facts garnered by others. Notwithstanding this, however, his method of dealing with the question is very interesting and throws an engaging light upon the mental attitude of many men of the time.

Moro's line of argument is as follows: In 1707 a submarine eruption took place in the Aegean Sea and the island of Santorin, one of the Cyclades, was thrown up. It is one of a circle of volcanic islands which form part of the crater of a partly submerged volcano. Moro says it was "Thrust out of the sea mountain high." In 1538 on a tract of level land in the Phlaegrean Fields near the town of Puzzuoli, a volcanic eruption broke out and in a single night Monte Nuovo was brought to birth. This is a conical hill 440 feet high and having a circumference of about a mile and a half. At the summit is a very well-defined crater, the bottom of which is 420 feet below the crater rim and it is now covered with luxuriant vegetation. The crater floor when seen by the writer some few years ago was being cultivated as a cabbage garden. This little volcano has attracted a great deal of attention because it came into existence suddenly in the middle of a well settled district and many persons were present at its birth. One account which has come down to us from an

[27] *De Crostacei e degli altri marini Corpi che si truovano su Monti*, Venice, 1740.
    There is a German translation of this work by Erhardt, bearing the title *Neue Untersuchung der Veränderungen des Erdbodens nach Anteitung der Spuren von Meerthieren und Meergewachsen, die auf Bergen und in trockener Erde gefunden werden, aufgestellt von Anton Lazzaro Moro*, Leipzig, 1751. A second edition was published in 1775.

eye witness states that after some earthquakes had taken place, the flat surface of the earth swelled up like a gigantic bubble and then burst at the top, forming an orifice from which were discharged great clouds of stones and ashes.   It was difficult to see just what took place on account of the fact that the clouds of ashes blown out of the volcano obscured the view and the observers also were in a somewhat excited frame of mind. An examination of the surrounding plain and of the hill itself, however, shows that in all probability the ground did not swell up but that a fissure was formed in the plain and from this immense volumes of steam and gases were emitted which brought with them stones, lapilli and ashes which, falling back about the vent, built up a cone composed of successive layers of this ejected material, which formed the hill as it now appears.

These two eruptions made a deep impression on Moro's mind, and searching the literature of the subject, he found descriptions of many similar occurrences, recorded by Pliny, Strabo and various other authors, some of which were true and some were not, and on these he bases his exposition of the theory of the development not only of volcanoes but of the earth's surface as a whole.   He gathered from the ancient records, accounts of the origin of some sixteen or more islands scattered up and down the coasts of Greece and Italy, which had been thrust up by the fiery forces within the earth.   Among these was Delos, the smallest of the Cyclades (which as a matter of fact is composed of gneiss and granite and not of volcanic materials), whose name refers to the tradition that it rose suddenly out of the sea and of which legend records that it at first floated about upon the surface of the waters till Zeus fastened it to the bottom so that it might be a safe refuge for Leto, who there gave birth to Diana and Apollo.   But islands rise abruptly from the surface of the sea in all the oceans of the world with which we are acquainted, many of them forming veritable mountains, and Moro thought that there was every reason to believe that all these had originated in the same manner as Santorin and Delos.

Moro goes on to say that all sound philosophers hold that the only way to reach correct results in the study of nature is to observe her operations with great care and then to base their conclusions only upon facts which have been actually observed.   The laws and rules by which she works however are so uniform and so invariable that the way on which she has once passed she always follows again.   Sir Isaac Newton found that "In Nature like causes give rise to like effects" ("Effectuum naturalium eiusdem generis easdem sunt causae"). And to this may be added another proposition accepted by all the wisest and most trustworthy philosophers, namely, that Nature takes the shortest, simplest and most direct way in all she does. The following rule, Moro says, may be laid down, namely, that Nature in all her operations always does the same thing in the same way. So then, following the established principles of sound philosophy, it is safe to conclude that since some islands are of volcanic origin, all islands have been thrust up from the ocean waters by the action of the fiery forces of the earth's interior. No distinction, furthermore, can be made between small and large islands. The same fiery forces are at work everywhere beneath the earth's crust for, as has been stated, islands are found rising from the surface of all the seven seas. It follows therefore that great islands such as Sicily, Great Britain, Madagascar, Java, Borneo, with the mountain ranges upon their surfaces have been thus upheaved. Size makes no difference. When gunpowder was first invented people were astonished to see a ball weighing two or three pounds shot high into the air by the explosion of a small amount of the powder placed in a cannon, but they would have been convulsed with derisive laughter had anyone told them that but a few years would elapse before great houses, yes even whole cities would be thrown down and destroyed by this same agent. And so Nature by her explosive forces has unlimited power at her command for the upheaval of the earth's crust. As a matter of fact, the continents are nothing but large islands in their universal ocean. It follows therefore that they also have been raised up from the ocean bed by the same forces.

But if the small islands, which are nothing but the tops of mountains, have been raised up by these subterranean forces, and the larger islands of the earth have had this same origin, why are not mountains which rise from the level surface of the dry land produced in the same way, for the sea bottoms are nothing but *plains covered by water?* They must have had this origin according to the rules of sound philosophy, as set forth above. In fact all mountains have had one and the same origin, all have been raised out of the bosom of the earth in one and the same manner, and indeed when examined closely they in almost every case show evidence of a fiery origin.

But Moro then goes much further and contends that all the stratified rocks of the earth's crust whether they be those laid down about volcanic vents or those underlying the level lands far removed from volcanoes, or indeed those composing great mountain ranges such as the Alps, no matter what their composition may be, consist of material thrown out from volcanic vents. Neither in form nor in composition, he says, do any of these strata differ from those which are found about volcanoes and which are formed of volcanic lapilli and ashes. The strata found in places far distant from volcanoes, consist, he says, sometimes of sandy materials and sometimes of coarser materials. So also do those found about volcanic vents. Even fragments of limestone are at times thrown out of Vesuvius in great abundance which, by pressure exerted on the fragments softened by the great heat to which they had been subjected, might easily be compacted into beds of solid limestone.

Vallisnieri held that the rounded pebbles in the beds of conglomerate which often occur in these stratified formations in the Alps are evidence that they have been laid down by water, but Moro holds that the rounding and polishing which these pebbles exhibit could equally well have been caused by mutual attrition of the fragments during eruption and that those which show a flattened form have had this impressed upon them by the pressure to which they had been subjected by overlying materials when still in a softened condition due to heat.

Looking now at the *structure* of mountains, Moro says that two kinds of mountains can be recognized. First, those which he designates as mountains of the first class (or *Primary Mountains*) and secondly, those which he calls mountains of the second class (or *Secondary Mountains*). Those of the first class are composed exclusively of rocks which are *unstratified* those of the second class are built up of *successive beds lying one upon the other* and composed now of one material and now of another. A Primary Mountain may in some cases be overlaid by stratified rocks characteristic of Secondary Mountains, this indicating that the forces which made it operated at two different and successive periods. Thus when the fiery forces thrust up an island from the sea, it consists of rocks and rocky parts of the sea bottom often covered with sea shells. This then becomes more or less mantled over with uncompacted earthy material blown out from the earth's interior and which settles down upon the underlying rock in successive layers.

Moro here makes for the first time the distinction between Primary (see Plates X and XI) and Secondary Mountains, a distinction of which Vallisnieri seems to have had some indistinct idea but which he does not actually mention, and one which is again made at a later date by Arduino and Lehmann, to whom the discovery of this important distinction is generally attributed.

PLATE X

PRIMARY MOUNTAINS (WITH MOUNTAIN GOATS)

(From Moro: *De Crostacei e degli Marini Corpi*, Venice, 1740.) Moro took this, with some slight modifications, from Scheuchzer

Moro then passes to the consideration of the origin of the extremely complicated foldings and contortions frequently displayed by the strata composing the mountains of the second class.

He can see no possible explanation of these other than that they are the result of the action of tremendous subterranean forces set in motion by the central fires within the earth.

But here an objection presents itself. He states that perhaps some one will say, if the whole or at any rate the greater part of the earth's surface is composed of materials which have been thrown out from the earth's interior, there must be within the earth great open spaces formerly occupied by these ejected materials, and as no empty spaces can exist ("Nature abhors a vacuum"), the explanation which Moro gives of the origin of mountains cannot be a correct one. An Epicurean would not put forward this objection because in his opinion an empty space may exist. But a Peripatetic or other philosopher who is a sworn enemy of an empty space would glory in such a contention. Moro meets this objection by saying that this question, much debated among philosophers, of the possibility of the existence of an empty space is one over which one might rack one's brains forever. It is a philosophical question concerning which it is impossible to reach a positive conclusion, no matter how much subtlety may be infused into the arguments on either side. Whether, however, a vacuum can or cannot exist within the earth's crust is not a question of any importance in the present connection. The island of Santorin *was* thrown up, Vesuvius and Etna *do* pour out floods of lava which rise from the earth's interior, these facts are indisputable whether these erupted materials leave empty cavities or not. But nevertheless since some people will feel that this apparent anomaly calls for some explanation and will not feel satisfied until one is found, Moro offers two solutions to the difficulty, the first being that the cavities are filled with "burning fiery materials" and should these force their way to the surface through any opening in the crust, their place will be taken by air or other fluid or gaseous materials and the cavities will be filled (see figure 63). The second is that the views of Empedocles may be true, namely, that the earth consists of a thick crust and a hollow interior which is filled up with fire, or rather with fiery fluid, and so no empty space would result from the eruption of an island or a continent, because when one portion of the crust was raised to the surface by the fire, another elsewhere would sink down into the fire beneath and so no cavity would be developed (see figure 64).

St. Augustine is then brought in to settle the controversy with his pertinent question, used in another connection: "Shall we deny that which is clear and evident because we cannot understand that which is hidden and obscure?" Moro supplies the answer—Certainly not—and then presents the conclusion of the whole matter in the following words: "All mountains, islands and level lands have been raised up out of the bosom of the earth into the position which they now occupy by the

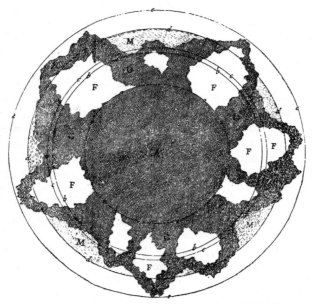

FIG. 63. Cross section of the earth, according to the first of Moro's hypotheses. *A.* Central portion of the earth composed of materials which are fixed and immovable. *G.* Exterior portion of *A* in which the fires are lighted. *H.* Materials torn from *G* and thrown out on to the surface of the earth by the action of fire. *F.* Open spaces left by the extrusion of the "burning fiery materials" and filled with air or gases. *M.* The ocean. (From Moro: *De Crustacei.* Tavola VII. Reduced one-half.)

action of subterranean fires," adding that there is no need to be disturbed by any spectre of open spaces below.

Moro's explanation of the origin of the central fire within the earth is however a distinctly medieval one which to the modern mind leaves much to be desired. It is as follows: "It pleased the great Creator of all things, when the dry land was to appear on the third day according to the sacred account in Genesis, that great subterranean fires should be kindled."

We have seen that Moro impressed upon his readers repeatedly that the way to reach a true explanation of the phenomena of Nature was, first to observe the facts most carefully and then to interpret them in the light of what he calls Newton's Law, namely that "Like causes give rise to like effects." This he endeavored to do, but reached conclusions widely at variance with what is now known to be the truth. The cause of this failure, as we look back upon his work, is seen to be the circum-

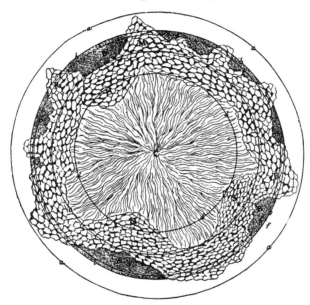

FIG. 64. Cross section of the earth according to the second of Moro's hypotheses (based on the views of Empedocles). *B.* The ocean. *C.* The central portion of the earth which is filled with fire. *M-R-P.* Mountains formed by the ejection of fragments of the earth's crust through the action of the central fire. *H-N-S.* Depressions on the surface developed in consequence of the elevation of *M-R-P.* (From Moro: *Crostacei*, Tavola VIII. Reduced one-half.)

stance that he did not observe correctly. His "facts" were not facts. All islands, for instance, are not alike, they are not all built up of volcanic rocks. The same origin cannot therefore be assigned to the island of Sanatorin and the island of Great Britain. Again, the "sands" which make up the sandstones of the Alpine succession are quite different from the "sand" i.e. volcanic ashes, which give rise to the tufas of volcanic islands. Nor yet again are the contorted strata of mountain ranges in any way analogous to the rope-like forms seen in the lavas poured out of many volcanoes, as he says they are. Furthermore, the opinions of Empedocles and St. Augustin contribute but little to a correct understanding of the subject in question, even though the former was such an enthusiastic student of volcanic phenomena that he lost his life by falling into the crater of Etna.

As has been mentioned, Moro was the first to recognize the distinction between *Primary* and *Secondary* mountains. This was a most important discovery and served as the initial step of a great advance toward a true understanding of the origin of mountain ranges.

It may here be noted that Antonio Vallisnieri, a contemporary of Moro, whose important studies on the origin of springs and rivers are referred to at some length in the chapter dealing with this subject (figure 74) was impressed by the complicated contortions which he had observed in the folded strata of the Alps. Scheuchzer[28] who had studied the contortions in his extended travels in these mountains gave Vallisnieri a number of diagramatic representations of certain sections in the Alps and Vallisnieri reproduces them in one of his books.[29] Elsewhere[30] Vallisnieri describes such contortions and mentions a number of different causes which might be put forward to account for their origin, but thinks it is impossible without further study to decide which, if any, of these is the true cause.

Arduino, Lehmann and Pallas—Mountain ranges are not produced by a single catastrophic act but are due to successive uplifts at different times

The next important step was the discovery that most of the great mountain ranges of the world were composite in structure and consisted of several formations widely differing from each other and widely separated in age. This led to the recognition of the fact that *mountain ranges were not as a general rule brought into existence at one period or by any one great catastrophic act, but that they had been built up by mountain making forces acting successively along the same lines of uplift, during two or more widely separated periods.* This fact made it possible to determine, within certain limits, the age of a mountain range. Its significance could not fail to be generally recognized as soon as the great mountain ranges of Europe were studied in some detail, and three men working in

[28] *Itinera per Helvetiae Alpinas Regiones*, Leyden, 1723.
[29] *Lezioni accademica intorno all' origine delle Fontane*, Venice, 1715.
[30] *De Corpi Marini che sù Monti si Trovano*, Venice, 1721.

three separate parts of Europe (all of whom may have been unacquainted with Moro's work) announced their discovery of it within a few years of one another.

These were Arduino in Italy in 1759, Lehmann in Germany in 1776 and Pallas in Russia in 1777.

But little is known of Giovanni Arduino. He was born at Caprino near Verona in 1714 of poor parents and died in 1795. He studied at Verona and then entered the "Vastissimi e spinose campi della geognosia."[31] Looking for guidance in this new field of work, he read the works of Woodward, Camerario, Whiston and Burnet but gained little from them; he then decided to seek information from Nature herself in mountain and mine. His first work was entitled *Osservazione sulla fisica constituzione delle Alpi Venete* which appeared in 1759. After holding an official position in the government mining service, he became professor of mineralogy and metallurgy in Venice. He also devoted much time and attention to agriculture, palaeontology, and other branches of natural science. Ronconi[32] calls him the "Father of Modern Italian Geology" and claims that he was the first to recognize that "The crystalline rocks were of volcanic origin," and to see the importance of the "application of palaeontology and chemistry to the chronology of the formations." Although Arduino wrote but little he exercised an important influence in his time.[33] In 1759 he sent two letters to Vallisnieri,[34] setting forth his views, reached after long and careful study of the surface characteristics of his own and other lands, concerning the origin of the materials of the mineral kingdom. The contents of these letters are practically identical with that of his memoir entitled *Saggio Fisico-Mineralogico di Lythogomia e Orognosia*.[35] He recognizes four "Divisions of the earth's crust," as follows:

1. *The Primary or Primitive Mountains:* These are composed of rocks frequently containing metallic ores and what are called by "practical" men "Monti Minerali."
2. *The Secondary Mountains:* These are built up of marble and bedded limestones,

[31] *Elogio di Giovanni Arduino*, by Catullo in his Prospetto, 1839, in which he says that the work of Arduino had not been properly appreciated and that he was writing this eulogy to revive his memory.

[32] *Giovanni Arduino e le Minerere della Toscana, Notizia Storica*, Padua, 1865.

[33] A list of papers by Arduino is given by Daubeny in his *Description of Active and Extinct Volcanoes*, p. 459.

[34] *Nuova Raccolta di Opuscoli Filologici &c. del Abb. Cologere*, T. VI, Venice, 1760.

[35] *Atti dell' Accademia delle Scienze di Siena, T. V.* This was reproduced with certain corrections and additions in 1778 under the title *Sammlung einiger Abhandlungen des Herrn Johann Arduino, aus dem Italienischen ubersetzt durch A. C. von F. Dresend*, 1778.

in which mineral deposits are rarely found, but in which "pietrificate organiche produzioni del Mare" are present in abundance.

3. *Thirdly,* certain low mountains and hills which are composed of gravel, sand, clay, marl, &c. which almost always hold a great abundance of marine debris. In this division are included those volcanic rocks which Arduino described in another publication.

4. *A Fourth Division* consists of earthy and rocky materials which repose upon these others. They are alluvial material washed down from the mountains by streams.

He states[36] that the earth as we now see it is not in its primitive state. It has undergone repeated upheavals and subsidences, many "revolutions" and "metamorphoses." These have been brought about by the injection of volcanic rocks, and to this same agency is due the remarkable contortion and twisting displayed by the strata in mountain regions.

The belief that "our mountains" which are composed of lava, tufa, lapilli, pumice and ashes owe their origin to fire is not, he says, a chimerical hypothesis, but rests on a solid basis of fact, even if there is no history of an actual eruption having taken place in the locality where they are found.

Arduino[37] in 1769 in a letter describing the extinct volcanoes near the village of Chiampo in northern Italy says that without the slightest doubt these were once active like Vesuvius or Etna. There is a footnote in the German translation, probably added by the translator, in which it is recorded how strange and unbelievable the contents of this letter were to geologists at the time, adding however, that Guettard had made similar observations with reference to the extinct volcanoes of central France in a paper which appeared in 1752.[38]

Seventeen years after Arduino's work appeared, Lehmann wrote his well-known volume in which the composite structure of mountains is described in much greater detail.

Johann Gottlob Lehmann[39] was a teacher of mineralogy and mining in

---

[36] *Giornale d'Italia,* Venice, T. 1.

[37] Schreiben des Herrn J. Arduino an Herrn Antonio Zanoni, die Wirkungen uralter feuerspeyen Berge betreffend welche von demselben in März 1769 auf denen Bergen bei Chiampo und andern dem Vizentinischen Gebiete nahe gelegenen Gegenden, beobachtet worden sind. In a volume entitled *Sammlung einiger Mineralogisch-Chemisch-Metallurgisch und Oryktographischer Abhandlungen des Herrn Johann Arduino und einiger Freunde desselben, aus dem Italianischen übersezt,* Dresden, 1778.

[38] See Academie Royale de Paris.

[39] Lehmann, *Versuch einer Geschichte von Flötzgebürgen,* Berlin, 1756. A French translation under the title *Essai d'une Histoire naturelle de la Terre,* forms Vol. III of an edition of Lehmann's works in 3 vols. in French published in Paris in the year 1759 under the title *Traités de Physique, d'Histoire naturelle, de Mineralogie et de Metallurgie.*

Berlin, from whence in 1761 he went to St. Petersburg, having been appointed by the Empress Catherine to the position of Professor of Chemistry and Director of the Imperial Museum. He died in 1767 as the result of the explosion of a retort filled with arsenic. His work on *Metallmüttern* is referred to in the chapter on the *Origin of Metals*. In connection with his researches on ore deposits he made extended studies of the mineral bearing regions of northern Europe and thus came to recognize that the earth's crust in which these deposits occurred was not a structureless mass but was built up of layers or "strata" which succeeded one another in definite order. He was the first in Germany to publish a geological section showing the succession of such "strata." His contemporary, Johann Christian Füchsel, a physician to the Prince of Rudolstadt in Thuringia, also observed this fact. Both of them worked out the succession as seen in areas which they examined, and presented the results of their observations in books,[40] the appearance of which, although attracting little attention at the time, really mark an important epoch in the history of geology. The geological succession which they thus established forms the basis of the more extended and complete column of formations and of the terminology connected with them, which was a few years later set forth by Werner, and was by him given such wide publicity that it is now associated with his name, (see Chap. VII).

It is not the intention, however, here to enlarge further upon the very important contributions made by Lehmann and Füchsel to the proper understanding of the stratigraphical succession in Europe (see Chapter VII), but rather to indicate the advance made by Lehmann toward the proper understanding of the origin of mountains.

Lehmann from his studies of the mountain ranges of Northern Europe was led to the belief that there were *three classes* of mountains. Those of the *first class*, which he calls the *Primitive Mountains*, include the oldest mountains. They were formed at the time of the creation of the world and differ in their structure from the other mountains in that they are composed of beds or strata which are thicker and show less variation in mineralogical character. Furthermore, having been deposited from waters which were in a state of violent agitation, the bedding is not even and well defined as in the mountains of the second class, whose constit-

---

[40] Füchsel, *Historia terrae et maris ex historia Thuringiae per montium descriptionen erecta.* Acta, Acad. Elect. moguntinae zu Erfuhrt, Vol. II, pp. 44–209, 1762 and Entwürfe der ältesten Erd und Völkergeschichte, 1773.

See also Keferstein, *Geschichte und Literatur der Geognosie,* Halle, 1840, pp. 54–58.

uent beds were deposited more slowly and from a more tranquil sea. Moreoever these beds are not horizontal but have a vertical or inclined attitude and pass into the depths of the earth so that their ultimate extension downward cannot be determined. The primitive mountains are also characterized by the occurrence in them of mineral deposits of a peculiar type. These differ from those found in the mountains of the second class in that they occur in the form of veins and irregular shaped masses rather than as bedded deposits. Certain metallic minerals are also peculiar to them. The veins and masses were not developed in the mountains when these latter were formed, but are later in age and are still being formed where fissures and cavities, due to the drying out of the materials of which the mountains were made, are now being filled by the deposition of ores and gangue stones in them through the action of vapors rising from the earth's interior.

The Primitive Mountains comprise all the great and high mountains in the world. They are occasionally found as isolated peaks but usually in the form of long mountain ranges, such as the Alps, Carpathians, and Apennines. The Hartz mountains and Erzgebirge also belong to this class.

Lehmann conceives of the earth as having been originally a great body of earthy matter mingled with water. "At the moment of the Creation" this earth was deposited and the water withdrew, a portion to form the ocean and lakes upon the earth's surface, while the remainder passed down into the great abyss which is in the center of the earth. The earth then dried out and on the surface of the world thus laid bare, the Primitive Mountains were present. These primitive mountains and valleys, he says, had upon their surface a mantle of earth.

Then came the Mosaic deluge, the physical cause of which Lehmann says will always present for us "An inexplicable enigma." The waters overtopped the highest mountains and from these waters earthy materials held in suspension were deposited. As the deluge subsided the retreating waters washed down from the slopes of the Primitive Mountains the loose earth which lay upon them, together with the remains of all animals and of any shell fish which were in the lakes on the mountain slopes, and redeposited these as a series of beds upon the adjacent plains. This is why the Primitive Mountains present the bare and barren surfaces which they now display. Thus also it comes about that the Primitive Mountains have, abutting against their lower slopes, a series of well bedded deposits which dip away from them at a low angle, or

FIG. 1. Portions of two Primitive Mountains, between which are bedded deposits laid down by the Mosaic deluge forming the materials from which a group of Secondary Mountains will be developed.

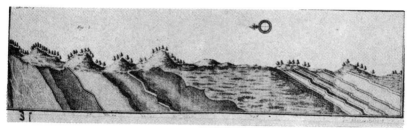

FIG. 2. A Primitive Mountain (31) which is overlain by a succession of thirty beds forming a group of Secondary Mountains.
Ilfeld—Rudigfdorf District.

PLATE XI

(From Lehman: *Essai d'une Histoire naturelle de la Terre*, Paris, 1759)

should two of the Primitive Mountains happen to be situated near one another, these bedded deposits form a basin-like deposit between them, the beds of which slope gently away from each mountain and toward the middle of the valley.

It is, Lehmann says, one of the characteristics of the Primitive Mountains that they are always surrounded by the Secondary Mountains composed or built up of a succession of well defined beds.[41]

While Lehmann sets forth in some detail the character of the beds composing these secondary mountains and gives the order of their succession in several of the regions which he had examined, describing also in detail the mineral deposits which are characteristic of this type of mountains, he does not state how it came about that these gently inclined or horizontal beds were actually fashioned into mountains.

In the accompanying Plate, reproduced from the French translation of his book, figure 1 is a diagram which shows two Primitive Mountains with the bedded deposits out of which the Secondary Mountains were to be developed, lying between them. Figure 2 presents an actual section, although not drawn to scale, of the geological succession which is seen in the Ilfeld-Rudigfdorf district near the Harz mountains. This shows a succession of thirty beds forming a group of Secondary Mountains, the lowest reposing on the slopes of a Primitive Mountain (number 31) traversed by its mineral veins. This is an almost complete section through what are now known as the Permian rocks of Thuringia. Together with other excellent sections showing the succession of the stratified rocks of various districts in Germany it represents a most important contribution made by Lehmann to the progress of geological science.

Of these beds 30–18, of coarser materials than those which succeed them, were deposited upon the slopes of the Primitive Mountain from the turbulent waters of the Deluge when these stood above the tops of the highest mountains. 18–1 which are shown as overlying this lower series unconformably, are those which were formed while the waters were abating rapidly and washing away the loose earth and debris which lay upon the mountain slopes as they were emerging from the waters of the retreating flood.

The mountains of the third class to which Lehmann refers as fewer in

[41] Delius, the Professor of Mining at Schemnitz, in his *Abhandlung von dem Ursprunge der Gebürge*, Leipzig, 1770, also expresses the opinion that the high central mountains of a range date back to the creation of the world, while the smaller hills flanking them were produced by the Mosaic deluge or by other great floods.

number and of less importance than those of the two classes just described, were formed, after the deposition of the materials of the Secondary Mountains from the waters of the Deluge, by various agencies operating in certain restricted areas.   Chief among these were volcanoes but also, in some cases, great floods.

Before passing on to the consideration of the work of Pallas a brief reference should here be made to a remarkable paper by the *Rev. John Michell*, Woodwardian Professor at Cambridge, which appeared a little later than Lehmann's work.[42]   The subject which Michell undertook to investigate obliged him to consider the structure of the earth, and in it he gives three sections across ideal mountain ranges showing them to consist of alternate anticlinals and synclinals.   These sections are reproduced in figure 67 (Chapter XI).   The folded structure, he says, is remarkable and "in most if not all large tracts of high and mountainous country, the strata lie in a situation more inclined to the horizon than the country itself."   This elevation, he thinks, is probably caused by a pressure exerted from below, due to vapors which are developed by the sudden access of the sea or other waters to the great fires which are present in the earth's interior and to which volcanoes and earthquakes are due as well as the elevation of mountain ranges, which is often accompanied by these latter phenomena.

Pallas[43] writing some twenty years after Lehmann, begins his excellent paper by observing that since the "Renouvellement des Sciences" there had been put forward a whole succession of hypotheses to account for the structure of the earth, the origin of mountains, the presence of marine fossils in the earth's strata and the other traces of the great catastrophes which have overtaken the globe.   These theories however were generally based on studies confined to a restricted area of the earth's surface, such as the particular country in which the author resided or even a very small corner of it, and regard the whole earth as built up on the same plan and as having the same structure as that which it displays in this particular locality.   Almost all of them start from the conception of a globe with a uniform and solid crust or surface, the submergence of the continents by transgression of the oceans for variously assigned reasons,

[42] *Conjectures concerning the Cause and observations upon the Phenomena of Earthquakes; particularly of that great Earthquake of the First of November 1755, which proved so fatal to the City of Lisbon and whose effects were felt as far as Africa and more or less throughout all Europe.* Philosophical Transactions of the Royal Society of London, Vol. LI, part 2, p. 582, 1761.

[43] *Observations sur la formation des montagnes et les changements arrivés au globe, particulièrement de l'Empire Russe,* St. Petersburg, 1777.

a complete disappearance of all land areas at the time of the Noachian deluge, then perhaps the impact of a comet, followed by a gradual retreat of the surface waters somehow and to somewhere; also a central fire whose existence is entirely conjectural, and which by a process of distillation gives rise to all springs or rivers, turning our planet into a great chemical and hydraulic laboratory, theories which, built up on speculation and hypothesis and often adorned with mathematical embroidery, are cruelly disrupted and destroyed by actual observation.

It is only in our days, Pallas says, that we have commenced to reach some generalizations based on an extended study of the great mountain ranges of the world.

Under the auspices of the Czarina Catherine II of Russia, Pallas traversed almost the whole of Asia and made a study of the greater part of two of its greatest mountain ranges, the Urals and the Altai. With the knowledge so gained, together with that garnered by other observers from their studies of the Alps, Apennines, Caucasus and other ranges, Pallas says that it has been shown that the highest mountain ranges of the world are composed of granite. This rock is massive and contains no organic remains. Accompanying and usually bordering it are a great variety of rocks (which geologists in modern times designate as crystalline schists), these are in the form of beds, standing vertical or inclined at high angles, and which often enclose ore deposits. Like the granite these rocks contain no fossils and were undoubtedly brought into being before there was any life on the earth.

Throughout the vast extent of Russia—as indeed throughout all Europe—careful observation, he goes on to say, demonstrates that in the mountain ranges the belts of crystalline schists above mentioned are succeeded and overlain by calcareous rocks forming another band or strip on either side of the axis of the range. These calcareous rocks constitute the Secondary Mountains and have a steep dip away from the axis of the range, which dip flattens out to a horizontal attitude in receding from the range and often gives rise to great plains, as is the case to the west of the Ural Mountains. Overlying the "Calcareous Border," which is of marine origin and often rich in fossils, is a great series of clays and marls which constitute the Tertiary Mountains, and which are excellently seen along the whole length of the Urals on their western side. These also are often filled with fossils.

His explanation of this succession is that in the original ocean the summits of the *Primitive Granite Mountains*, which were formed when the world was created, protruded above the waters as lines of islands. The disintegration of the granites under the attack of the ocean waters gave rise to accumulations of quartzose, feldspathic and micaceous sands which hardened into the crystalline schists which lie upon the slopes of the granite ranges. Later there were deposited from the ocean the calcareous rocks of the *Secondary Mountains* and still later the clays and marls of the *Tertiary Mountains*.

While all these rocks of the threefold succession were being laid down, there were accumulated in many places on the sea bottom great bodies of iron pyrites formed from ferruginous materials washed into the sea and which there became mingled with great quantities of decomposing remains of various animals which lived in the sea. These accumulations of pyrites taking fire gave rise to the volcanoes which broke out in various parts of the world and are still active in so many places. It was the explosive forces developed in these volcanoes that causes the fracturing and upheaval of the strata seen in the series of crystalline schists as well as in many parts of the Secondary and Tertiary Mountains, in fact, Pallas says, the explosions from these bodies of pyrites were probably the cause of the elevation of the whole chain of the Alpine Mountains, composed largely of calcareous rocks formed from the remains of corals and shells like those which are accumulating on the sea bottom of the present ocean, and where the associated clays commonly contain a great abundance of pyrite.

He says that the theory put forward by some, which explains the presence of the fossiliferous rocks at very high levels by supposing the sea to have originally covered the summits of the highest mountains, cannot be entertained, for there would not be sufficient vacant space within the earth, even were it traversed by caverns in every direction, to accommodate the enormous volume of water which would have to retreat into the interior of the earth's interior in order to lay bare the mountains as we now see them.

In his opinion, the sea never ascended high enough to cover more than the little limestone hills which rise from the plains—that is to say, not to an elevation of more than 100 toises (i.e., 640 feet) above sea level; all the limestones in the Alps which occur at a greater elevation than this have undoubtedly been upheaved by the force of subterranean eruptions.

Edward D. Clarke, who was subsequently appointed the first professor of mineralogy in the University of Cambridge, while he knew but little about mineralogy, was a great traveller and a sprightly soul who Adam Sedgwick says "kept them awake" at that seat of learning. He mentions having visited Pallas in Russia in 1800:

Luckily, just at this critical season we met with the best of friends, Professor Pallas, to whom the late Empress had given an estate in the Crimea and who received us into his house and was in benevolence a father to us. With him we remained the last summer, till we arranged matters, so as to enable us to quit the empire. . . . In all virtues of hospitality, humanity and the whole chapter of what men should be, he is a Samaritan.[44]

Leopold von Buch—Craters of Elevation

In Chapter VII reference has been made to Werner's views concerning the origin of mountains and to the very important contributions made by his pupil von Buch to the development of the science of geology. One outstanding portion of von Buch's work has however been reserved for consideration here, namely, the theory of "Craters of Elevation" which is always associated with his name. This is a theory to explain the *form* of volcanic mountains. Werner held that this was due to certain peculiar conditions of sedimentation when rocks composing such mountains were deposited from the waters of the primeval ocean, or else to superficial erosion by these waters as they subsided. Von Buch saw plainly that the materials composing these mountains were not of aqueous origin and deposited from above. They had come from below, and must, he thought, have been formed by the upward push of a soft pasty mass. He explained this by his "Erhebungs theorie."

When von Buch, deeply imbued with Werner's belief that volcanoes and volcanic phenomena were sporadic in nature and played a very minor rôle in the evolution of the earth's history, visited Italy, the Canary Islands and the volcanic region of Central France, he was more and more deeply impressed, in fact almost overawed, by the widespread development of volcanic action and by the tremendous forces which were or had been at work in these great centers of volcanic activity. He describes in detail the enormous streams of lava poured out by the volcanoes which he visited, the great blocks of rock hurled high into the air from their craters, the deafening explosions which accompanied their ejection and the dense clouds of ashes which fell covering the surface of the country in all directions for miles about the volcanic vents. He

[44] Otter, W. *The Life and Remains of Edward Daniel Clarke*, Vol. II, p. 78, London, 1825.

noted that the volcanoes were conical in form and frequently displayed a distinctly stratified character, the layers of material of which they were composed conforming to the slope of the mountain sides and dipping away from the crater at its summit. He held that this structure could not be due to the pouring out of successive floods of lava from the crater, because the lava streams were not continuous over the whole surface of the volcano. There were no sheets of lava completely enveloping the whole surface of the mountain, nor was the lava sufficiently fluid to spread out widely in a lateral direction. Furthermore, volcanic islands could not be formed by lava flowing out from a vent on the sea bottom because the molten rock would be solidified immediately on coming in contact with the sea water.

When he came to study the Puy de Dôme and certain other quartz trachyte or "Domite" mountains in the Auvergne district, he found that they consisted of dome-shaped masses of this rock, rising abruptly from a plateau of granite, and apparently thrust upward out of the underlying granite. These remarkable dome-shaped mountains displayed a smooth and uniform outline and *had no crater*. Associated with the dome-shaped elevations and forming members of the same line of isolated hills were many extinct volcanoes composed of dark basic lavas and having the perfect conical form of a typical volcano, with a normal crater at the summit. Of these the Puy de Pariou and the Puy de la Nugère were typical examples.

After many months of careful observation and mature consideration of what he had seen in the Canary Islands and in these remarkable fields of volcanic activity in Central France, culminating in his study of Mount Dore, he put forward as an explanation of the phenomena which they displayed, his theory of Craters of Elevation in which he held that each of these hills was formed of an extrusion of molten material from the interior of the earth as a dome-shaped mass forced upward through the agency of highly heated steam and vapor. If this extruded mass solidified retaining its form, a mountain of the Puy de Dôme type resulted. If on the other hand the bubble-like mass burst at the summit, the plastic rocky material collapsing and falling inward toward the center thus forming a crater, a typical volcano came into being.

Von Buch considered the very narrow steep-walled valleys or "Barrancas," which are observed cutting the slopes of the volcanic cone radially and often extending from this crater to the foot of the mountain, as evidence of a violent upward thrust which had brought these

mountains into existence and which had in so doing rent the whole structure from top to bottom, since he says running water could not cut such
knife-like gashes in the mountain side.[45] This "Erhebungs Theorie"
was fully set forth in the account of his travels in Italy.[46]

It is confusing when von Buch calls, as he does, these dome-shaped
protrusions of igneous rock *craters* of elevation, for a crater is by the very
derivation of the word, a bowl-shaped depression, and these domes have
no craters. But he states that in many cases the summit of the dome-
shaped mass collapses and through lack of support from below an actual
crater is thus formed. This also he calls a Crater of Elevation. Within this crater volcanic action may develop and thus a volcano is formed
within a Crater of Elevation. He holds that the Peak of Teneriffe is an
example of this. Also that "The whole cones of Etna, Vesuvius, Volcano
and Stromboli owe their first elevation to a sudden projection above the
surface." Craters of Elevation associated with volcanic activity often
develop in the sea giving rise to islands. "It has frequently been conjectured that all coral islands of the South Sea which contain a shallow
lake (lagune) in the middle may be regarded as islands of elevation.". . .

These quotations are from an article which he wrote 27 years after his
theory was first announced, and in which, after having visited many
islands and studied their volcanoes, he reiterates these views for which
he says he had now obtained "complete proof." "The object of this
paper,[47] he says, "is to show that Elevation Craters are not volcanoes,
that the distinction between the two is well grounded and important"
but that "Even the cones of volcanoes can be formed only by a sudden
elevation and never by the building up of streams of lava." He supported the theory vigorously to the time of his death.

But craters of elevation and volcanic mountains are sometimes arranged in a linear series and von Buch was of the opinion that if active
along a belt of country instead of merely at isolated localities, the
elevating process would give rise to mountain chains.

The question as to the manner in which the huge mass of a volcano,
upheaved by the subterranean forces, could be supported over the
underlying subterranean cavity formed by its elevation was one which

---

[45] *Ueber die Zusammensetzung der Basaltischen Inseln und über Erhebungs Cratere.* A
lecture delivered before the Prussian Academy of Science, May 1818, Berlin, 1820.

[46] *Geognostische Beobachtungen auf Reisen durch Deutschland und Italien,* Berlin, 1809.

[47] *Volcanoes and Craters of Elevation,* Edin. *New Phil. Journal,* Vol. XXI, p. 206, and in
Poggendorfs Annalen XXXVII, 1836.

of course presented itself to von Buch's critical mind. "There is not," he says, "one vast open space beneath the mountain but a series of great cavernous openings situated one above the other and separated by walls and arches of the lava. Vesuvius is probably built up in this way. The upraised lava when it sinks back after an eruption probably encloses many lakes of "gasförmigen Flussigkeiten." These upon cooling leave great vaulted spaces supported by their own strength and that of the adjacent wall surrounding the whole structure."[48] This problem indeed was one which had forced itself upon the attention of many other observers in times before and after von Buch. Thus Lucretius,[49] although knowing nothing about Craters of Elevation, speaking of Etna says: "The whole mountain is hollow beneath, being supported upon caverns in the rock," and De Saussure[50] long afterwards again refers to the "Vast caverns beneath Mount Etna."

Leopold von Buch was gradually, with the greatest regret, relinquishing the Neptunian teaching of his great master Werner and he was eventually forced by the accumulating evidence to adopt, in large measure at least, the views put forward a few years earlier by James Hutton, on the elevation of tracts of land and ranges of mountains, although apparently he never formally acknowledged that he had done so. As Vogelsang[51] says:

Auf der skandinavischen Reise (1806–1808) überzeugte sich L. v. Buch zuerst, dass die Englander Recht hatten, wenn sie erklärten, der Granit sei keineswegs immer das älteste unterste Grundgebirge, sondern er durchsetze die Schichten und überlagere sie als eruptive Masse. Durch die Beobachtungen auf dieser Reise, verbunden ohne Zweifel mit dem Studium englischer Schriften, besonders aber bei dem späteren Aufenthalt in England (1815) zur vorbereitung auf die Reise nach den Canarien, vollzog sich bei von Buch die folgenreiche Wendung, die man übrigens mit dem grossten Rechte einfach als einen Abfall von Werner und Uebergang zu Hutton bezeichnen kann, wenngleich des Letzteren Name, soviel ich mich erinnere, in Buch'schen Schriften nirgendwo genannt wird. . . .

Die Hebung der Gebirge, in Verbindung mit dem Hervorbrechen eruptiver Massen, bildet nun ein Hauptthema, welches sich in wechselnden, aber immer sehr allgemein gehaltenen Sätzen durch die theoretischen Betrachtungen von Buch's hindurchzieht. Endlich wird jede Insel, jeder Berg zum mehr oder weniger ausgebildeten Hebungskrater, und schliesslich thut Elie de Beaumont seinem Freunde den Gefallen, den Gedanken zu einem geologisch-mathematischen Chaos zu verarbeiten.

[48] *Geognostische Beobachtungen auf Reisen durch Deutschland und Italien*, Berlin, 1809, Bd. II, p. 208.
[49] *De Rerum Natura*, Book VI, 680.
[50] *Voyages dans les Alpes*, Vol. I, Discours Préliminaire, p. VIII.
[51] *Philosophie der Geologie*. Bonn. 1867. p. 88.

An interesting glimpse of von Buch himself is given by Murchison who chanced to meet him at St. Cassian in 1847 when both were engaged independently in their studies of the geology of the Alps. Von Buch at that time has passed his three score years and ten, but Murchison was impressed by his stubborn energy, perseverance and contempt for physical privation even in rough weather and still rougher country. Von Buch admired Murchison and thought highly of his work, but he did not hesitate to express in round terms his dissent from the conclusions of any geologist with whom he did not agree.

Von Buch ridiculed the idea that glaciers had the power to transport large rock masses and when accompanying Murchison in the Alps would halt here and there and plant his staff triumphantly on a big eratic boulder and ask, "Where is the glacier that could have transported that block and left it sticking here?" Mourning over the spread of such heresies and looking back with regret to the creed of the great pioneer of Alpine research he wrote in the book of the little inn on Mount St. Gotthard:

"O Sancte De Saussure
Ora pro nobis"

Murchison thought himself a true and thorough disciple of the school that preached the doctrine of convulsion and cataclysm as the origin of the present irregularities of the earth's surface, but in von Buch he found a far more thorough-going disciple of convulsion for, Murchison writes "When von Buch says that the granite blocks on the top of the Jura were shot across the valley of Geneva like cannon balls by the great power of the explosive forces of elevation I feel the impossibility of adhering to him."[52]

On December 20th, 1850, only two years before his death, in a letter addressed to Murchison, von Buch writes—in his quaint English:

You do approach nature to lift her veil with due reverence and attention to her, and then she speaks to you graciously. Others come hastily with spurs and boots and gross hands draw the veil, as it were a curtain, and they discover behind not the flying Nature, but a phantom they have constructed themselves. Such are the makers of coral islands dancing up and down on the sea, the builders of volcanic cones by successive lava threads and so many ingenious explainers of Nature.

"This dear old geological Tory" as Geikie calls von Buch, was clearly not a believer in Darwin's theory of the origin of coral reefs, and it was Darwin[53] whose observations supplied some of the most telling evidence

[52] Geikie, Archibald, *Life of Sir Roderick I. Murchison*, London, 1875, Vol. II, p. 79.
[53] Darwin, Charles, *Geological Observations on Volcanic Islands and parts of South America*, with critical introduction by Prof. John W. Judd, Minerva Library of Famous Books, London, 1890, pp. 161–227.

against von Buch's theory of Craters of Elevation and in support of the opposite view which now prevails.

Hutton—Elevation of mountains through plutonic intrusion

Hutton's[54] style is ponderous and often rather obscure, which makes it difficult to give a clear and concise presentation of his views by quotations from his works.   These views however are briefly as follows: The loose materials which accumulate on the ocean bottom are consolidated and compacted into solid rocks by the action of subterranean heat.   The power of heat for the expansion of bodies is, so far as we know, unlimited.   "By the expansion of bodies placed under the strata at the bottom of the sea the elevation of the strata may be affected."   This elevation is accompanied by every species of fracture, dislocation and contortion of the strata in question.   The spaces opened up by these fractures and dislocations are filled by "mineral veins."   In this connection, it must be borne in mind that this term, as employed in Hutton's time, included not merely veins in the present sense of the word, but dykes and intrusions of igneous rock.   These were the disruptive agents. Hutton makes use of comparison to enforce his view:

> If, for example, a tree or a rock shall be found simply split asunder, although there be no doubt with regard to some power having been applied in order to produce the effect, yet we are left merely to conjecture at the power.   But when wedges of wood or iron, or frozen water, should be found lodged in the cleft, we might be enabled, from this appearance, to form a certain judgment with regard to the nature of the power which has been applied.   This is the case with mineral veins.

These intrusions have been driven into the positions they now occupy by the power of tremendous fires, which has been sufficient to carry the great bodies of molten matter up to the highest mountain ranges, such as the Andes and the Alps.   It is by the force of such intrusive masses that the sea bottom in many places is elevated to form dry land, and by the same force in other places mountain ranges are thrown up.   If this be true, volcanoes might be expected to occur in mountain ranges at points where the molten materials chance to find egress to the earth's surface.   Such as a matter of fact is the case:

> A volcano is not made on purpose to frighten superstitious people into fits of piety and devotion; nor to overwhelm devoted cities with destruction; a volcano should be

---

[54] *Theory of the Earth or an Investigation of the Laws observable in the Composition, Dissolution and Restoration of Land upon the Globe.*   (Read March 7 and April 4, 1785).   Transactions of the Royal Society of Edinburgh, Vol. I, Edinburgh, 1788, pp. 261–285.

considered as a spiracle to the subterranean furnace, in order to prevent the unnecessary elevation of land, and fatal effects of earthquakes; and we may rest assured that they, in general, wisely answer the end of their intention, without being in themselves an end, for which nature has exerted such amazing power and excellent contrivance.

In conclusion, Hutton says:

We only know that the land is raised by a power which has for principle subterraneous heat; but how that land is preserved in its elevated station, is a subject in which we have not even the means to form conjecture.

The next contribution to our knowledge of mountain ranges was derived from studies of the structure of the Alps and Pyrenees made by a number of geologists of whom De Saussure may be selected as the outstanding representative, and ever since De Saussure led the way, the Alps being, as they are, so easy of access, have been the subject of continuous study and research by a host of geologists from all parts of Europe, and will continue to be so until the fascinating but difficult problem of the origin of mountain ranges reaches its final solution.

De Saussure and von Buch—Mountain Ranges due to folding through compression

De Saussure was the first geologist who made a prolonged study of the Alps.[55] His great work, giving a detailed account of his researches in this great mountain range of central Europe, was published in four volumes, which appeared successively between the years 1779 and 1796. During these years he crossed the whole chain of the Alps no less than fourteen times and made in addition sixteen other traverses from the plains flanking the range to its central axis, and this at a time when there were but few roads in that part of Europe, which was then shunned by all travellers and where the passage of the mountains was not only difficult but often dangerous.

De Saussure was born in Geneva in the year 1740 and belonged to a patrician family, many of whose members had already achieved renown in the world of learning. In the altogether admirable *Discours Préliminaire* in Volume I of the above-mentioned work, he speaks of the enthusiasm and delight awakened in him from his earliest years by the mountains of his native land. At the age of 18 he had already made repeated excursions to all the mountains about Geneva. These ex-

[55] *Voyages dans les Alpes*, 4 Vols, Neuchatel.

periences aroused in him an intense desire to explore the remoter portions of the Central Alps whose towering peaks were seen in the distance. Every year from 1760 onward he visited some portion of the range and for purposes of comparison also studied the structure of the Vosges, Jura and most of the mountain ranges of other European countries. He was an indefatigable field geologist, making all these voyages, he tells us, hammer in hand. He climbed all the accessible summits, collected specimens of all rocks which seemed to be worthy of more detailed examination and study, and it was his invariable rule to jot down the results of his observations on the spot and to write his notes *in extenso* within the next 24 hours, so far as this was at all possible. He also adopted what he found to be a very useful habit of writing out, before he started on any of his expeditions, a list or "agenda" as he calls it, setting forth in order the subjects which he intended to investigate, and the various questions for which he hoped to obtain an answer on the excursion which he was about to undertake. This concentrated his attention on the immediate objects of his expedition and prevented his attention from being diverted from them by any chance and distracting incidents of travel.

De Saussure states that in his studies of the Alps he paid especial attention to the Primitive Mountains and particularly to those composed of granite, because in them lies hidden the mystery of the "first origin of things." He was a contemporary of Werner and with Werner believed that this granite had crystallized out of the waters of the primitive ocean. The great mountain peaks, composed of massive granite, he considered to represent very thick beds formed in the intervals of the "stagnation du fluide." He was impressed with the fact that many of the great peaks, such as that of Mont Blanc, were distinctly pyramidal in form, and in contemplating these he asks himself the question whether they should be regarded as enormous crystals or as the results of denudation. The former theory, surely one of the most fantastic that ever entered the mind of man, was held by J. C. Delamétherie[56] who contended that the materials which crystallized out from the waters of the primitive ocean were not laid down as a unifrom layer over the whole surface of the globe, but as when common salt or alum is crystallized out of solution in a flat bottomed vessel, little ridges and elevations are seen

[56] *Suite à mes vues sur l'action galvanique comme cause principal des commotions soutérraines et des Volcans   Jour. de Physique de chimie et d'histoire naturelle*, Vol. 81, Paris, 1815, p. 280.

rising here and there from the approximately level surface of the solid mass, so from the rocky surface of the ocean bottom of the primitive earth on which were being deposited successive layers of crystalline material, here and there in isolated areas or along lines having no constant direction, similar elevations were developed and these constituted the primitive mountains and mountain ranges.

De Saussure observed that in the Alps the primitive rocks associated with the granite showed evidences of stratification and even the granite itself in places showed indications of the same structure. He also observed that in certain mountains when rocks which belong to the Secondary Mountains repose upon the slopes of these Primitive Mountains, there is evidence of a transition from one into the other. Thus mica, which is one of the ordinary constituents of the primitive rocks, is found in these intermediate strata, as well as in the limestones which form the base of the secondary rocks. It was not known in De Saussure's time that this mica was of metamorphic origin. While the strata forming both the Primitive and Secondary Mountains in many places are horizontal in attitude, in very many other places they have a steep dip and indeed are often vertical in position. In some mountains the strata can be observed to rise gradually in the form of a great arch, in others again they plunge downward forming great synclines, in still others the strata are seen to be bent into the form of the letter S or to display irregular and complicated contortions. These phenomena impressed De Saussure as having a very important bearing on the question of the origin of mountains. He gives a drawing of the great fold seen on the flank of the Mountain of Nant d'Arpenaz in *Plate IV in Volume I (and in Plate I (at end) of Vol. III)*. Such steeply inclined beds, he says, were either laid down from the waters of the primitive ocean on some steeply inclined portion of the ocean bed, or they were deposited in a horizontal position and raised up by some force which was brought to bear upon them subsequently. If, he goes on to say, the former were the case, the materials of which the beds are composed would tend to slide down the steep slope and the beds would thus be thicker at the foot of the slope than in its upper parts. But by careful examination of such steeply inclined beds in many different places in the Alps, De Saussure found that no such thickening had taken place. In fact in one case he found that a bed which formed part of a great syncline occupied a horizontal position in the valley but became more and more steeply

inclined as the formation of which it was a member curved upward to
form the mountain which made one side of the valley. At the summit
of the mountain it had assumed a vertical attitude, yet the bed was
identical in composition and uniform in thickness throughout. Again
he points out that if these beds had been deposited originally on a steep
slope their inclination would progressively decrease until in the valley
at the foot of the mountain they would become horizontal, as is seen in
the case of alluvial deposits washed down from the mountain sides at
the present day; but this was not the case. It is therefore evident,
De Saussure says, that these curved strata were not deposited on a
steeply sloping surface but were originally laid down in a horizontal
position and were later thrown up into the great folds which now form
the Alps, by some mightly force which was subsequently brought to bear
upon them. "But," he asks, "what was this force?" In the first
volume of his *Voyages dans les Alpes* which appeared in 1779, De Saussure
says (p. 334) that in his opinion these bizarre forms can be explained by
the vagaries of crystallization, and compares them to the contortions
seen in the laminated deposits of alabaster laid down in caverns and
other places in various parts of the world. By the time he came to
write the last volume of this great work seventeen years later, he had
evidently revised his views on this matter, for in describing some of the
remarkable contortions displayed by the mountains bordering the Lake
of Lucerne, he says[57] the compact limestones which constitute these
mountains must have been formed by mechanical deposition and not by
crystallization from the waters of the primitive ocean and therefore the
great anticlinal folds which they display could not have been the forms
in which they were originally laid down. They must owe their origin
either (1) to some force acting from below upward or (2) to the folding
of the original horizontal strata. He regards the latter explanation as
more probable then that they had been forced upward by an "explosion."
We must suppose, he goes on to say, that when these contortions took
place, the rocks were in a soft and plastic state.

When one looks for some source whence the tremendous forces required to bring about such gigantic folding can come, the first suggestion
that presents itself is that of subterranean fires, but he hesitates to call
in agencies "presque surnaturels," especially as he can see no trace of
the action of fire in the rocks which display these puzzling phenomena.[58]
In conclusion he says, 'I can offer no general theory to account for the

[57] *Voyages dans les Alpes*, Vol. IV, 1796, p. QQR.
[58] *Voyages dans les Alpes*, Vol. I, p. 333.

origin of these mountains, for this I await the results of further observations.[59]

De Saussure's *Voyages dans les Alpes* presents the results of the first great geological reconnaissance of the whole area of the Swiss Alps. It was a magnificent piece of work which, while presenting an immense body of new facts, did not contribute much to our knowledge of the structure of the range or to the general theory of the origin of mountains. This is not to be wondered at in view of the enormous complexity of the Alps as revealed by the work of the generations of geologists who have devoted their lives to unravelling their complex structure since De Saussure's time.

While De Saussure's book deals chiefly with the geology of the Alps, it contains also many references to the natural history and meteorology of this part of Switzerland, and being written in a fine humanistic spirit contains many references to the inhabitants of the mountains, their hard condition of life and their sterling traits of character. He also speaks in the most glowing terms of the pleasure of mountaineering and Alpine exploration and of the exhilaration which he experienced while prosecuting his work in the keen atmosphere of the lofty mountains.

What has been called the False Dawn of Modern Mountaineering came in the sixteenth century when Conrad Gesner[60] wrote "I have determined for the future so long as the life divinely granted to me shall continue, each year to ascend a few mountains or at least one—for the sake of suitable bodily exercise and the delight of the spirit. For how great the pleasure, how great think you are the joys of the spirit touched, as is fit it should be, in wondering at the mighty mass of mountains when gazing at their immensity and, as it were, in lifting one's head among the clouds. In some way or other the mind is caught up in contemplation of the Supreme Architect. Those to be sure whose spirits are sluggish wonder at nothing; they remain idly at home, do not enter the theatre of the universe, hide in a corner like dormice—let such wallow in the mire, let them be stupied amid gains and sordid pursuits." Some 200 years later in the eighteenth century a great awakening to the delights of mountaineering took place in the Western Alps. The work of De Saussure contributed very largely to this and within two generations later devotees to mountain climbing were to be found all over the world.

While, then, De Saussure made an admirable geological reconnaissance

---

[59] *Voyages dans les Alpes*, Vol. III, p. VII.
[60] *On the Admiration of Mountains*, Zurich, 1543. Translated from the Latin by H. B. D. Soule, San Francisco (the Grabhorn Press) 1937.

of the whole chain of the Swiss Alps, and found that they owed their origin to upheaval by some force whose nature he could not determine, and while he was able to ascertain the structure of certain individual mountains, he did not succeed in solving the problem of the origin and structure of the range itself.

Shortly after De Saussure had brought his work to a close, von Buch, to whose work on volcanoes reference has already been made, having completed his academic studies under Werner, commenced his investigation of the mountains of Silesia and of certain parts of Austria and then of the Swiss Alps, an account of which has been given in Chapter VII, dealing with the Rise and Fall of the Neptunian Theory. The remarkable folding of the strata in all these mountain ranges impressed him deeply, and he sought to account for it by Werner's theory of an irregular settling of sediments deposited from a universal ocean, upon the uneven surface of the primitive crust of the earth. Apparently however he was not altogether satisfied with Werner's explanation, but remarks that if the forces which it called into action "were aided by some exterior mechanical force, the wonderful hieroglyphic-like forms displayed by the strata might be produced."[61]

If any general law governed the structure of the Alps he thought that it would be possible to discover this, if two geological sections were made across the range along different lines more or less parallel to one another but separated by a considerable interval, since local variations in structure could be eliminated by a comparison of the two sections. In this way he thought it should be possible to decide whether the range owed its origin, (1) as De Saussure believed, to a great upheaval of the Primitive Mountains above the valleys or, (2) as De Luc held, that the original surface of the earth had sunk down leaving the Alpine peaks as upstanding remnants of the original crust, or (3) as Werner assumed, that the mountains rose above the plains in virtue of compression exerted within the mountains themselves by the squeezing together of the sediments as they settled down.

With this interesting problem in view, von Buch made two geological sections across the Alps, one on the line of Mount Cenis and the other on that of the Brenner Pass.[62] He found that to the north of the central

---

[61] *Geognotische Uebersicht des Osterreichischen Salzkammerguths*, in his *Geognostische Beobachtungen auf Reisen durch Deutschland und Italien*, Bd. I, p. 146, 1802.

[62] *Vergleichung des Passes über den Mont-Cenis mit dem über den Brenner.* Ibid, Bd. I. p. 297.

axis of the range there was a general similarity in the two sections, but to the south of the axis, there was a great and outstanding difference, since in the section along the Brenner Pass, enormous bodies of porphyry and limestone occur which were absent in the corresponding portion of the Mount Cenis section. The comparison of the two sections, he says, did not supply a solution to the problem, additional data had to be secured before a final answer could be obtained, but the Brenner section to the south of the central axis afforded distinct evidence that this portion of the range at any rate had been elevated by external forces of compression.

Élie de Beaumont—Mountain ranges arranged in a number of parallel series each of a different age—Réseau Pentagonal

The next noteworthy attempt to explain the origin of mountain ranges was that of Élie de Beaumont[63] which, although it attracted widespread attention at the time, has now little more than an historic interest. Bischof and St. Claire Deville had measured the amount of contraction in volume which took place in fused rock upon solidification and Élie de Beaumont based his theory on the result of their experiments. If the earth had been at one time in a state of fusion, as it cooled a crust would be formed, and as the cooling progressed the crust, owing to the resultant radial contraction, would become distorted and develop a gently undulating surface. As the cooling proceeded still further the increasing tension would cause the crust to break suddenly along a certain line or lines. Along this line one side would sink down, while the other, relieved of the tension, might rise, the high projecting edge would form a mountain range, while the subsiding side, through intense lateral compression as it moved downward, would be thrown into a series of folds parallel to the fissure, which folds would gradually die out as the distance from the line of fracture increased. The as yet unsolidified magma beneath the crust, being suddenly subjected to the pressure exerted by the subsiding portion of the crust, would be forced up, in places at least, along the line of fracture. In this way the granite axes seen in many mountain ranges were formed, on whose sloping sides are displayed a succession of bedded rocks, dipping away from the central axis. If, as is frequently the case, a series of later rocks in flat undisturbed strata are found reposing on the

[63] *Extrait d'une Série de Recherches sur quelques-unes des Révolutions de la Surface du Globe.* Annales des Sciences naturelles, T. XVIII & XIX, 1829 & 1830.

uplifted straia, the age at which the mountain range was formed could be determined.   It would be later than the age of the highest stratum of the uptilted series and older than that of the undisturbed beds.

A very interesting colored diagram presenting an ideal section through the earth's crust showing how the ages of his respective *systems* of mountains can be determined by the age and attitude of the sedimentary formations on either side, is given by Élie de Beaumont in a memoir entitled *Recherches sur quelques-unes des Revolutions de la Surface du Globe* which appeared in the Revue Française, No. XV, for May, 1830. Élie de Beaumont held that every mountain range came into existence suddenly, it was catastrophic in origin, and marked a "Révolution."[64]

It has been in vain attempted to explain the geological facts observable in high mountain chains by the action of the slow and continuous causes now in force on the surface of the globe.   No satisfactory result has been obtained by these means.   In fact everything shows that the instantaneous elevation of the beds of a whole mountain chain is an event of a different order from those which we daily witness.

Werner in his study of mineral veins had shown that very frequently if a fissure which, when filled, gave rise to a vein, ran in any direction across the country rock, other fissures parallel to it would be produced at the same time.   Élie de Beaumont tried to prove that the same was true of mountain ranges, and that all mountain ranges of the same age were parallel to one another or at least cut the meridian at the same angle and belonged to one system.   He at first recognized 12 such systems, the number of which he later (1852) increased to 21, and he believed that the complicated series of intersections produced by the crossing of these systems had as its basis a pentagonal network,—his celebrated "Réseau Pentagonal," producing what Vogelsang[65] refers to as a "Geologisch-Mathematisches Chaos."

As already stated, this theory has now been abandoned.   The orographic map of the world displayed no such symmetry as it demanded, the mountain ranges did not run in parallel lines across the surface of the globe but usually took their course in great sweeping curves.   Nor was each mountain range thrown up at one time, but many of the great ranges were found to have come into existence through several successive uplifts at widely separated periods.

[64] *Researches on some of the Revolutions which have taken place on the surface of the Globe Phil. Mag.*, New Series, Vol. 10, p. 241.   (No. 58) 1831.

[65] *Philosophie der Geologie*, Bonn, 1867, p. 88.

A BRIEF REFERENCE TO LATER WORK

The successive attempts which have been made from the beginning of history down to the close of the first quarter of the nineteenth century, that is to say to the years 1829–30, to discover the causes which have given rise to mountains have now been reviewed.

On looking back over the path which has been travelled it will be noted that all the explanations which were offered during classical and medieval times, as set forth by a long succession of distinguished men and summarized by Faventies have now been rejected. In fact many of them are now seen to be little short of ridiculous. The only exceptions which should be made to this general statement are the explanations given of the origin of certain more or less isolated mountains or mountain ridges which have been brought into being by processes of erosion. The true origin of these was indicated in a general way by Avicenna in 1021 and later by Leonardo da Vinci and set forth in detail by Agricola in 1546. It was also recognized that volcanoes owed their origin in one way or another to the agency of heat or fire, a fact indeed that was evident to all. In the next period, extending from the close of the Middle Ages to the end of the first quarter of the nineteenth century, a period when observation in the field replaced speculation, much was learned concerning the structure of mountain ranges and a basis was laid down for a true understanding of their nature. During this time another series of theories was put forward many of which have since been abandoned in the light of a wider knowledge. Among these may be mentioned that which held that mountain ranges were ridges of materials crystallized out of the waters of a primitive ocean, or that they were caused by an irregular settling down of sediments deposited in such an ocean or perhaps in a later universal ocean due to the Noachian deluge. Also those theories which held that the Volcanic fires were produced by the combustion of coal, bitumen, sulphur, or other similar inflammable bodies which were constituents of the earth's crust, or the later theory of Davy which attributed these fires to the oxidation of masses of metallic potassium or sodium present in the earth and ignited through the agency of water, which gained access to them from the sea.

The years 1820–30 ushered in a period which has been called "The Golden Age of Geological Science."[66] Historical geology came into being and the stratigraphical succession was studied in every land.

[66] Kober, Leopold, *Gestaltungsgechichte der Erde*, Berlin, (1925) p. 14.

Geological surveys were established in all civilized countries and greatly increased attention was devoted to the study of the structure and origin of the mountain ranges of the world.

It is not within the scope of this present work to follow out the further development of those studies in this later time.[67] A brief mention, however, may be made of some of the chief, among many, theories which have been advanced to account for the origin of mountain ranges in more recent times.

The next important developments of the theory of mountain building came from North America. The earliest of these was that of W. B. Rogers and his brother, H. D. Rogers,[68] based on a prolonged investigation of the geological structure of the Appalachian mountain chain. These able investigators reached the conclusion that the wave-like structure displayed by the Appalachians is due to:

An actual undulation of the supposed flexible crust of the earth exerted in parellel lines and propagated in the manner of a horizontal pulsation from the liquid interior of the globe. We suppose the strata of such a region to have been subjected to excessive upward tension, arising from the expansion of the molten matter and gaseous vapours, the tension relieved by linear fissures, through which much elastic vapour escaped, the sudden release of pressure adjacent to the lines of fracture, producing violent pulsations on the surface of the liquid below. This oscillating movement of the fluid mass below would communicate a series of temporary flexures to the overlying crust and those flexures would be rendered permanent (or keyed into the forms which they present) by the intrusion of molten matter. If, during this oscillation, we conceive the whole heaving tract to have been shoved (or floated) bodily forward in the direction of the advancing waves, the union of the tangential with the vertical wave-like movement, will explain the peculiar steepening of the front side of each flexure, while a repetition of similar operations would occasion the folding under, or inversion, visible in the more compressed districts.

It is interesting to note that the opinion of the Rogers brothers that movements in the earth's crust were due to a great lake of molten rock beneath, was in part at least shared by Darwin, who states, as the result of his researches in South America, that all earthquakes on the west coast of that continent are caused by the injection of liquified rock be-

[67] The student who desires to do so should consult Kober (op. cit.); Bucher, *The Deformation of the Earth's Crust* (Princeton, 1933) or other works.

[68] Rogers, W. B., and Rogers, H. D. *On the Physical Structure of the Appalachian Chain, as exemplifying the Laws which have regulated the Elevation of Mountain Chains Generally.* Brit. Assoc. Advancement of Science, 1842. *On the Laws of Structure of the More Disturbed Zones of the Earth's Crust.* Trans. Royal Society of Edinburgh, 1856.

tween the strata and that the line of volcanoes in Chile, 800 miles long, rests upon a sheet of fluid matter.

The *Geosynclinal theory* of Hall and Dana, held that the major mountain ranges of the world originated in belts of sediments, narrow but very thick, which were deposited along the slowly subsiding margins of continental areas and subsequently elevated into mountain ranges by the expansive action of the heat to which these sediments became subjected as they sank into the deeper parts of the earth's crust. This theory underwent a number of modifications by other and later investigators.

The *Decken or Nappe theory* of Heim,[69] Suess,[70] Lugeon and many others was put forward to account for the structure of the Alps, this range being interpreted as compounded of a great number of folds lying upon and carried over one another for great distances from their original position.[71]

Another important theory is that of *Isostacy*. In a letter written by Herschel to Lyell in the year 1837 there is to be found the first suggestion of the theory of isostacy as a mountain making process. Airy had in mind the same idea in his suggestion concerning the cause of the gravity anomalies in the Himalaya region in 1855. The term Isostacy[72] was, however, introduced by Dutton, who developed the idea in a paper in which he criticized the Contractional theory, and the theory of Isostacy is in consequence generally associated with his name.

The *Drift theory*, associated with the name of Wegener, holds that portions of the original continent or contients broke away from the parent mass and slowly moved through a lower plastic zone of the earth's crust. The movements of these upper "Sial" blocks through the lower "Sima" is assumed to have led to the formation of mountain ranges.

CONCLUSION

In all probability it will be found eventually that the complete truth does not lie in any or in all of these theories. The primary cause of the

[69] Heim, *Untersuchungen über den Mechanismus der Gebirgsbildungen*, Basel, 1878.

[70] Suess, *Die Entstehung der Alpen, 1875;* also *Das Antlitz der Erde*, Vol. I, 1885, Vol. II, 1888.

[71] Heritsch, Franz, *The Nappe Theory of the Alps*, translated by P. G. H. Boswell, London, 1929.

[72] Isostacy was once wittily defined by Professor B. K. Emerson as "A sort of hydrostatic equilibrium with the water left out and the equilibrium somewhat doubtful."

great variety of structures presented by the mountain ranges, which these theories attempt to explain, is not yet established. The general opinion is that it is to be found in radial contraction of the earth's crust, giving rise to a compression and folding of the surface along certain lines.[73]   Others, however, assert that this is contrary to all modern geophysical conceptions.[74]

The reason of the great diversity of opinion which still prevails concerning the origin of mountain ranges, is that, while modern research has gathered an enormous volume of knowledge and has definitely established many facts concerning these ranges, much with reference to their character and structure still remains unknown or is a mere matter of conjecture.   Although there have been great advances since his time we may still say as Collini did some 150 years ago:

> Il y a longtemps que les minéralogistes et les géologues observent les montagnes mais leurs observations ne sont point encore suffisantes pour applanir les doutes et les difficultés qui naissent des différents phénomènes que nous présentent ces parties élevées de notre globe. . . . A chaqu'instant l'on remarque dans les montagnes de nouvelles circonstances et de nouveaux phénomènes, qui renversent les principes qu'on avait adoptés et qui nous jettent derechef dans l'embaras.[75]

More facts are required and must be gathered by long continued observation in many different and widely separated parts of the globe, and furthermore since the source of the movements which give rise to mountain ranges is to be found ultimately in forces which have their origin within the earth, until a much more complete knowledge of the nature of the interior of our planet is obtained many problems concerning the origin of mountain ranges and also of volcanoes will remain unsolved. We cannot indeed but recognize that not only is the problem of the origin of mountain ranges still unsolved, but that toward the final elucidation of this subject geological science has made a less satisfactory advance than in many, if not in most, other directions.

[73] Jeffreys, Harold, *Earthquakes and Mountains*, London, 1935, p. 162.
[74] Kober, Op. cit, p. 177.
[75] *Considérations sur les Montagnes volcaniques*, Mannheim, 1781, pp. 1 and 2.

CHAPTER XI

EARTHQUAKES AND THE CHARACTER OF THE INTERIOR OF THE EARTH

1. VIEWS CONCERNING THE ORIGIN OF EARTHQUAKES IN CLASSICAL TIMES

It is strange to reflect that while mankind has for many thousand years lived and pursued its manifold activities on the earth's surface, men have remained in almost complete ignorance concerning the character of the interior of the earth, even for a few miles beneath their feet. The ancients knew nothing about it, and even at the present time with all the resources for investigation which have been acquired by science within recent years but little is yet known with certainty. The reason for this is of course that the interior of the earth is not accessible to us.

The only facts which the ancients had on which to base their conjectures or generalizations were the following:

1. The earth is not always stable but from time to time it trembles or shakes violently.

2. These *earthquakes* are especially frequent in certain localities.

3. The movements are often accompanied by the elevation or subsidence of portions of the surface.

4. Fissures sometimes appear in the earth's crust.

5. When an earthquake takes place great sea waves often roll in upon the coasts.

6. Low noises or heavy rumblings are at times heard within the earth.

7. Floods of hot or cold water may issue from the earth.

8. Earthquakes are often associated with volcanoes or "burning mountains," from which tremendous blasts of hot air or mephitic winds are given off, with steam and sulphurous vapors, clouds of dust, frag-

ments of rocks which are hurled into the air and streams of molten rock which are poured out accompanied by the appearance of fire.

All the early writers on this subject were deeply impressed, as they well might be, with the tremendous and awful nature of the forces at work in the production of these phenomena. They felt that these must be "Elemental Forces." The very elements themselves (fire, air, earth and water) were here seen in action.

While the early writers were impressed by the elemental character of the forces at work, they were far from agreeing as to which of the four elements was the moving cause. Each of these in turn was held, by one writer or another, to play this rôle.

Water was the cause alleged by Thales, the earliest of all the Greek writers. Coming from the sea girt island of Miletus, he was impressed with the force displayed by this element and the devastation wrought by it when the sea waves in time of great storms, beat upon the coast line. They seemed at times to shake the very land. According to Thales, the globe itself was like some great lumbering vessel floating on the surface of a great body of water, and the movements of the water gave rise to earthquakes. As Seneca observed later, "No lengthened consideration is needed to prove the falsity of this view."

Fire was by Anaxagoras believed to be the cause of some earthquakes at least. Thick clouds of vapor in the caverns within the earth coming into rapid collision strike out fire, as lightning is developed by the impact of one dark cloud against another in the stormy sky. And as it is the nature of fire ever to rise up, the fire thus produced within the earth rises rapidly toward the surface and, should it meet with any obstructions, bursts through these with great violence, shaking and shattering the earth. Some later adherents of this view believed that the fire burned away everything with which it came in contact and thus the supports sustaining the roofs of the earth-caverns being destroyed these crash down giving rise to tremendous reverberating shocks.

Earth itself is the cause according to Anaximenes. This writer is somewhat obscure in his statement but believes that earthquakes are caused by masses of rock within the earth falling for some cause and striking others with great violence. Seneca again observes that all things "fall through age," even the most solid buildings, and so it is with the rocky cliffs of the earth's interior.

Air (or vapor) according to Archelaus and his followers gives rise to earthquakes. This philosopher of the Ionian school was one of the most distinguished followers of Thales. The air finds its way into the earth's interior through passages leading downward from the surface. It is forced into the underground caverns and these become filled to their utmost capacity. If then additional supplies of the element are forced down from the surface the air becomes highly compressed and violent storms result, which blast away everything in their path and cause the movements of the earth and the uproar manifested in times of earthquake. Sometimes the air thus compressed breaks out at the surface of the earth, causing widespread devastation. Callisthenes, a follower of Archelaus, says that the "Air enters the earth by hidden openings under the sea just as everywhere else," and if the water cuts off its retreat in the rear, the air is borne hither and thither and undermines the earth. This is the reason why the land near the sea is more frequently shaken by earthquakes than elsewhere, and hence Neptune is called the Earthshaker.

The latest deliverance of the classical world on the subject of physical speculation was that of Seneca[1] in his *Quaestiones Naturales*, which may best be translated *Physical Inquiries or Investigations in the Domain of the Physical Sciences*, and one of the best chapters in this book is that on Earthquakes.

This chapter was inspired by the occurrence of the great earthquake in the Campania on February 5, A. D. 63. The shock caused great damage in Pompeii, but the city was quickly restored, only to be completely destroyed and buried by the first eruption of Vesuvius in historic times, that of the year A.D. 79, in which the elder Pliny lost his life.

"We have just had the news, my esteemed Lucilius," says Seneca, "that Pompeii, the celebrated city in the Campania, has been overwhelmed in an earthquake which shook all the surrounding district as well."

Having reviewed the opinions of his predecessors Seneca states his own opinion as to the origin of earthquakes. He takes his place with Archelaus:

The chief cause of earthquake is air, an element naturally swift and shifting from place to place. As long as it is not stirred but lurks in a vacant space it reposes innocently, giving no trouble to objects around it. But any cause coming upon it from without rouses it, or compresses it, and drives it into a narrow space . . . and when

---

[1] John Clarke, M.A., *Physical Science at the time of Nero, being a translation of the Quaestiones Naturales of Seneca, with notes by Sir Archibald Geikie*, (Macmillan & Co.), London, 1910.

opportunity of escape is cut off, then "With deep murmur of the Mountain it roars around the barriers" which after long battering it dislodges and tosses on high, growing more fierce the stronger the obstacle with which it has contended.

Seneca also sets forth the conception of the character of the earth's interior which was undoubtedly prevalent long before his time and continued to be commonly held until the rise of modern science, namely, that the interior of the earth presented in a general way at least, a reproduction of the conditions found on the earth's surface:

Be assured that there exists below, everything that you see above. There are antres vast, immense recesses and vacant spaces, with mountains overhanging on either hand. There are yawning gulfs stretching down into the abyss which have often swallowed up cities that have fallen into them. These retreats are filled with air, for nowhere is there a vacuum in nature; through their ample spaces stretch marshes over which darkness ever broods. Animals are also produced in them, but they are slow paced and shapeless; the air that conceived them is dark and clammy, the waters are torpid through inaction.

It was generally believed indeed that throughout all parts of the earth's interior there ran rivers of water or fire, fissures or passages through which passed whirling currents of air, all arising in or communicating with cavernous spaces filled with fire, air or water, whose fortuitous contact gave rise to every manner of destructive or constructive action which was believed to be in operation in the dark and mysterious depths of mother earth.

The earth is indeed compared by Seneca, as it was by a host of writers in later times, to the human body with its veins and arteries and its circulatory system through which the blood and other fluids circulate quietly in normal life, but which, when the body processes become disturbed give rise to all manner of diseases and troubles.

Lucretius[2] in his splendid poem *De Rerum Natura*, written rather more than one hundred years earlier than the work of Seneca to which reference has just been made, gives a graphic description of the manner in which earthquakes take their origin, his views being essentially those set forth later by Seneca:

Now attend and learn the reason for earthquakes. And in the first place be sure to consider the earth below as above to be everywhere full of windy caverns, bearing many lakes and many pools in her bosom with rocks and steep cliffs; and we must suppose that many a hidden stream beneath the earth's back violently rolls its waves

[2] Book VI, 535 et seq.   Trans. by W. H. D. Raise, D. Litt., in the Loeb Classical Library.

and submerged boulders; for nature itself demands that she be everywhere like herself. Since therefore she has these things attached beneath her and ranged beneath, the upper earth trembles under the shock of some great collapse when time undermines those huge caverns beneath; for whole mountains fall, and with the great shock the tremblings creep abroad from the place,—and with good reason; since when a waggon of no great weight passes, whole buildings hard by the road tremble with the shock, nor less do they dance also when a stone in the road jolts up the iron tires of the wheels on this side and that. Sometimes also, when from lapse of time a huge mass is rolled forwards from the earth into some great and wide pool of water, the earth also is moved and shaken by the wave of water: even as a vessel often cannot remain still unless the water within it ceases to be moved about in waves to and fro.

Besides, when a wind gathering together from some one quarter through the hollow places beneath the earth throws itself forward, and bears hard, thrusting with great force into the lofty caverns, the earth leans over in the direction of the wind's headlong force. Then those buildings which are built up above the earth, and each all the more, the more they tower up towards heaven, lean suspended, pushing forward in the same direction, and the beams dragged forward hang over ready to go. . . .

There is also another cause of the same trembling, when wind or a very great force of air, either from without or arising within the earth itself, has thrown itself suddenly into the hollow places of the earth, and there in the great caverns first growls tumultuously and is carried whirling about, afterwards the force thus excited and driven outwards bursts forth, and at the same time cleaving the earth asunder makes a great chasm. This befell at Syrian Sidon, and came to pass at Aegium in the Peloponnese, when such an issue of air overthrew those cities with the earthquake that followed. Many another city wall has fallen by great quakings in the earth, many cities have sunk down to destruction in the sea along with their inhabitants. But if there is no breaking forth, yet the impetuous air itself and the furious force of wind is distributed abroad through the many interstices of the earth like an ague, and thus transmits the trembling; even as when cold penetrates deep, into our limbs it shakes them, making them tremble and quake against our will. Therefore men shiver in their cities with a twofold terror: they fear the houses above, they dread the caverns below, lest the earth's nature loosen all asunder in a moment, or torn asunder open abroad her own gaping jaws, and in confusion seek to gorge it with her own ruins.

The Venerable Bede[3] about the year 700 refers to another and ancient explanation of the origin of earthquakes, when, after various conjectures which had been put forward concerning the origin of earthquakes, he says that others attribute earthquakes to that sea beast "Which lies

---

[3] *De Mundi Coelestis Terrestrisque Constitutione Liber.* *Migne, Patrologiae Latinae Tomus X C Venerabilis Bedae,* Vol. 1, 1862. (p. 187). This is attributed to Bede but is placed in the class "Dubia et Spuria."

in the ocean wrapped around the world holding its tail according to its custom in its mouth, but this Leviathan from time to time when scorched by the sun struggles to seize the sun and in its titanic struggles shakes the very earth."

2. VIEWS IN MEDIEVAL TIMES AND DOWN TO THE MIDDLE OF THE EIGHTEENTH CENTURY

Conrad von Megenberg (died 1374) in his *Buch der Natur* sets forth the opinions concerning earthquakes commonly held in his time. The book is written in popular style and his account shows incidentally the influence of astrology at this period. He says:

It often happens in one place or another that the earth shakes so violently that cities are thrown down and even that one mountain is hurled against another mountain. The common people do not understand why this happens and so a lot of old women who claim to be very wise, say that the earth rests on a great fish called Celebrant, which grasps its tail in its mouth. When this fish moves or turns the earth trembles. This is a ridiculous fable and of course not true but reminds us of the Jewish story of the Behemoth. We shall therefore explain what earthquakes really are and what remarkable consequences result from them. Earthquakes arise from the fact that in subterranean caverns and especially those within hollow mountains, earthy vapours collect and these sometimes gather in such enormous volumes that the caverns can no longer hold them. They batter the walls of the cavern in which they are and force their way out into another and still another cavern until they fill every open space in the mountain. This unrest of these vapours is brought about by the mighty power of the Stars, especially by that of the God of War as Mars is called, or of Jupiter and also Saturn when they are in constellation. When then these vapours have for a long time roared through the caverns, the pressure becomes so great that they break a passage through to the surface and throw the mountains against one another. If they cannot reach the surface they give rise to great earthquakes.

He then goes on to refer to the different kinds of earthquakes, of the subterranean noises which frequently accompany them and of the effect of the escaping vapor and dust in darkening the heavens and producing sunglows. Megenberg states that these subterranean vapors when they remain within the caverns for long periods become highly poisonous and escaping to the surface in certain localities have led to the death of a great number of persons. This poisonous effect is often experienced in mines, and vapor absorbed by the waters of springs also renders them poisonous. He attributes the Black Death, which in 1348 and subsequent years devastated Europe and which it has been estimated caused the death of approximately one-third of the whole population, to the escape of such posionous vapor from the interior of the earth during the earthquakes which at this time were experienced in various localities. He also states that these vapors escaping from the earth in certain places turned men into stone and in the vicinity of some salt mines turned men and cattle into masses of rock salt.

As good examples of later medieval treatises on earthquakes those of Galesius,[4] Maggio[5] and Zuccalo[6] may be cited. A devastating earthquake was experienced at Ferrara and the adjacent parts of Italy in the month of November in the year 1570 A.D. This had been preceded in

Fig. 65. Results of an earthquake. (From Reisch: *Margarita Philosophica*, book IX, chapter 17.)

the spring of the same year by very heavy earthquake shocks which did great damage further south in Italy in the district about Naples and Pozzuoli. These earthquakes led to the preparation of three treatises by these authors which were all published in the same year, 1571, in the city of Bologna. They have all essentially the same content and that of Galesius, being perhaps the best of the three, may be taken as representing opinions of the other two.

Galesius first gives a list of the authors which he had consulted in the preparation of his treatise and it is characteristic of the time, that of the seventy one authors in his list thirty seven or over half are commentators or epitomists of Aristotle's *Meteorologica*, of the others, many are poets,

[4] August Galesii, *de Terraemotu Liber*, Bologna, 1571.
[5] *Del Terremoto Dialogo del Signor Lucio Maggio*, Bologna, 1571.
[6] *Del Terremoto Trattato di Gregorio Zuccolo*, Bologna, 1571.

among them—Virgil, Ovid and Lucretius, while of the remainder the
most important are Seneca, Agricola, Cardan and Scaliger.

He then sets forth in tabular form all the causes which had been assigned as the origin of earthquakes by former writers, and proceeds to
examine and discuss these in turn.   This table is here reproduced since

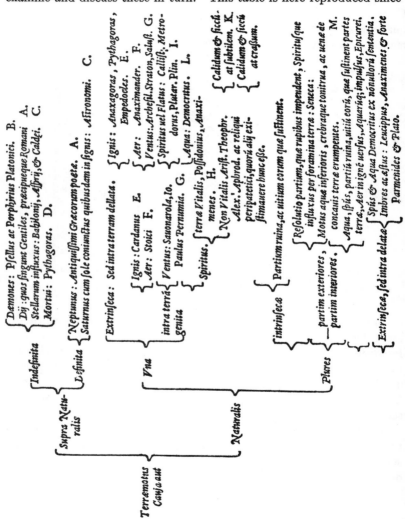

it is interesting as presenting a compendium of ancient opinions, or perhaps it would be more correct to say, ancient *conjectures*, on this subject. As will be seen these are first divided into Supernatural and Natural Causes. The Supernatural explanations are divided into Indefinite and Definite. The former class includes Daemons, who by Porphyry and others were believed to have their abode within the earth and when aroused to fury for any reason shook the earth. Others thought the Gods shook the earth to indicate their displeasure or to punish men for their sins. Another of these "indefinite" supernatural causes was the influence exerted by the stars upon the earth while the fourth, "Mortui," is found in an obscure passage of Pythagoras which explains earthquakes as caused by disturbances in assemblies of the dead within the earth so violent in character that the earth itself was shaken.

Under the heading of Natural Causes there is a subdivision into those opinions which attribute earthquakes to a single cause and those which hold that there are several contributing causes. The large capital letters following certain subdivisions in the table refer to paragraphs in the text where these several causes are explained and then refuted. His own conclusion is that most earthquakes are due to the generation of exhalations in the deeper parts of the earth by the interaction of fire, air, earth and water, which exhalations, in the form of winds forcing themselves outward toward the earth's surface and meeting obstructions of various kinds, shake the earth with violence and in a manner depending on the character of the resistance which they experience. This explanation he probably took from Aristotle who taught in his *Meteorologica* that earthquakes were caused either by wind rushing violently into or out of the earth or moving tumultuously within it.

Galesius passes on to enumerate the various types of earthquakes— "Vibratio," or movement from side to side, "Tremor," a shaking in various directions, "Elatio," an upward movement, "Depressio," a downward movement, and "Pulsus," an up and down movement. He then discusses the question as to which portions of the earth's surface are most subject to earthquake shocks. Those parts of the world which are either very cold, as the polar region, or very hot, as the torrid zone, are seldom visited by these devastating occurrences because the intense cold of the former prevents the development of exhalations from the earth and the intense heat of the latter rarifies the exhalations so that they pass out into the air, meeting with but little resistance.

The temperate zones therefore are those in which earthquakes are most frequently experienced. Again, places far distant from the sea are seldom visited by earthquakes; these are more abundant near the seashore, for the waves of the sea beating upon the shore drive back and compress the escaping exhalation and prevent its free escape. Again, islands far out to sea are less exposed to earthquakes than those off shore, because the great sea cools the exhalations, overpowers and forces them down by its weight. Places where there is an abundance of sulphur, nitre, and bitumen are more subject to earthquakes because these materials are favorable for the production of exhalations, which may also give rise to conflagrations.

Then after treating of the hours of the day and the seasons of the year in which earthquakes are most common, he passes to the signs of the approach of earthquakes. These are periods of great dryness, times of heavy rains, times of the appearance of comets, or sword shaped stars, fiery columns, violent winds and frequent lightning, or of great stillness in the atmosphere, a haziness of the sun, fearful sounds within the earth, disturbances in the sea, lakes and other bodies of water, birds flying in a confused and restless manner.

He goes on to refer to earthquakes as portents of coming disasters and finally gives the "remedies" for earthquakes. These are as follows:

1. Placing statues of Mercury and Saturn on the four containing walls of the building.

2. Bearing patiently these troubles which cannot be avoided.

3. Places facing to the north are less shaken.

4. Caves and subterranean passages and drains whence exhalations can escape easily from the earth are believed to be safer.

5. Vaulted brick passages and rooms are safer.

6. Dwellings should not be made too high—not over 60 feet.

7. Houses should be supported by propping them up with beams.

8. During the occurrence of the earthquake people should move out of their houses and live in tents or huts made of light boards.

9. Pray to God and ask his mercy.

While the Aristotelian view of the origin of earthquakes through mighty winds which had the sun as their primary source of motion, was that still commonly held at the close of the sixteenth century, there were a number of "Philosophers" who were beginning to reject this view or accept it only as a partial explanation of the observed phenomena.

Thus Agricola[7] states that in his opinion earthquakes are indeed due

[7] *De Ortu et Causis Subterraneorum*, Basel, 1546, Book 2.

to the development of vapors within the earth but that these are caused by the action of the central fire on moisture present in the earth and that the vapors under great compression rend the rocky crust as they force their passage outward toward the earth's surface. If fire appears with the escaping vapor this is due either to the fact that the vapors have become ignited by internal friction or that the district through which they are passing is one which has been formerly the scene of volcanic activity.

Cardan[8] writing only a few years later put forward another view, namely, that fiery vapors charged with nitre, bitumen and sulphur from the earth's interior are the active agents in the production of earthquakes, a view adopted by Milton[9] when writing about one hundred years later:

> As when the force
> Of Subterranean wind transports a hill
> Torn from Peloris, or shatter'd side
> Of thund'ring Etna, whose combustible
> And fuel'd entrails thence conceiving force,
> Sublim'd with mineral fury, aid the winds
> And leave a singed bottom, all involv'd
> With Stench and Smoke.

This view gained ground rapidly so that in the latter half of the seventeenth century it was generally accepted.

Varenius[10] sets it forth in detail as follows:

Those countries which yield great store of Sulphur and Nitre, or where Sulphur is sublimed from the Pyrites, are by far the most injured and incommoded by Earthquakes; for where there are such Mines they must send up Exhalations, which meeting with subterraneous Caverns, they much stick to the Arches of them, as Soot does to the Sides of our Chimnies, where they mix themselves with the Nitre or Saltpetre, which comes out of these Arches, in like manner as we see it come out of the inside of the Arch of a Bridge, and so makes a kind of Crust, which will very easily take Fire. There are several ways by which this Crust may take Fire, viz., (1) By the inflammable Breath of the Pyrites, which is a kind of Sulphur that naturally takes Fire of itself. (2) By a Fermentation of Vapours to a degree of Heat, equal to that of Fire and Flame. (3) To the falling of some great Stone, which is undermined by Water, and striking against another, produces some Sparks which set Fire to the combustible Matter that is near; which, being a kind of Natural Gun-powder,

---

[8] *De Subtilitate*, Nuremberg, 1550, p. 67.

[9] *Paradise Lost*, Book I, vv 230–237.

[10] *A Complete System of General Geography*. Originally written in Latin by Bernard Varenius since improved and illustrated by Sir Isaac Newton and Dr. Jurin and translated into English by Mr. Dugdale, 4th Edition, Vol. I, p. 155.

at the Appulse of the Fire, goes off (if I may so say) with a sudden Blast or violent Explosion, rumbling in the Bowels of the Earth, and lifting up the Ground above it, so as sometimes to make miserable havock and Destruction 'till it gets Vent or a Discharge. Burning Mountains and Vulcanos are only so many Spiracles serving for the discharge of this subterranean Fire, when it is thus preternaturally assembled. And where there happens to be such a Structure and Conformation of the interior Parts of the Earth, that the fire may pass freely and without Impediment from the Caverns therein, it assembles unto these Spiracles, and then readily and easily gets out from time to time without shaking or disturbing the Earth. But where such Communication is wanting, or the Passages not sufficiently large and open, so that it cannot come at the said Spiracles without first forcing and removing all Obstacles, it heaves up and shocks the Earth, till it hath made its Way to the Mouth of the Vulcano; where it rusheth forth, sometimes in mighty Flames, with great Velocity, and a terrible bellowing Noise.

Lemery,[11] in one of the very earliest essays in experimental geology, gave what he considered to be a most conclusive demonstration that earthquakes and also volcanoes and thunder storms *could* originate "If iron and sulphur happened to meet together." His account of the experiment, which is frequently referred to in works of this period, is as follows:

I take a mixture of equal parts of iron and sulphur powdered; this I form into a paste with water, and leave it to digest two or three hours, without fire, in which time it ferments and swells with a considerable heat; the fermentation cracks the paste in divers places, and through the crevices there issue vapours, which indeed are but barely warm if the mass be small, but when it is considerable, as thirty or forty pounds, an actual flame comes forth.

The fermentation accompanied with heat, and even fire, which happens in this operation, proceeds from the penetration and violent friction which the acid points of the sulphur exert upon the particles of the iron.

This single experiment seems, to me, fully sufficient for explaining after what manner fermentations, shocks and conflagrations are excited in the bowels of the earth, as happens in Vesuvius, Ætna and divers other places: for if iron and sulphur happen to meet together, and are intimately united and penetrate each other, a violent fermentation must ensue, which will produce fire, as in our operation.

Kircher who wrote at the time when gunpowder had come to be largely used in warfare, was impressed with the similarity of the sounds heard and the concussions experienced in earthquake shocks, with those produced by the explosion of gunpowder, especially when used in cannon.

[11] *Explication Physique & Chemique des Feux souterrains, des Tremblemens de Terre, des Ouragans, des Eclairs et du Tonneur.* Histoire de l'acad Roy. des Sciences. Paris. 1700. p. 101. This is one of the papers included in the volume edited by Bevis.

This led him, as well as many other writers who noted this resemblance, to attribute earthquakes to the explosion of a mixture of sulphurous and nitrous vapors within the earth.

The raging of the pent up vapors was thought by some ancient writers to bring about within the earth's interior a condition which they compared to fever in the human body, induced in their opinion by an unnatural accumulation or excessive movement of "humours" in the body. It was this view that Shakespeare[12] had in mind when he wrote:

> Diseased nature oftentimes breaks forth
> In strange eruptions: oft the teeming earth
> Is with a kind of colic pinch'd and vex'd
> By the imprisoning of unruly wind
> Within her womb which for enlargement striving,
> Shakes the old bedlam earth and topples down
> Steeples and moss grown towers.

About the middle of the eighteenth century, a new force may be said to have been discovered when the strange, mysterious, and powerful effects of electricity attracted widespread attention. The Leyden Jar was first constructed in 1745, while Benjamin Franklin's experiment of drawing electricity from the clouds was carried out in 1752.

A group of writers saw in this new force the cause of earthquakes. Stukeley[13] seems to have been the first to advocate this hypothesis. Isnard[14] elaborated the theory that the "electric fluid" was only a more tenuous form of the "mineral exhalation" or "mineral spirit," to which extended reference has already been made in Chapter IX. The heat of the earth's interior developed this exhalation, a portion of which was condensed to form veins of metallic ores, but another part of which spread throughout the whole body of the earth, penetrating it everywhere.

This exhalation, as well as the electricity which accompanied it and was associated with it everywhere, was composed of minute "particles of fire" mixed with minute saline and sulphurous particles. The electricity was identical in character with the exhalation except that the particles of the constituent materials were still more minute and it therefore possessed a still more subtile character than the exhalation with which

---

[12] *King Henry IV*, pt. 1, Act. III, Scene i.
[13] W. Stukeley, *On the Causes of Earthquakes*, Phil. Trans. Roy. Soc. London, 1750. (See fuller reference on p. 414)
[14] *Mémoire sur les Tremblemens de Terre*, Paris, 1758.

it was mingled. Evidence of the nature of its constituent particles was found in the resemblance of the odor developed by the passage of an electric discharge through the air, to the smell of arsenic, phosphorus, or to the fumes given off when iron is dissolved in nitric acid.

The exhalation gradually accumulated in caverns within the earth's crust and eventually, becoming greatly compressed by the continual forcing of new bodies of the exhalation into these large, yet confined, spaces, within which it was often in violent motion, sometimes exploded, these explosions shattering the roof and walls of the caverns and giving rise to disastrous earthquakes. The earth furthermore being traversed in all directions by mineral veins which act as conductors of this electric exhalation, if it may be so termed, just as metals at the earth's surface are good conductors of electricity, these in many cases carry the exhalation from the sources where it originates to distant parts of the earth where it may accumulate. Isnard states that mineral veins are most common in the roots of the mountain chains, which is the reason why earthquakes are frequently experienced in such situations. This electric exhalation is also present in the atmosphere. The tremendous lightning flashes seen in times of storm, as well as other electric phenomena in the air, are manifestations of the same force as that which gives rise to earthquakes within the earth.

Vannucci[15] describes the great earthquake which shook Rimini and the surrounding country in 1786 and contends at great length that this had its origin in "showers of electric fire" which fell upon the unfortunate city, at intervals during several days, from dark clouds which continually overhung it. His discourse gave rise to many hostile and impolite criticisms to which he replies in his paper. Vivenzio, whose work in Calabria will be referred to further on, also held the view that earthquakes were of electric origin. Jean-Claude Delametherie,[16] a prolific writer who expressed fantastic opinions on many geological questions, goes so far as to compare earthquakes to the electric shocks which are given by the Torpedo or Electric Ray and the strata of the earth's crust to the muscles of this fish. The fish moves its muscles and an electric shock results. All earthquakes, he says, are accompanied by a "forte électricité."

The widespread belief that earthquakes were due to winds, vapors or

[15] *Discorso Istorico-Filosofico sopra il Tremuoto...*, Cesena, 1789.
[16] *Sur les causes des Commotions souterranes par l'action galvanique, Journal de Physique, de Chemie et d'Historie Naturelle*, Vol. 81, 1815, p. 393.

exhalations imprisoned within the earth suggested to many of those who thought upon these subjects, what seemed to them an obvious and at the same time certain method of protecting any building or even any district against earthquake shocks, namely, that of sinking deep pits or shafts in order to allow such vapors free passage for their escape. Pliny[17] advocates this and Giovanni Villani in his *History of Florence* states that when Arnolfo di Lapo was about to construct the cathedral church of Santa Maria del Fiori in that city, he caused a number of deep wells to be dug at intervals all around the building in order that any vapors arising from the earth's interior might escape through them and the building be thus preserved from destruction by earthquakes, which from time to time were experienced in Florence and the adjacent parts of Italy.

Isnard[18] concludes his book by pointing out how by this simple expedient, known even to the ancient Persians, the inhabitants in earthquake districts may ensure their safety. If the Romans, he says, had by this simple operation made free passage for the exhalations from Vesuvius, Herculaneum and Pompeii might have been preserved from destruction, and had the Portuguese also taken the same precaution, the Lisbon earthquake might have been prevented as well as the loss of more than 40,000 lives to say nothing of property of untold value; unless indeed, he prudently adds, the subterranean furnaces which give rise to these dreadful phenomena are so vast and so deeply seated that it is beyond all human power to bridle their impetuosity.

And here may be mentioned an interesting work by Bonito[19] which appeared in 1691. It is a stout octavo volume, in which the author, after a general discussion of the nature and origin of earthquakes, gives an annotated list of all the earthquakes of which he could find any record that had taken place in the preceding 3456 years, that is to say, from the "Creation of the world" to the year when his book was written. While, as he remarks, there is a certain difference of opinion among the various authorities who have attempted to determine the date of the creation of the world, he adopts that given by the great chronologist Girolamo Bardi, namely 3967 years before the birth of Christ. In the year 1765 after the creation, Bonito tells us, it is recorded by several of the ancient

[17] *Natural History*, Book II, chapter 82.
[18] *Mémoir sur les Tremblements de Terre*, Paris, 1758.
[19] D. Marcello Bonito, *Terra Tremante overo continuatione dé Terremoti dalla creatione del Mondo sino al tempo presente*, Naples, 1691, 822 pp.

historians that Noah the "Restorer of the human race," bearing the name of Giano, came to Italy after the flood subsided, and founded Rome, of which he became the first king, "establishing his royal seat on Mount Gianicolo." In that year, he says, the first earthquake is recorded to have taken place and this was in Italy. He then goes on to mention in their order the recorded earthquake shocks down to the time of that which took place at Ancona, in the year 1690 A.D. These are over 1400 in number.

During the decade between 1750 and 1760 Western Europe was repeatedly visited by heavy shocks, the great Lisbon earthquake in 1755 being one of the most disastrous catastrophes recorded in history. The careful study of these by a number of able observers brought to light many new facts the study of which opened a new chapter in the history of seismology. These new facts with the arguments and conclusions based upon them are excellently set forth in a series of ten papers[20] which appeared in various publications in different countries, several being taken from the *Philosophical Transactions of the Royal Society of London,* and which were collected by John Bevis or Bevans and reprinted, some of them in a somewhat abridged form, without mention of Bevis' name, under the title *The History and Philosophy of Earthquakes, collected from the best writers on the subject by a member of the Royal Academy of Berlin. London, 1757.* A perusal of these Papers will enable the reader who wishes to follow the subject further, to gain a clear idea of the views concerning earthquakes and their origin which were held in the middle of the eighteenth century. Another important work which appeared about this time is that by Élié Bertrand,[21] in the first half of

[20] 1. Sturmius, J. C. *De Terrae-Motibus,* etc. Altdorff, 1670, 32 pp.

2. Lister, M. Phil. Trans. vol. 13, 1683, pp. 512–519.

3. Hooke, R. *Posthumous Works,* 1705, pp. 277–450.

4. Woodward, J. *An Essay Towards a Natural History of the Earth,* etc., 3rd ed. 1723, pp. 149–160.

5. Lemery, N. Paris, Ac. Sci. Hist. Mem.

6. Bouguer, P. *La Figure de la Terre,* 1749, pp. lxiv–lxxviii.

7. Buffon, G. L. L. *Histoire Naturelle,* 3rd ed. 1750, vol. 1, pp. 502–533.

8. Ray, J. *Three Physico-Theological Discourses* (second discourse, *Consequences of the Deluge*) 3rd ed. 1713, pp. 289–291, 294.

9. Hales, S. Phil. Trans. vol. 46, 1752, pp. 669–681.

10. Stukeley, W. *The Philosophy of Earthquakes,* Natural and Religious a pamphlet of 139 pp. 3rd ed. 1756, based on three articles in Phil. Trans. vol. 46, 1752, pp. 641–645, 657–669, 731–750. The second edition, a pamphlet of 39 pp., was published in London in 1750.

[21] *Mémoires Historiques et Physiques sur les Tremblements de Terre,* La Haye, 1757 (Reprinted in his *Reçueil de Divers Traités sur l'Histoire Naturelle,* Avignon, 1766).

which he deals with the general subject of earthquakes, while in the second he gives a valuable account of the earthquakes which had taken place in Switzerland.

3. JOHN MICHELL AND THE GREAT LISBON EARTHQUAKE OF 1755

Still another very important paper which contributed largely to the further advance of seismological knowledge, is that by John Michell[22] Woodwardian Professor in the University of Cambridge, concerning which Geikie goes so far as to say that "it stands out conspicuously as by far the most important contribution on this branch of science that had yet appeared in any language or country."[23]

Michell was a man of wonderful versatility, being an excellent classical scholar and a mathematician, as well as an astronomer of distinction. He was not only the author of important contributions to the science of astronomy but he made the 10 foot reflector which was later bought and used by Herschel. He also invented the torsion balance subsequently employed by Cavendish in his determination of the density of the earth. His outstanding contributions to a true understanding of the stratified succession of the sedimentary formations and the origin of mountains have been referred to in Chapter X. He was one of those men who "seemed by gentle persuasion to have penetrated that reserve of nature which baffles smaller men."[24] Furthermore, like so many English men of science in those earlier times when the stream of life ran more quietly than at present, he was a clergyman. Michell in his treatment of earthquakes differs from his predecessors in that he has freed himself from the shackles of ancient views and traditions. His paper is based wholly upon the work of later observers, largely, as he himself states, on the papers in the volume edited by Bevis referred to above, and it breathes the spirit of the modern time.

Michell first looks for the cause of earthquakes and says:

We need not go far in search of a cause, whose real existence in nature we have certain evidence of, and which is capable of producing all the appearances of these extraordinary motions. The cause I mean is subterraneous fires. These fires if a large quantity of water should be let in upon them suddenly, may produce a vapour whose quantity and elastic force may be fully sufficient for that purpose.

[22] *Conjectures concerning the cause, and Observations upon the Phenomena of Earthquakes, particularly of that great Earthquake of the First of November 1755, which proved so fatal to the city of Lisbon, and whose Effects were felt as far as Africa and more or less throughout almost all Europe.* Philosophical Transactions of the Royal Society of London, Vol. LI, part II. 1761.
[23] *The Founders of Geology*, London, 1905, p. 274.
[24] Davidson, Charles. *The Founders of Seismology*, Cambridge, 1927.

The fires he thinks are the same as those which give rise to volcanoes; "Those places which are near volcanoes are always subject to earthquakes." He however falls into the old error, which survived down to the time of Werner, of attributing these fires to the combustion of beds of coal or pyritous shale occurring interstratified with other beds within the earth's crust.

When such a set of strata is thrown into folds and the pyritous beds happen to be brought near the surface at the summits of the folds, the steam and vapors resulting from their combustion being under great compression, may at the tops of mountain ridges force their way to the surface, and this is why volcanoes are found on the summits of hills or mountains (see figure 66, nos. 1 and 3.)

Michell's conception of the manner in which an earthquake may originate at some particular point in the burning stratum is shown in figure 66, number 4. He says:

Both the tremulous and wave-like motions observed in earthquakes, may be very well accounted for by such a vapour. In order to trace a little more particularly the manner in which these two motions will be brought about, let us suppose the roof over some subterraneous fire to fall in. If this should be the case, the earth, stones, &c., of which it is composed would immediately sink in the melted matter of the fire below: hence all the water contained in the fissures and cavities of the part falling in would come in contact with the fire and be almost instantly raised into vapour. From the first effort of this vapour, a cavity would be formed (between the melted matter and the superincumbent earth) filled with vapour only before any motion would be perceived at the surface of the earth: this must necessarily happen, on account of the compressibility of all kinds of earth, stones, &c., but as the compression of the materials immediately over the cavity would be more than sufficient to make them bear the weight of the superincumbent matter, this compression must be propagated on account of the elasticity of the earth, in the same manner as a pulse is propagated through the air; and again the materials immediately over the cavity, restoring themselves beyond their natural bounds, a dilatation will succeed to the compression; and these two following each other alternately, for some time, a vibratory motion will be produced at the surface of the earth.

As a small quantity of vapour almost instantly generated at some considerable depth below the surface of the earth will produce a vibratory motion, so a very large quantity (whether it be generated almost instantly, or in any small portion of time) will produce a wave-like motion. The manner in which this wave-like motion will be propagated, may, in some measure, be represented by the following experiment. Suppose a large cloth, or carpet, (spread upon a floor) to be raised at one edge, and then suddenly brought down again to the floor, the air under it, being by this means propelled, will pass along, till it escapes at the opposite side, raising the cloth in a wave all the way as it goes. In like manner, a large quantity of vapour may be conceived to raise the earth in a wave, as it passes along between the strata, which it

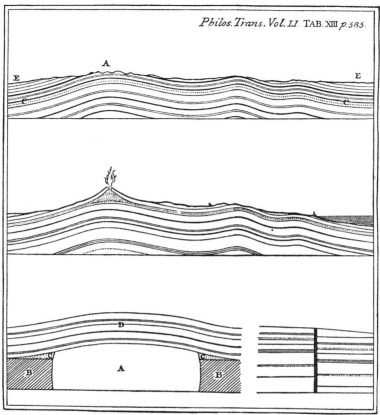

Philos. Trans. Vol. LI. TAB. XIII *p. 585.*

FIG. 66. Plate showing the origin of volcanoes and earthquakes, according to Michell.

Nos. 1 and 3. Vapors (dotted stratum) passing between beds in a series of strata and in figure 3 finding access to the surface at the summit of an anticline giving rise to a volcano. These vapors when confined beneath lower land where there is a greater thickness of overlying strata or beneath the waters of the sea, would be under a higher pressure, might give rise to an earthquake. This is the reason (according to Michell) why many earthquakes originate beneath the bed of the ocean.

No. 2. A fault. Michell says "This is what is usually called by miners the trapping down of the strata—it may have a great effect in producing some of the singularities of particular earthquakes."

No. 4. *A* represents a vertical section of the matter on fire, *BB* parts of the same stratum as yet unkindled. The overlying strata *D* have been arched up by the expansion induced in them through the heat from the combustion of *A*. This separates *D* from *B*, leaving an empty space *CC* around *A*. This space becomes filled with water coming in from adjacent cracks and fissures in the country rock. The burning material in *A* will not pass into the space *CC* on account of lack of fluidity, but the waters in the earth, will on account of their cooling action form a wall or crust of hard rocky material separating *CC* from *A*. As *D* continues to rise the wall between *CC* and *A* becomes higher. This wall will eventually give way, precipitating the water into the burning mass of *A*, and the steam and vapor thus produced will find an exit by passing, often for long distances between certain strata, causing the elevation of the overlying country.

may easily separate in an horizontal direction, there being, as I have said before, little or no cohesion between one stratum and another. The part of the earth that is first raised, being bent from its natural form, will endeavour to restore itself by its elasticity, and the parts next to it beginning to have their weight supported by the vapour, which will insinuate itself under them, will be raised in their turn, till it either finds some vent, or is again condensed by the cold into water, and by that means is prevented from proceeding any farther.

If a large quantity of vapour should continue to be generated for some time, several waves might be produced by it.

The vibratory motion occasioned by the first impulse of the vapour, will be propagated through the solid parts of the earth, and therefore, it will much sooner become too weak to be perceived, than the wave-like motion; for this latter, being occasioned by the vapour insinuating itself between the strata, may be propagated to very great distances; and even after it has ceased to be perceived by the senses, it may still discover itself by the appearance before-mentioned.

The compressibility and elasticity of the earth, are qualities which don't show themselves in any great degree in common instances, and therefore are not commonly attended to. On this account it is, that few people are aware of the great extent of them, or the effects that may arise from them, where exceeding large quantities of matter are concerned, and where the compressive force is immensely great. The compressibility and elasticity of the earth may be collected in some measure, from the vibration of the walls of houses, occasioned by the passing of carriages in the streets next to them. Another instance to the same purpose, may be taken from the vibration of steeples, occasioned by the ringing of bells, or by gusts of wind: not only spires are moved very considerably by this means, but even strong towers will, sometimes, be made to vibrate several inches, without any disjointing of the mortar, or rubbing of the stones against one another. Now, it is manifest, that this could not happen, without a considerable degree of compressibility and elasticity in the materials, of which they are composed: . . . There are several things that seem to argue *a considerably greater density in the internal, than the external part of the earth*; and why may not this greater density be owing to the compression of the superincumbent matter, since it is probable, that the matter, of which the earth is composed, is pretty much the same kind throughout?

Michell seems to have been the first to estimate the velocity of an earthquake. Referring to the shocks which succeeded the first great one at Lisbon in 1755, he says:

The velocity with which they were all propagated was the same, being at least equal to that of sound, for they all followed immediately after the noise that preceded them, or rather the noise and the earthquake came together: and this velocity agrees very well with the intervals between the time when the first shock was felt at Lisbon and the time when it was felt at other distant places, from the comparison of which it seems to have travelled at the rate of more than twenty miles per minute.

Lastly, Michell takes up the question of the means of determining the position of the place where an earthquake originates:

If we would inquire into the place of origin of any particular earthquake, we have the following grounds to go upon:

First, The different directions, in which it arrives at several distant places: If lines be drawn in these directions, the place of their common intersection must be nearly the place sought: but this is liable to great difficulties; for there must necessarily be great uncertainty in observations, which cannot, at best, be made with any great precision, and which are generally made by minds too little at ease to be nice observers of what passes; moreover, the directions themselves may be somewhat varied, by the inequalities in the weight of the superincumbent matter, under which the vapour passes, as well as by other causes.

Secondly, We may form some judgment concerning the place of the origin of a particular earthquake, from the time of its arrival at different places; but this also is liable to great difficulties. In both these methods, however, we may come to a much greater degree of exactness, by taking a medium amongst a variety of accounts, as they are related by different observers.

Thirdly, We may come to the greatest degree of exactness in those cases where earthquakes have their source from under the ocean; for, in these instances, the proportional distance of different places from that source may be very nearly ascertained, by the interval between the earthquake, i.e. the vibrating motion, and the succeeding wave: and this is more to be depended on, as people are much less likely to be mistaken in determining the time interval, than in observing the precise time of the happening of some single event.

If we would inquire into the depth, at which the cause lies that occasions any particular earthquake, I know of no method of determining it, which does not require observations not yet to be had; but if such could be procured, and they were made with sufficient accuracy, I think some kind of guess might be made concerning it. . . .

For the Lisbon earthquake the observations made were "too gross":

but if I might be allowed to form a random guess about it I should suppose (upon a comparison of all circumstances) that it could not be much less than a mile or a mile and a half and I think it is probable that it did not exceed three miles.

The first method was that which Mallet adopted in his study of the Neapolitan earthquake of 1857. The greater accuracy to be obtained by measuring intervals rather than absolute times, referred to in connection with the third method, lies at the root of the method depending on the duration of the preliminary tremor now so widely used. Davidson observes that Michell's "random guess" as to the depth of the focus, like Newton's guess with regard to the density of the earth, is one of those intuitions which only occur to the ablest minds.

One of the important contributions which Michell made to the advancement of seismology was in pointing out that the vibratory motion in earthquakes was due to the propagation of elastic waves in the earth's crust and that such waves would be transmitted outward from the source of their origin for long distances and would gradually die away.

4. GRIMALDI, VIVENZIO, AND THE CALABRIAN EARTHQUAKES OF 1783

After the publication of Michell's paper no other very important contribution was made to the science of seismology for the next quarter of a century, when Milne's work appeared.   During this period however the science did make a distinct advance through the contributions of a number of persons whose labors, however, for the most part lay in other fields.   The great Calabrian earthquakes which in 1783 brought about widespread destruction in this part of Italy with the loss of 35,000 lives, directed the attention of a number of able investigators to the study of earthquake action in this area.   *Francesco Grimaldi*, the Neapolitan Secretary of War, a man whose contributions to knowledge had been in the sphere of literature and history, was the first to study this devasted area.   He did so at the wish of his king but died in 1784 before his report was published.   In it he gave an account of the successive earthquake shocks and the damage which they wrought, mentions the relief measures which were taken and gives a list of the destructive earthquakes which were experienced in Calabria between 1181 and 1756.[25]

In the same year in which the earthquake took place, Giovanni Vivenzio, Chief Physician in the court of Naples, also wrote his *Istoria*, a second and enlarged edition of which was published four years later. The first edition[26] of this work, while giving an account of the character and extent of the devastation wrought at different places in Calabria, is devoted largely to an attempt to demonstrate that earthquakes are due to electricity, a fact, he says, which is now "coming to be recognized by all the greater physicists and by those persons who are informed concerning the operations of Nature."   The second edition[27] is in two quarto volumes and is composed largely of irrelevant matter.   It however contains a valuable table, compiled by Domenico Pignataro, of 1186 earthquake shocks which took place in this region in the years 1783 to 1786, in which he makes the earliest known attempt to devise an "In-

[25] *Descrizione de Tremuoti accaduti nelle Calabrie nel 1783*, Naples, 1784.

[26] *Istoria e teoria de' Tremuoti in generale ed in particolare di quello della Calabria e di Messina del 1783*, Naples, 1783.

[27] *Istoria de' tremuoti avvenuti nella Provincia della Calabria ulteriore e nella Città di Messina nell' anno 1783*, Naples, 1788.

tensity Scale," classifying the earthquake shocks as, Slight, Moderate, Strong, and Very Strong, denoting these intensities in his table by the symbols F^1, F^{11}, F^{111}, F^{1111}, with a fifth class in which he places the five most violent earthquakes, indicating these in the tables by a maltese cross. P. N. G. Egen[28] in 1828 also devised and made use of another similar scale in which the six classes were rather more sharply defined. Mallet introduced two other scales in 1859 and 1862 respectively, while still others were put forward by De Rossi[29] in 1879 and Forel[30] in 1883, each of these having 10 classes. Others were adopted by later investigators.

5. LYELL AND DAVID MILNE

Next should be mentioned the contributions made by Lyell in his *Principles of Geology* which appeared in 1830. Up to the middle of the seventeenth century the writers on earthquakes confined their attention to the question of the origin or when describing the effects dwelt almost exclusively on the number of cities destroyed, the great loss of life which resulted and the various atmospheric disturbances which accompanied the shocks. It was not, however, until Hooke in his *Discourse on Earthquakes* published in 1668 and in a number of papers by him which appeared in the succeeding years, chiefly in the Transaction of the Royal Society of London, pointed out and emphasized the striking results brought about by earthquakes in the elevation and depression of land areas and in modifying the character of the earth's surface that this aspect of earthquake action came to be fully recognized.

It was to this aspect of earthquake action that Lyell chiefly devoted his attention in his *Principles*, drawing his instances from a number of earthquakes, most of which took place within the 150 years before he wrote. His account is illustrated by a number of cuts showing changes in the configuration of the earth's surface brought about by faults, landslides, fissures and surface depressions due to earthquake shocks.

Important investigations were also carried out by David Milne, working at times in association with James David Forbes, on the earthquakes of Great Britain, the results appearing in a series of papers published in various British scientific journals between 1841 and 1844. Milne describes the transmission of vibrations passing outward from the

[28] Ann. Phys. Chem., 1828, pp. 153–163.
[29] *Bullettino del Vulcanismo Itialiano*, 1874, p. 1.
[30] *Bullettino del Vulcanismo Itialiano*, 1883, p. 67.

source or origin in spherical waves and shows that at the epicentre the movement was vertical and that it became more nearly horizontal as the distance from the epicentre increased. He also remarks that if instruments could be invented which would record not merely the intensity of earthquake shocks but the direction from which they came, by placing such instruments in different localities it would be possible to determine the point within the earth's interior at which the shocks originated.[31]

6. Humbolt

Humbolt writing about the same time (1844) in referring to "the very considerable number of earthquakes which I have experienced in both hemispheres, alike on land and at sea" says:

The propagation is most generally effected by undulations in a linear direction, with a velocity of from twenty to twenty-eight miles in a minute, but partly in circles of commotion or large ellipses, in which the vibrations are propagated with decreasing intensity from a centre towards the circumference. There are districts exposed to the action of two intersecting circles of commotion. In northern Asia, where the Father of History, and subsequently Theophylactus Simocatta, described the districts of Scythia as free from earthquakes, I have observed the metalliferous portion of the Altai mountains under the influence of a twofold focus of commotion, the lake of Baikal, and the volcano of the Celestial Mountain (Thianshan). When the circles of commotion intersect one another, when, for instance, an elevated plain lies between two volcanoes simultaneously in a state of eruption, several wave-systems may exist together as in fluids, and not mutually disturb one another. We may even suppose *interference* to exist here as in the intersecting waves of sound. The extent of the propagated waves of commotion will be increased on the upper surface of the earth, according to the general law of mechanics, by which on the transmission of motion in elastic bodies, the stratum lying free on the one side endeavours to separate itself from the other strata.[32]

In the last volume of Cosmos,[33] which did not appear until 1858, Humbolt says:

Since the appearance of volume I of this work in 1845, the obscurity in which the seat and the causes of the phenomena are involved has but little diminished, but the excellent work of Mallet (1846) and Hopkins (1847) have thrown some light

[31] *Edinburgh New Philosophical Journal*, Vol. 31, 1841, p. 276.

[32] *Cosmos*, Vol. I, (Translated from the German by E. C. Otté, Bohn's Standard Library), London, 1849.

[33] *Cosmos*, Vol. V, (Translated from the German by E. C. Otté and W. S. Dallas.), London 1858, p. 166 & p. 180.

upon the nature of the concussions. . . . The analogies between the oscillations of solid bodies and sound waves in the atmosphere to which Thomas Young[34] had already called attention, lead to simpler and more satisfactory views upon the dynamics of earthquakes.

It is furthermore interesting to note that Humbolt emphasizes the distinction, which he says has been "lately made" (although he does not mention by whom) between *Volcanic* and *Plutonic* earthquakes. Many volcanoes, he observes, go into violent eruption without any earthquakes taking place even in their immediate vicinity. In fact,

The most destructive earthquakes recorded in history and which have passed through many thousand square miles, if we may judge from what is observable at the surface, stand in no connection with the activity of volcanoes. These have lately been called Plutonic, in opposition to true Volcanic earthquakes which are usually limited to smaller districts. By far the greater number of earthquakes upon our planet must be called Plutonic.

7. The Great Achievements of Seismology in Recent Years and the Many Problems Which Still Await Solution

To trace out the great and very rapid forward strides which have been made by seismology during more recent times does not lie within the scope of the present work. It is admirably set forth by Davidson[35] in his *Founders of Seismology* to which work the reader who wishes to familiarize himself with the later advances made by investigators in America, Italy, Central Europe and Japan, is referred. While these have been striking and most significant, demonstrating as they do that the study of earthquake phenomena affords a means of exploring the interior of the earth, a region which had previously been merely a playground for human imagination and speculation, and of ascertaining its true character and structure.

Recent seismic studies have shown that the wonderful, romantic and age-long ideas and concepts concerning the nature of the earth's interior have been completely dissipated. The earth does not resemble a great sponge containing innumerable caverns and "antres vast," into and out of which the violent winds rush to and fro, shaking the whole globe. Nor does it enclose a multitude of caverns filled with fire or water. The central portion of the earth is not a hollow space in which

[34] *Lectures in Natural Philosophy*, 1807, Vol. I, p. 717.
[35] Charles Davidson, *The Founders of Seismology*, Cambridge University Press, 1927.

rage the fires of an enormous furnace, the source and origin of all manner of valuable products and strange phenomena which make themselves manifest at the surface.

The results which have been attained by the investigations of seismologists in recent years have been set forth by one of their most able investigators:

Is the interior of the earth solid or fluid?   We shall not enter into a discussion of the meaning of these terms further than to say that, at least in the popular sense, a gas is a body which will fill space in which it is placed, a liquid is one which occupies a definite volume and shows a negligible resistance to change of form, and a solid has both a definite volume and a fixed form so that it resists any force tending to change its size or shape.   Hence, we may take form-elasticity or rigidity as a criterion of solidity.   Form-elasticity is necessary for the transmission of waves of elastic shear. Now shear waves are observed not only in the earth's crust but in the deeper mantle. Their speed increases from about four kilometers a second at the base of the crust to about seven kilometers a second at a depth of 1,200 kilometers and reaches a maximum value of about seven and one-half kilometers a second between there and the boundary of the core.   They practically disappear as do the compression-dilation waves at a little over one hundred degrees distance, where they are cut off by the core. We must conclude from the fact that shear waves traverse the earth down to the core that the earth is solid, at least to that depth, and furthermore, we must conclude from the high value of the shear-wave velocity that these outer shells are highly rigid.[36]

It has been established that the earth consists of a heavy inner core probably metallic in character enclosed in outer envelope or "mantle" of rock, which latter is made up of a series of subordinate concentric shells each of which transmits earthquake shocks more rapidly than the one enclosing it.

Mr. Macelwane however in a recent address[37] shows that the story which earthquakes have yet to tell is only beginning to unfold itself:

The most accessible part of our earth is its outer crust; yet this very outer crust bristles with unsolved problems.   Geologists have been accustomed to speak of the zone of fracture near the surface of the earth and of the underlying zone of flow. Now it has been found that the depth of first yielding in most destructive earthquakes is not, as might be expected, near the top of the zone fracture but is at or somewhat below its base.   The *normal* depth of focus of earthquakes or depth of first significant

[36] *Our Present Knowledge Concerning the Interior of the Earth*, James B. Macelwane, Bull. Seis. Soc. of Amer., Vol. 21, No. 4, Dec. 1931, pp. 245, 247.

[37] *Problems and Progress on the Geologico-Seismological Frontier*, Address of the retiring Vice-president and chairman of the Section of Geology & Geography of the American Association for the Advancement of Science, Dec. 30, 1935, Science, Feb. 28, 1936.

radiation of earthquake waves seems to be between ten and fifteen kilometers. Does this first sudden failure take place by faulting properly so-called? ... Wood presented evidence that volcanic earthquakes as well as *volcanic eruptivity* itself may be largely controlled by tectonic factors. The problem is complicated by the necessity of explaining both the truly plutonic earthquakes to which we shall refer later in this address and also the large-scale surface faulting that has clearly been caused by certain great earthquakes.

As evidence accumulates in regard to crustal structure the impression given is one of increasing complexity, almost of confusion. Before any certain conclusions can be drawn we must have a much greater body of evidence for a large number of regions. Such study can be prosecuted with great promise of fruit where there are a number of stations close together equipped with sensitive seismographs and recording with an open-time scale; and then only when earthquakes occur within the proper range of distance.

Furthermore, this mysterious, intriguing mantle is the seat of *plutonic earthquakes* which have occurred at a wide variety of depths down to 700 kilometers and are frequently violent. In some of the deepest of these earthquakes the total energy suddenly released is enormous. This energy must have been stored as potential energy in some form or other. Two possibilities suggest themselves: the energy might have been stored chemically or it might have been stored as potential energy of elastic strain.

In conclusion, it would seem that the entire picture of positive geological results attained by seismological methods, of the new problems realized and formulated, and of the better understanding and readiness to coöperate which is evidenced in the ranks of geologists and of seismologists, is a most encouraging one and foreshadows a new and exceedingly interesting era of geologico-seismological research.

CHAPTER XII

THE ORIGIN OF SPRINGS AND RIVERS

There is one problem—that of the origin of Springs and Rivers—which was the subject of continuous speculation and controversy throughout the ages, but which may now be said to have been definitely and finally solved, in all its chief features at least, and it is the aim of the present chapter, commencing from the earliest times of which we have records in Europe, to trace in outline the opinions which have been held on this subject and the changes which have taken place in these opinions from age to age throughout the centuries.

That streams of water, in some cases indeed rushing torrents of water, issued from cracks and fissures in the earth's crust in certain places, giving rise to rivers or at least contributing to their flow, is a fact which made a deep impression on men's minds from very ancient times. The question as to whence all this water came at once presented itself.[1]

1. VIEWS OF ARISTOTLE AND OTHER CLASSICAL WRITERS

The earliest writers were of the opinion that these streams of water issued from one or many great lakes, or caverns, which existed within

[1] See also M. N. Baker and R. E. Horton, *Historical Development of Ideas regarding the Origin of Springs and Ground-Water, Trans. American Geophysical Union*, Washington, D. C., 1936, Pt. II, p. 395.

the earth. This was taught by Anaxagoras, (500–428 B. C.) and Plato (429–347 B. C.) imagined that there is within the earth an immeasurable cavern filled with water in continual motion, from which all rivers issued and to which their waters returned. Later Virgil in his *Aeneid* sings of a great subterranean abyss of waters:

> Hinc ira Tartarei quaefert Acherontis ad cendas,
> Turbidus hic caeno vastaque voragine gurges,
> Aestuat atque omnem Cocyto eructat arenam.

Also in the *Georgics* he refers to rivers which issue from caverns and are the homes of Nymphs, which fair creatures even took up their residence sometimes in the great subterranean caverns from which the rivers came.

That there was a great body of subterranean waters seems to have been the thought in the minds of the writers of the books of *Genesis* and *Exodus*. In the story of the deluge it is stated that "The fountains of the Great Deep were broken up," that is to say cleft open, and from these the waters which covered the whole earth gushed forth, the "Waters under the Earth" are also referred to in the fourth verse of the 20th chapter of Exodus and in other passages.

The chief objection to the view that springs and rivers issued from great subterranean reservoirs, was that no matter how large such reservoirs might be they would eventually be drained dry by the continuous flow of the rivers, unless there were some paths by which the waters might return to the central reservoirs from which they came.

The foremost man of science in early classical times was undoubtedly Aristotle (B. C. 384–323). His views on the origin of springs and rivers are to be found in his *Meteorologica* (Book I, chiefly in chapter 13). He commences his discussion of the subject by referring to the views held by Anaxagoras[2] and the reasons why he holds these to be erroneous. Summing up the opinions of this writer he says:

It is thought that the water is raised by the sun and descends in rain and gathers below the earth and so flows from a *great reservoir*, all the rivers from one, or each from a different one. *No water at all is generated*, but the volume of the rivers consists of the water that is gathered into such reservoirs in winter. Hence rivers are always fuller in winter than in summer, and some are perennial, others not. Rivers are perennial where the reservoir is large and so enough water has collected in it to last out and not be used up before the winter rain returns. Where the reservoirs

[2] *Meteorologica*, Book I, Chapter 13, 249^b 2 to 20 (Webster's translation). The references given follow those of Webster's text.) Clarendon Press, Oxford.

are smaller there is less water in the rivers, and they are dried up and their vessels empty before the fresh rain comes on.

But if any one will picture to himself a reservoir adequate to the water that is continuously flowing day by day, and consider the amount of the water, it is obvious that a receptacle that is to contain all the water that flows in the year would be larger than the earth, or, at any rate, not much smaller.

In order to set forth clearly Aristotle's views, these will be given in his own words, not indeed in the original Greek which might present certain difficulties to many modern readers but from the critical translation of the *Meteorologica* by E. W. Webster of Oxford, an excellent Greek scholar, who made this work a subject of special study, the translation in question being published by the Clarendon Press of the University of Oxford.

Aristotle opens Book I, chapter 13 which treats of the origin of springs and rivers with the words:—

Let us explain the nature of winds and all windy vapours, also of rivers and of the sea . . . in other matters so in this, no theory has been handed down to us that the most ordinary man could not have thought of.[3]

In chapter 2, Aristotle has already stated that there are three zones surrounding the earth, those of fire, air and water, in descending order.

Let us then[4] go on to treat of that region which immediately surrounds the earth. It is the region common to water and air and the processes attending the formation of water above,

that is, in the atmosphere.

The efficient, chief and first cause is the circle in which the sun moves. For the sun as it approaches or recedes causes dissipation and condensation and so gives rise to generation and destruction . . . the moisture surrounding the earth is made to evaporate by the sun's rays and other heat from above and rises. But when the heat which was raising it leaves it, in part dispersing to the higher region, in part quenched through rising so far into the upper air, then the vapour cools because its heat is gone and because the place is cold and condenses again and turns from air into water and after the water has formed it falls down again to the earth. The exhalation of water is vapour: air condensing into water is cloud, so we get a circular process that follows the course of the sun. For according as the sun moves to this side or that, the moisture in this process rises or falls. We must think of it as a river flowing up and down in a circle and made up partly of air and partly of water. When the sun is near, the

---

[3] *Meteorologica*, 349[a], 12–16.
[4] *Meteorologica*, Book I, Chapter 9, 346[b], 16–19.

stream of vapour flows upwards; when it recedes the stream of water flows down . . .
when the water falls in small drops it is called a drizzle; when the drops are larger it
is rain.[5]

Reference has been made elsewhere in this volume to Aristotle's views
concerning the nature of his four "elements"—fire, air, earth and water.
His differentiation of air and water is much less sharp than that which
we now make between these two substances, in fact in his work *De
Generatione et Corruptione* Book II. 3, he says:

Air, being hot and moist, is a sort of aqueous vapor.

Aristotle then proceeds:

Just as above the earth, small drops form and these join others, till finally water
descends in a body as rain, so too we must suppose that in the earth the water at first
trickles together little by little and that the sources of rivers drip, as it were, out
of the earth and then unite.   Hence too the headwaters of rivers are found to flow
from mountains and from the greatest mountains there flow the most numerous and
greatest rivers.   Again most springs are in the neighbourhood of mountains and of
high ground.   For mountains and high ground, suspended over the country like a
saturated sponge, make the water ooze out and trickle together in minute quantities
but in many places.[6]   Even rivers when they flow from marshes, these latter lie
below mountains or on gradually rising ground.[7]

But while Aristotle believes that a portion of the water which flows
out of the earth in the form of springs, is derived from rain which having
been formed by the condensation of air into water in the high atmosphere
falls upon the earth's surface as rain and sinks into the earth which then
holds it as a sponge might do, he also states that:

It is unreasonable for anyone to refuse to admit that air becomes water in the earth
for the same reason that it does above it.   If the cold causes the vaporous air to
condense into water above the earth, we must suppose the cold *in the earth* to produce
this same effect, and recognize that there not only exists in it and flows out of it,
actually formed water, but that water *is continually forming in it too.*[8]

Furthermore there is a passage in the thirteenth chapter of the *Meteorologica* in which Aristotle states that there is a third source from which
the water within the earth is derived.   This is a continuation of the passage quoted above, in which reference is made to the mountains and high

[5] *Meteorologica*, Book I, Chapter 9, 346[b], 20–36 & 347[a], 1–13.

[6] *Meteorologica*, Book I, Chapter 13, 349[b], 31–36 & 350[a].

[7] *Meteorologica*, Book I, Chapter 13, 350[b], 2–23.

[8] *Meteorologica*, Book I, Chapter 13, 349[b], 21–27.

ground being suspended over the lower lands like a saturated sponge. It reads as follows:

> They (the mountains & higher lands) receive a great deal of water falling as rain and they *also* cool the vapour that *rises* and condense it back into water.[9]

This passage is obscure and there was much discussion and controversy[10] with reference to Aristotle's meaning by many writers throughout the Middle Ages and in later times, for Aristotle does not state where this rising vapor, which is condensed into water, comes from. In fact it is interesting to note that very diverse views were attributed to Aristotle by these various writers on the basis of their interpretation of this and the other passages quoted above.

As has been already mentioned (Chapter II, p. 16), most of the works of Aristotle which have come down to us are in the nature of notes and memoranda for lectures. Of these the *Meteorologica* is one, a fact which accounts for the obscurities and repetitions which are found in places, but which would have been cleared up and explained when the lecture was being given.

It will thus be seen that Aristotle held that the water which flowed out of the earth in the form of springs consisted in part 1) of rain water which had percolated into the earth's crust, and 2) in part of water which was formed within the earth's crust by the condensation of air which made its way into the earth apparently from the external atmosphere, and 3) in part from the condensation of vapors which "*rose*" from some source which is not stated; and that all this water was not gathered together in subterranean reservoirs but was held within the mountains and the higher lands as it were in a sponge.

Another interesting and curious fact may here be noted, namely, that while it is now recognized that rivers derive most of their water from rains or from the melting of the snow on high mountains which runs off on the surface of the land, such streams according to Aristotle would not be *rivers* at all, for he says:

> We do not call water that flows anyhow a river, even if there is a great quantity of it, but only if the flow comes from a spring.[11]

This belief that springs were the source of all or of most rivers, may have been due in part to the fact that he lived for the greater part of his life in

---

[9] *Meteorologica*, Book I, Chapter 13, 350ᵃ, 9–14.
[10] See Kircher, *Mundus Subterraneus, Amsterdam,* 1678, p. 248.
[11] *Meteorologica*, Book II, Chapter 4.

Greece and other countries where limestone is the usual country rock and where consequently springs are abundant.

Seneca[12] (3 B.C.–A.D. 65) also like Aristotle found it impossible to believe that it was rain water alone that issued as springs. This is seen from the following passage:

> You may be quite sure that it is not mere rain water that is carried down in our greatest rivers, navigable by large vessels from their very source, as is proved by the fact that the flow from the fountain-head is uniform winter and summer. Rainfall may cause a torrent, but it cannot maintain the steady, constant flow of a full river. Rains cannot produce, they can only enlarge and quicken a river.

He then deals with the question as to the true source of the water which flows out of the earth and gives rise to springs and rivers:

> We Stoics[13] are satisfied that the earth is interchangeable in its elements. . . . Air passes into moisture but nevertheless contains moisture. Earth yields both air and water and is never at any time devoid of water any more than it is of air. The mutual transition is the easier, because there is already an admixture of the element to which the transition is to be made. So then, the earth contains moisture which it forces out. It contains air, which the darkness of its wintry cold condenses to form water. By nature, too, earth has itself the power of changing into water: this power it habitually exerts.

Seneca's solution of the problem is then that the earth itself was changed into water and thus provided a continuous source of supply to the rivers "As long as the earth endureth," but this water in the course of time changes back into earth, thus all the elements maintain their due proportions and nature preserves its equilibrium.

One other classical author may be mentioned, who, while not treating of the subject at any length, touches upon the question of the origin of springs and rivers and in part anticipates the true explanation which was not to be established until some sixteen centuries later. This is Vitruvius, the Roman engineer and architect who, in Book VIII of his treatise *On Architecture* says:

> Now vapour and clouds and moisture seem to rise from the earth, for this reason that the earth contains both fervid heat and huge blasts of air, coldness and a large amount of water. Hence when the earth is cooled at night, and the rising sun touches it by its force, and gusts of wind rise up through the darkness, the clouds rise on high from the damp places . . . Whithersoever the wind carries the massed moisture from

---

[12] Seneca, *Quaestiones Naturales*, translated by John Clarke and published under the title *Physical Science in the time of Nero*, London, 1910, p. 123.

[13] Seneca, *Quaestiones Naturales*, pp. 120–121.

springs, from rivers and marshes and the sea, the moisture under the sun's influence is collected and drawn forth, and the clouds are raised on high. Then the clouds, supported on the waves of air, meet the resistance of the mountains and becoming liquid in rain storms, by their fullness and weight, break and their water is poured over the fields.[14] . . . Valleys between mountains are *subject* to *much* rain and because of the dense forests, snow stands there longer under the shadow of the trees and the hills. Then it melts and percolates through the interstices of the earth and so reaches to the lowest spurs of the mountains, from which the product of the springs flows and bursts forth. But on the plains one cannot get supplies of water.[15]

2. VIEWS HELD IN THE MIDDLE AGES AND IN THE RENAISSANCE

In the long interval of the Dark Ages between the classical times and those of the middle ages but little attention was directed to this subject, although Isidor of Seville[16] (A.D. 570–636) says that springs and rivers issue from a great abyss of subterranean waters to which all eventually return through hidden passages. With the coming of the middle ages, however, this subject of the origin of springs and rivers again attracted general attention. But now the Christian Church had come to occupy a very prominent place in the intellectual life of Europe: indeed almost every writer and scholar was connected with it in some capacity. The authority of Holy Writ was regarded as paramount on all questions, and certain passages found in it were regarded as indicating and indeed definitely proving, that the ocean was the source of all rivers and other bodies of moving water. And chief among these was that passage from the book of *Ecclesiastes*, Chapter I, verse 7, which reads:

All rivers run into the sea, yet the sea is not full: unto the place from which the rivers come, thither they return again.

This is quoted with tiresome reiteration, by almost every writer on this subject throughout the middle ages and even later. The conception of a continual circulation of water passing from the sea bottom by hidden channels into the earth and then ascending to the surface, or indeed to the mountain tops and thence running back again into the sea, was already set forth by Ristoro d'Arezzo, one of the earliest writers on the phenomena of nature in the middle ages. The power by which the ocean water was drawn up through the porous body of the earth is stated by

[14] Op. cit. Chapter II, p. 145. From the translation by Dr. Frank Granger, The Loeb Classical Library.

[15] Op. cit. Chapter I, p. 143.

[16] *Etymologiae*, XIII, 20.

him to be "The virtue of the heavens," which draws up the water in the same way as a magnet attracts iron.

Reisch[17] on the other hand says:

> Within the earth as we have shown there are many open spaces and passages, into which (since there can be no such thing as a vacuum) vapours are drawn up from the earth and condensed into drops of water which unite to form rills, which running down to lower levels issue into the open air as springs,

by which condensation, he goes on to say, a new vacuum is formed which in its turn brings up more vapor and so the process goes on continually and a permanent outflow of water results.

Among the many writers who made reference to this subject Kircher is especially worthy of mention, and it will be well to consider his contribution in some little detail. Athanasius Kircher was a member of the Society of Jesus and the author of a number of great folio volumes dealing with a great variety of widely different subjects. Some of these, such as that on Magnetism embody important contributions to knowledge while others, as for instance those on the Tower of Babel and Noah's Ark are interesting displays of a brilliant and fertile imagination. His best known work is that entitled *Mundus Subterraneus*,[18] which appeared in 1664.

In this volume he treats of all things that are within the earth and certain chapters deal with the origin of springs, rivers and mineral waters. After a general review of the question he presents a theory which with minor modifications was very generally accepted during the later middle ages and in the earlier years of the Renaissance.

In the preface of this great tome Kircher gives a brief account of the manner in which he gathered the material for writing it. He relates that he was making a journey through Calabria, Sicily and the Lipari Islands in March 1636 when the great Calabrian earthquake wrought widespread destruction in this region. This display of the tremendous power of the forces of nature and their terrible results made a very deep impression upon him. This was intensified by a subsequent visit to the volcano Vesuvius and the surrounding district. These experiences aroused in him an intense desire to penetrate into the mysteries of the

---

[17] *Margarita Philosophica*, Strasburg, 1504, Book IX, Chapter XV.

[18] The edition from which the quotations are taken is the third and was published in Amsterdam in 1678.

hidden interior of the earth, in which these devastating forces had their origin and if possible to unlock its secrets. He was unable, he says, to visit every part of the earth's surface where such manifestations of these forces of nature were to be studied, indeed a whole lifetime would not be sufficient to attain this end, but he carried on an extended correspondence with the members of his order who were stationed in every country in the world, and from these he obtained much of the information which he sought. He also, through letters of introduction from influential persons, got into communication with many men in various lands from whom such information as he required could be obtained, and finally he made a thorough study of everything that had been written on the subject by ancient authors. He goes on to say that he has a constitutional distrust of what he reads or is told concerning the wonders of nature and the virtues of natural objects, unless the information comes from someone whose word is known to be thoroughly reliable or until the statements made to him have been verified by actual experience. Notwithstanding all this, the book, while containing much that is true, is made up largely of mere speculations concerning the character of the earth's interior and many other subjects, which have since been dissipated in the light of our wider knowledge, while his account of the animals, men and demons which inhabit subterranean places, as well as the "figures" which, he says, are found impressed upon the strata of the earth's crust and their symbolic, mystical or theological significance, belong to an age which has long since passed away and which can never return.

Aristotle ("Peripatus"), Kircher says, opposes the teaching of Holy Writ as set forth in the book of *Ecclesiastes*, denying that springs and rivers take their origin in the sea, and asserting that they are produced only by the condensation of air into vapor. This, after a discussion of the subject, Kircher states to be an erroneous conclusion. He holds that there are, in many different places, holes in the sea bottom which communicate with subterranean passages leading to the mountain summits, and through these the ocean waters pass into great caverns within the mountains or sometimes even into lakes which are open to the sky, upon the sides or tops of the mountains. From these the waters flow out again as springs or rivers and make their way over the surface of the land, back into the sea. But, he says, the opponents of this view ask how can water rise up from the sea level to the mountain tops, this being contrary to its nature. If once I overcome this difficulty, Kircher goes on to say, I certainly feel that there is no man with mind so dull, as not to be ready to follow on hands and feet, as the saying is, my trains of thought.[19]   There are, he says, several ways in which water may be made to "run uphill" and he then proceeds to give a series of diagrams showing various instruments or pieces of machinery by which this effect may be accomplished.   The first of these is a pair of double bellows arranged to deliver water instead of air, the power to work the machine being supplied by a water wheel.   Another is a large U tube with one arm longer than the other.   This is filled with water up to the level of the top of the shorter arm, which is fitted with a flexible diaphragm over this end, so that when this is pressed down by the hand, the water is carried to a much higher level in the longer arm.   A third piece of apparatus demonstrates that water may be sucked up to a high level by means of a vacuum developed above it.   These and a number of other instruments Kircher states that he had caused to be constructed and that he found in every case they worked satisfactorily, as undoubtedly they would.   He then points out that there is reason to believe there are in nature, conditions which parallel those which exist in the case of these machines.   Thus for instance when the tide in the ocean rises, owing to the attraction of the moon, the ocean water over one portion of the sea bed piles up and exerts an increased pressure upon it, which causes the sea water, in any holes at the bottom of the sea which communicate with subterranean passages, to be forced up into the interior of the mountains along these channels and thus the high tides which keep sweeping around the world act as the bellows do, or like the increased pressure exerted on the short arm of the U tube in the apparatus just described.   He believes also that the pressure of high winds upon the ocean surface adds its share to the pressure upon the adjacent waters and aids in forcing these up into the interior of the mountains.   The interior of the earth, he says, is traversed by a multitude of these subterranean passages which the Divine Wisdom has placed there for His own purpose, just as in the body of man, the microcosmos, He has ordained the innumerable veins and arteries which carry the blood everywhere throughout the body, to acccomplish the purposes for which it and the body itself were created.

Many wonder and are not able to understand how it is that in the Alps of France, Germany and Italy so many streams take their rise and have flowed without cessation since the beginning of the world to the present day.   From such streams the Danube, Rhine, Moselle, Meuse, Rhône, Saone, Po, Tessino and Oenus and even certain lakes

[19] *Mundus Subterraneus*, p. 250.

originate.   To say, as the Peripatetics assert to be the case, that all this water has
formed by the condensation of drops of water by distillation in caverns within the
mountains is to assert something which common opinion holds to be ridiculous.   All
the waters which collect in the caverns of these mountains would be scarcely sufficient
to form a single river.   There is some hidden secret of nature which must here be
brought to light.   The reader will remember (he goes on to say) that in Book III of
this work it was shown that as in the Alps so in all other great mountain ranges God
decreed at the time of the Creation of the world that great hydrophylacia (i.e. caverns
containing water) should be formed, from which in the various parts of the world
rivers should flow, some for the irrigation of the adjacent regions and others which
might be used by men for the purpose of navigation.   But so many and such great
rivers which had a perennial flow could not be drawn from such caverns unless these
were being continually furnished with new supplies of water, which would not be
possible through the condensation of vapours in caves within the mountains.   For as
already stated, if all the vaporous air was condensed into water a sufficient and
continuous supply of water would not be afforded, nor would a perpetual flow be
secured by the accession to it of supplies of water from rain and the melting snow, for
such contributions would be intermittent and the flow supply of water to the rivers
would be irregular, which is not the case.[20]

He then goes on to give as a case in point: (see figure 67)

The lake at the summit of Mount St. Gotthard from which the river Tessino with
many windings precipitates itself with a continuous flow into a succession of deep
valleys, affording the most striking evidence that it is continuously fed with new
supplies of water from some source or other, and this source cannot be the condensation of vapours, since here there is no covered vault, the lake being open to the sky.[21]

He gives a picture of an ideal section of a mountainous region and the
adjacent margin of the sea.   In this a number of whirlpools are shown
marking the position of openings in the sea bottom, into which the sea
water is rushing down, each of which communicates with a subterranean
channel leading to a cavern at the summit or on the slope of a mountain,
from the floor of which the water is carried away by a stream or a river.
But according to Kircher there is another force which is at work in
the elevation of the waters of the sea to the tops of the mountains.   This
is fire.   For, of those caverns great and small which exist within the
earth, some, as has been shown, are occupied by water, but others are
filled with fire.   There is a great fire at the center of the earth and this
through branching fissures and passages communicates with all these
fire-filled caverns, some of which lie deep down within the earth and

[20] *Mundus Subterraneus*, Book V, p. 254.
[21] *Mundus Subterraneus*, Book V, p. 255.

PLATE XII

An explanation of the Origin of Rivers given by Kircher. The whirlpools mark the positions of openings in the sea bottom into which the sea water rushes. These communicate
with subterranean channels through which it rises into caverns in the mountain tops, and
issuing as springs gives rise to rivers, returning again to the sea. (From his "Mundus Subterraneus".)

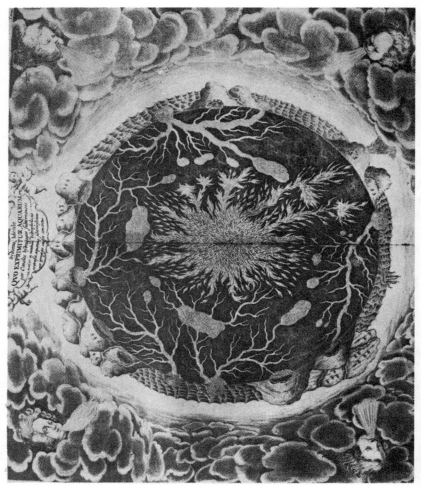

PLATE XIII

A section through the earth showing the great fire at its center, which communicates with many smaller bodies of fire. Also many caverns within the earth filled with water which has passed into them from openings in the sea bottom (whirlpools). From these caverns it rises through other channels to mountains on the earth's surface. Volcanoes, Fumaroles, as well as springs of hot water take their rise through the action of the central fires. (From Kircher: *Mundus Subterraneus*.)

others nearer to the surface. The fire in some of those which lie nearest the surface sometimes actually breaks through to the surface, producing volcanoes or burning mountains. But in others the fire never reaches the surface but, remaining deeply buried within the earth, serves to heat the water in adjacent caverns. Or the fire in the passages leading from the central fires to the fire-filled caverns may pass near caverns filled with water with the result, in all such cases, that this water being heated or vaporized, comes to the surface, and gives rise to hot springs or fuma-

FIG. 67. A Hydrophylacium or great cavern within a mountain, filled with water and feeding springs in the vicinity. Shown as it would appear if a portion of the roof were removed. (Herbinius, *Dissertationes de Admirandis Mundi Cataractis*, Amsterdam, 1678.)

roles; but on the other hand if these vapors fail to gain immediate access to the surface, they may pass on through cooler portions of the earth's crust and condensing again to water, eventually reach the surface and appear as springs of cold water. These processes are displayed in full operation in a large and striking section through the earth given by Kircher which is reproduced on a reduced scale in Plate XII. The underground apparatus which gives rise to hot and cold springs occurring in close proximity to one another on the surface of the earth is illustrated in figure 68. Figure 69, also taken from his book shows how it comes

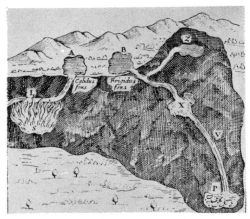

FIG. 68. Two springs, one hot (A) and the other cold (B) occur close to one another. The underground passage L passes directly over the fire S and the water becomes heated. B on the other hand comes from some hydrophacium remote from the fire—P, X or Z—and the water remains cold. (From Kircher: *Mundus Subterraneus*.)

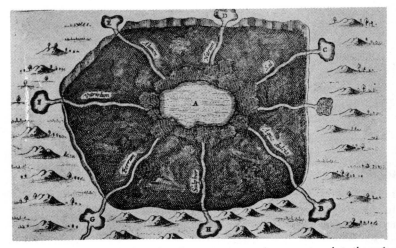

FIG. 69. The water issuing from the subterranean cavern A passes outwards to the surface by several different channels and issues as springs marked B, C, D, E, F, G, H. That coming by the passage A–B passes through rocks containing no soluble matter and issues as pure water. That reaching the surface by A–C passes through a country rock which is impregnated with saline material and issues as a salt spring. The others take up sulphurous, ferruginous and other materials from the rocks which they traverse and come to the surface as mineral springs of various kinds. (Kircher, *Mundus Subterraneous*.)

about that mineral springs displaying very diverse characters often occur in the same district.

Thus through the balanced interplay of fire and water so wonderfully designed by the "Naturae Opifex" each element tempers the other. Had the former won the mastery, the earth would have been reduced to a mass of glowing cinders, had the latter preponderated unduly the earth would have been a frozen waste, but with the nice balance which is maintained there results the smiling earth, so admirably adapted in its various parts to the needs of various races of men. The abundant rivers and the springs of water, some hot and others cold, watering the earth supply the needs of man and beast. The numerous springs of mineral waters of the most varied kinds bestow on him their healing properties, while still others deposit metallic ores and a great variety of other minerals for human needs.[22]

Reference has already been made, in the chapter dealing with the origin of Ore Deposits, to a theory put forward by Becher to account not only for the origin of ores and metals but at the same time for the origin of springs and rivers. Sea water passing through the ocean floor gains access to a great fiery cavity occupying the center of the earth and, being vaporized, passes outward in all directions toward the surface. It rises most easily into the interior of the mountains, these being more cavernous than the rest of the earth's crust, and is there condensed into water by the intense cold of the high mountains, covered with snow and ice, and comes to the surface as streams of water which run down the mountain slopes. Becher,[23] quoting Fromondi, compares every mountain to an *alembic*, a piece of apparatus much used by the alchemists for purposes of distillation, the fluid which was to be distilled being placed in the lower part of the vessel which was heated, the rising vapor being condensed in the cold "head" of the alembic, from whence it was carried off by a pipe into the receiver.

Barattieri[24] also holds this theory of alembics, although he believes that melting snow as well as rain makes its contribution to the development of springs and rivers.

And so in nature the vaporized sea water rising into icy cold caverns within the high mountains was there condensed again to water—as in a mighty alembic—and flowed out on the floors of the caverns to the outer

[22] *Mundus Subterraneus*, p. 186.
[23] *Chemisches Laboratorium*, Frankfort, 1653, p. 92.
[24] *Architettura d'acque*, Piacenza, 1699, p. 18.

air through crevices in the mountain sides in the form of springs.   Many writers observe that springs are found in mountainous places rather than on the plains, because the intense cold of the former readily condenses the rising vapor, while in the latter, the temperature being higher, condensation is not so readily effected.

Even the mysterious Nile which has such a long uninterrupted flow through the level land is now known, so these writers tell us, to take its rise on the slopes of a great mountain in a far distant land known to them as Mount Argento.

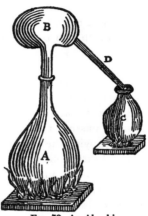

Fig. 70. An Alembic.

The opinion that the sea water in some way found its way into the bowels of the earth and was distilled up through the mountains, in whose cold caverns it was condensed, every mountain range thus representing a *series of giant alembics*, was very widespread and general and appears with monotonous repetition in the writings of almost all the authors who deal with these subjects in the sixteenth and early seventeenth centuries.[25]

Herbinius[26] is another writer who has left a very interesting work on this subject.   He defines "Cataractes" as bodies of fire, air or water in violent motion, situated either on the surface or within the earth, and his book is a storehouse of references to early writers on springs and rivers.   Many of these believing that springs and rivers had their origin in the sea, tormented their ingenuity to find some plausible explanation

[25] Giovan. Sarava, *Della Filosofia Naturale*, Venice, 1557, p. 92.

René Descartes, *Principorum Philosophiae*, Amsterdam, 1678, p. 164.

[26] *Dissertationes de Admirandis Mundi Cataractis*, Amsterdam, 1678.

for the rise of the waters of the sea to the tops of the mountains. Schottus[27] expresses himself feelingly as follows:

> According to this theory, sea water may be carried through subterranean canals to the surface of the earth and quite frequently to the top of the highest mountains. How this takes place is something which hitherto has baffled the minds of all and has led to an almost interminable amount of conjecturing.

The causes which were put forward to account for the rise of these waters into the mountains were many in number and often sound very strange to modern ears. Herbinius himself says that the true causes are (1) God, the originating cause and (2) as a secondary cause, the unceasing movements of the waters in the great subterranean abyss, which by their violence drive these waters up to the surface of the earth.

The following are among the other causes offered as explanations of this remarkable phenomenon—

1. The Surface of the Ocean is higher than the Land. This belief had a wide acceptance in the Middle Ages. The convexity of the ocean surface was believed to represent a semi-globe superimposed upon the globe of the earth. Many facts which now meet with quite a different explanation were adduced to support this view. It is stated, in several texts in the Bible (*Job* XXXVIII–10, *Jeremiah* V–22, *Proverbs* VIII–29), that the Lord had fixed the boundaries of the sea so that it should not overflow the land. Such being the case it was evident that the waters, standing as they did above the earth's surface, were held up miraculously by the Divine power. Since however the texts state only that the Lord had fixed the boundaries of the sea so that it would not overflow the land, there was no reason to suppose that the ocean waters had been forbidden to pass through subterranean fissures into the land and rise to the mountain tops where they could issue as springs for the use and support of mankind.[28] Papin,[29] holding the view that the surface of the ocean is higher than that of the land, explains this fact as due to an "esprit concrétif" which binds and holds together all things in which it is present, and this in the case of fluids causes them to assume a globular form, as seen in drops of water, and to this is due the convexity of the ocean's surface.

[27] *Anatomia Physico-Hydrostatica Fontium ac Fluminum explicata.* Würtzburg. 1663. p. 174.

[28] This is set forth in detail with quotations from several of the church fathers by Schottus in his *Anatomia Physico-Hydrostatica*, p. 181.
See also Neckam, *De Naturis Rerum*, London, 1863, p. 159.

[29] *Raisonnemens Philosophiqes*, (twelfth century) Blois, 1647, p. 85.

2. To the attractive power of the earth through which the water rises. This explanation was probably based on the fact that capillary attraction will cause water to rise to a certain height in a mass of dry earth.

3. Sennertus[30] and others being ignorant of the principle of hydrostatic equilibrium held that the ocean, by the enormous weight of its waters increased at certain seasons by high tides or heavy winds, was able to force these, through fissures in the earth, up to the tops of the highest mountains.

4. Ristoro d'Arezzo,[31] as already mentioned, says it is raised up by the influence of the heavens, as a magnet attracts iron.

5. Fontana[32] says that the movement of waters, whether it be the continuous flow of rivers fed by the rain from the clouds and from waters rising through passages from the abyss, or of fountains issuing from the mountain sides, are due to one and the same cause. The waters rise up to the mountain top actuated by the same cause as that by which they flow down hill, namely, the "anima" of the "Geocosmos," owing to the fact that these movements are part of the economy of the "Geocosmos"— just as in the human body the blood circulates and fluids rise and fall moved by the "anima" or vital principle of the "Microcosmos."

6. Owing to the air enclosed in the bowels of the earth.

7. Theodoretus[33] goes back directly and at once to the Great First Cause and says that they rise to the mountain tops in obedience to the word of God.

The Maelström on the coast of Norway attracted the attention of a long succession of writers from Aristotle onward. By many it was supposed to be a whirlpool caused by the ocean waters running down into a great hole in the bottom of the ocean. Later, however, this explanation seemed less acceptable, since the action of the Maelström varied with the tides. Ortelius[34] describes it as:

No less than 40 miles round, and upon the tides coming in, it swallowes in a manner the whole sea with an incredible noise, drawing Ships, Whales or whatever else comes within its compass and dashing them to pieces against the sharp rocks, that there are in the descent of this dreadful *Hiatus;* and then upon the ebb throwing them out again with as prodigious a violence in so much that some have attributed the whole flux and reflux of the sea (and not without some reason) to this vast *Vorago.*

[30] *Epitome Naturalis Scientiae*, Chapter X, p. 64, Venice, 1651.
[31] *La Compozione del Mondo*, edition published in Rome in 1859, p. 85.
[32] P. D. Cajetano Fontana, *Instituto Physico-Astronomica*, Modena, 1695, p. 198.
[33] Sermon 2, *De Providentia*.
[34] From Robert Plot, *The Natural History of Staffordshire*, Oxford, 1686, p. 73.

One of the most complete and at the same time most interesting treatises on the origin of springs and rivers which appeared in these ancient times is that of P. Gasparus Schottus.[35]   Schottus like Kircher belonged to the Society of Jesus, and he has a chapter on Kircher's views

Fig. 71. The Maelström near the island of Mosken (*c*) on the coast of Norway, which according to Herbinius is a whirlpool due to a hole in the sea bottom, through which the water of the ocean is alternately sucked down into and with great violence forced out from the Abyss of Waters within the earth.   (From Herbinius: *Dissertationes Admirandis Mundi Cataractis.*)

as these are set forth in a book by the latter.[36]   Schottus dissents from Kircher's views and critically examines those of all the writers of any importance who had previously discussed this subject, so that his work is a most valuable compendium of the history and literature of the subject

[35] Op. cit.

[36] *Iter Extaticum Coeleste*, Würtzburg, 1660, pp. 529–541.   In this Kircher sets forth his conclusions concerning the origin of springs and rivers which are given in more extended form later, in his *Mundus Subterraneus.*

up to his time.   Having set forth the views of former writers with the
reasons which may be urged for and against them, he summarizes his own
opinions and conclusions in the following words:[37]

We are of the opinion that some springs and rivers have their origin from subterranean air and vapours which have been condensed into water.   Others from rain
and snow which have soaked into the earth, the greatest number and the most important rivers, however, from sea water rising through subterranean passages and
issuing as springs which flow continuously.   And so the sea is not the only source, at
least it does not distribute its water through underground passages to all these springs
and rivers.   But this statement would seem to run contrary to the clear teaching of
Holy Writ found in Ecclesiastes, chapter 1 and verse 7, *All rivers run to the sea; yet
the sea is not full; unto the place whence the rivers come, thither they return again.*   The
real meaning of these words however seems to be: All rivers run into the sea, from
the place out of which they come, to it they flow back again.   Consequently those
which enter the sea have issued from the sea, and those which have issued from the
sea return to it and enter it that they may flow out of it again.   But all enter it and
all return to it, therefore all have issued from it.   But it does not follow that some, as
we believe, have not come out of the sea by another road than that just mentioned.
I am, therefore firmly of the opinion and again repeat, all rivers do not issue from the
sea—at least all do not make their exit directly out of the ocean into the depths of the
earth and from there rise through subterranean channels to their fountain heads.
This is held to be true not only by recent authorities as, "Conimbricensium" Fromondus, Cabeus, Cornelius, Magnanus, as seen in their words above referred to,
but also by Albertus Magnus, (Duns) Scotus and a multitude of others who believed
it to be consonant with the teaching of Scripture.[38]

But there was another question which thrust itself forward in all these
discussions concerning the sea as the source and origin of rivers, through
the rise of sea water to the summits of the mountains, namely, how is it
that while the water of the ocean is salt, the water of the rivers is fresh?
The answer generally given to this question was that the sea water in
rising through the earth's crust, traversed fissures which were so narrow
that they permitted only the pure fresh water to pass through them,
while the grosser particles of salt were retained or filtered out.   When it
came to be recognized, however, that salt cannot be filtered out of water
but that it may be removed by a process of distillation, the theory of
Fromondus and Becher, to which reference has already been made, took
its place.   This theory held that the sea water made its way into the
interior of the earth and there encountering the central fire was evaporated, and thus steamed up through the earth's crust to the mountain

tops, leaving the salt entangled in the mesh of minute fissures in the rock through which the steam passed.

Agricola[39] who was the leader along so many lines in the great advance which was made by the geological sciences at the opening of the Renaissance, does not formally discuss the question of the origin of rivers and the movement of subterranean waters, but confines his attention to the question of the origin of those waters which occur within the earth. He agrees with Aristotle that these waters take their origin in part from rain soaking in from the earth's surface and in part from the condensation of vapors within the earth, but he believes that a contribution to them is also made by water from the ocean and from rivers flowing on the earth's surface, which makes its way into the earth's interior. He goes on to say that the vapors which are found within the earth originate from the waters which penetrate into the earth from these several sources, and that the springs which break out on the earth's surface usually derive their water from rain. His belief that the water of the ocean does, to some extent at least, percolate into the land, was due to the fact that wells dug on the sea coast near the margin of the ocean are often found to yield saline waters.

Agricola's explanation of the origin of the vapors which by their condensation give rise to water within the earth's crust is very unsatisfactory, since the waters turn into vapor and the vapor condenses again into water without any apparent cause. The same is true of his treatment of those springs of hot water which issue from the earth in certain places, since, after reviewing the various sources from which their heat might conceivably be derived, he concludes that it probably comes from the combustion of supplies of carbonaceous materials within the earth.

3. THE TRUE EXPLANATION

Leonardo da Vinci (1452–1519) that universal genius, the most outstanding figure of the Renaissance in Italy, who possessed to such a remarkable extent the rare gift of freeing himself from the entanglements of prevailing opinion and of approaching all questions from the standpoint of direct personal observation and independent thought, was one of the earliest to see the true explanation of the origin of rivers. He wrote but little on the subject, yet from his observations in the Alps he recognized the important rôle played by the more pervious beds in the synclinal

[39] *De Ortu et Causis Subterraneorum*, Basel, 1546, Book I. p. 12.

folds of great mountain ranges, especially when they are dipping at a high angle and lie between impervious beds, in carrying the rain and snow waters deep down into the crust of the earth, whence they may be brought again to the surface at some distant point or perhaps onward into the ocean without again reaching the surface at all.   Indeed he seems to have reached the same conclusions as those arrived at by Vallisnieri when he observed the same phenomena many years later.[40]

Among the very earliest, if not indeed the first writer to recognize and insist that rain and the melting snows on the earth's surface were not only one but the only source from which springs and rivers derived their waters, was Bernard Palissy the celebrated potter (c. 1514–1589). Palissy, who was a true son of the Renaissance, was born of humble parents in Aquitaine.   Being endowed with marked natural ability and force of character, as well as with great determination and courage, overcoming all obstacles which presented themselves, he made outstanding contributions to both science and art and occupies an important position in the history of both.   He first studied drawing and modelling, inspired by the works of Raphael, Albert Durer and Leonardo da Vinci. Chemistry had not at that time developed into a science.   Palissy therefore had recourse to the alchemists and apothecaries in order to learn the nature and properties of materials, and while gaining from them much information that was of great value to him later, found at the same time that in their teaching and practice was a large element of imposture and fraud.   From all this he turned aside and became an advocate of sound philosophy and a stout opponent of charlatanism in all its forms. Thus in education he followed quite a different path from that of the schools.   His opponents pointed the finger of scorn at him because he knew neither Latin nor Greek, a fact which he did not attempt to deny, and what was worse did not even seem to regret seriously, pointing out that for the study of nature, it was much more important to observe and investigate at first hand the actual phenomena of nature, than to acquire a knowledge of the speculations concerning them which had been formulated by the ancient philosophers.   Palissy breaks away completely from the speculative attitude of the medieval writers and bases his whole treatment of the subject on the facts of nature which he had observed and which can be observed by all who will take the trouble to study nature in the field and in the laboratory.   After a long period of poverty and disappointment he achieved success.   He was the first to make

[40] Giuseppi de Lorenzo, *Leonardo da Vinci e la Geologia*, Bologna, 1920, pp. 109–111.

faience in France, and his pseudo-chemical knowledge gained from the alchemists, aided him in the preparation of fine glazes for its adornment. Further, he was the first to tell the "Doctors" in Paris that marine fossils were true animal remains left by the sea in the places where they are now found. Later, when in danger of his life because he had embraced the reformed religion, he was befriended by the Constable Montmorency and taken under the protection of the King and Court. He lectured to large audiences in Paris on natural history, illustrating his remarks by specimens from collections which he had made and insisting that the direct study of nature and experiment was the true path to knowledge. Palissy was eventually thrown into the bastille because he was a Calvinist, and died in prison.

His views are set forth in his work entitled *Discours Admirables de la Nature des Eaux et Fonteines, tant naturelles qu'artificielles, des metaux, des sels & salines, des pierres, des terres, du feu & des emaux*, the first edition of which was published at Paris in 1580. It is a very interesting little work, written in old French. In it he deals first with waters, rivers, springs and fountains, and then passes on to the discussion of alchemy, the nature and "generation" of metals, potable gold, salts, minerals, clays and other allied subjects. The book is written in the form of a dialogue between two persons "Practice" and "Theory," and it is always "Practice" that instructs "Theory." Palissy's views on the origin of springs and rivers may best be set forth in the following abstract from his work:

It is necessary that you realize distinctly the fact that all the waters which there are, or which there have been or will be on the earth, were created when the world was made: and God not willing that anything should remain in idleness commanded them to come and go and be productive. As I have pointed out the sea is always in motion, its waters are continually rising and falling; so it is with rain water. It falls in winter and rises again in summer, only to fall again next winter. The sun's heat, dryness and the heavy winds blowing toward the land cause the evaporation of great quantities of water from the ocean, which water gathers in the heavens in the form of clouds which speed from one coast to another like heralds sent by God, and when it pleases God that the clouds (which are nothing more than bodies of water) should become dissipated, these let their waters fall all over the earth in the form of rain.[41]

Having made the question of the origin of springs the subject of close and long continued study, I have learned definitely that these take their origin in and are fed by rain and by rain alone.[42]

And these rain waters rushing down the mountain sides pass over earthy slopes and

[41] *Discours Admirables* p. 43.
[42] Op. cit. p. 34.

fissured rocks and sinking into the earth's crust, follow a downward course till they reach some solid and impervious rock surface over which they flow until they meet some opening to the surface of the earth, and through this they issue as springs, brooks or rivers according to the size of the exit, and the volume of water to be discharged.   Thence they continue their downward course into the valley, and as they flow through this they are augmented by the waters from other springs in the mountains on either side.   This in a few words is the way in which streams and rivers originate.[43]

It may here be mentioned that Conrad von Megenberg in his popular work *Das Buch der Natur*, a printed edition of which appeared as early as 1475, gives nearly the same explanation as Palissy, although not treating of the subject in detail.

The time had now come when close and critical observation was to replace conjecture and ingenious speculation, and when a true basis was thus to be laid for a final solution of many, if not all the problems concerning the origin of springs and rivers, with reference to which there had been such widespread differences of opinion down through the centuries. Pierre Perrault in his book entitled *De l'Origine des Fontaines* which was published anonymously at Paris in the year 1674 presents the results of the first serious attempt to actually measure the rainfall and determine its relation to the amount of water carried off by the rivers.[44]   Perrault selected and measured the drainage area of the Seine from its source to Aignay-le-Duc.   He found this area to have a superficies of 121.50 square kilometres.   He then determined by means of a rain gauge the average amount of water which fell as rain upon this area annually during

[43] Op. cit. p. 48.

[44] In the second abridged reprint of the early volumes of the *Philosophical Transactions of the Royal Society of London*, the first volume of which appeared in 1809, Perrault's book is by mistake attributed to Papin.   Another point which may be worthy of mention in this connection is that in the abstract of Perrault's book which appears in the original edition of the *Philosophical Transactions* and which is reproduced without substantial change in both sets of the abstracts which appeared later, certain figures of rain-fall and run-off in the district of the Upper Seine taken from Perrault are presented, upon which he based his conclusion that the rainfall of that district was six times as great as the run-off.   These figures, however, do not bear out this result.   This fact remained unnoticed until the publication of the second set of abstracts, when the editor remarked that some mistake must have been made in the numbers.   On tracing this back to Perrault's book it is found that the abstractor for the original edition of the *Philosophical Transactions*, and whose abstract was copied by the others, confused the words "vingt quatre" (24) and "quatre vingt" (80), the amounts in the book being given in words and not in numerals.   The amount of rainfall on the area selected for study is actually given by Perrault as 224,899,942 "muids"—and not "a little over 280 million" as stated by the abstractor.   With this alteration the result given by Perrault—namely that the run-off is one sixth of the rainfall—is seen to be correct.

the years 1668, 1669 and 1670 and found it to be 520 millimetres.[45] Taking this as 500 millimetres, if all the rain which fell upon this area during the course of a year remained in place and if no portion of it was lost by evaporation or otherwise, on the last day of the year the area would be covered with water to a depth of 50 centimetres, which would represent a volume of 60,750,000 cubic metres. He then found the amount of water which was carried off the area by measuring the amount which passed through the Seine canal at Aignay-le-Duc each year and found it to amount to 10,000,000 cubic metres, that is to say, about one-sixth of the total rainfall.

A few years later Mariotte,[46] the Prior of St. Martin-sous-Beaune in Burgundy, a well known physicist, made a more extended investigation embracing the whole drainage basin of the Seine above Paris. This he found to have an area of 60,356 square kilometres. Taking the rainfall at only 40 centimetres per annum, this gave a total of 24,142,400,000 cubic metres of annual precipitation, he determined the amount of water carried past Paris annually and found it to be 3,553,056,000 cubic metres showing that the rainfall was over six times, in fact nearly seven times greater than the discharge of the river. If the rainfall is taken at 50 centimetres, the precipitation would be nearly eight and a half times greater than the discharge.[47]

An interesting book was written in 1691 by Bernardino Ramazzini,[48] professor in the medical school of Modena, in which he gives an account of the "Wonderful Springs of Modena," artesian waters concerning the nature and origin of which there was much discussion about this time. His investigations and conclusions may best be given in a translation of his own words:

[45] These figures are given by Perrault in "poulces" and "muids," the equivalents of these in the metric system as given here are taken from Paramelle. (See Paramelle: *L'Art de Decouvrir les Sources*) Paris & Liege, (6th edition), 1926, p. 106).

[46] *Traité du movement des Eaux par feu M. Mariotte de l'Academie Royale des Sciences, mis en lumiere par les soins de M. de la Hire*, Paris 1686, pp. 31–33. (This work was issued by la Hire after Mariotte's death). See also *Memoires de l'Académie Royale*, Paris, 1703, and Paramelle, *L'Art de Decouvrir les Sources*, pp. 106 & Q 107.

[47] See also John Dalton, *Experiments & Observations to determine whether the quantity of Rains & Dew is equal to the quantity of Water carried off by the Rivers and raised by Evaporation, with an Enquiry into the Origin of Springs*. Literary and Philosophical Society of Manchester, March 1, 1799. As a result of these experiments Dalton says: "I think we may fairly conclude that the rain and dew of this country are equivalent to the quantity of water carried off by *evaporation* and *by the rivers.*"

[48] *De Fontium Mutinensium admiranda scaturigine tractatus physio-hydrostaticus*, Modena, 1691.

We may be bold to say that *Modena*, a most ancient City, which *Tully* has of old dignified with the title of the Most Noble Colony of the *Romans*, has been well situated by its first Founders: For seeing it stands in a great Plain, ten Miles distant from the Foot of the Rising Hills, it has such a Situation, that, with the wholsom Temper of the Air, and a fruitful Soil, it has a great abundance of most pure Water, which neither can cease through length of Time, nor be ever vitiated or diverted by the Craft of Enemies: For this City has under its very Foundations a great Repository of Waters, or whatever else it may be called, out of which it draws an inexhaustible Stock of Waters; and, which is very rare, is got at a very small Charge; seeing for the getting of this treasure (for Water, according to the Testimony of Pindarus, is the best of all things) there is no need of great stir, in digging through Mountains, or keeping a great many Workmen, as is usual elsewhere, and such as *Rome*, formerly had divided, as *Frontinus* says, into Searchers, Water-Finders, Water-Bayliffs, Conveyors, Distributors and many other Workmen.

But that I may not keep the Reader longer in Suspense, you must know for a certain Truth, which many Thousands of Experiments have already confirmed,

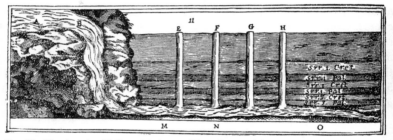

FIG. 72. The Artesian Wells at Modena.   (From Ramazzini: *De Fontium Mutinensium.*)

That in any place within, or without the City, for some Miles round, one may see a Spring which shall constantly send forth most pure Water.

And seeing every Citizen may take out of this great Stock, as much Water for his private Uses as he pleases; without fear of wronging the Publick, or being Fin'd for it: Therefore when any will have a Spring in his own House, he calls some Workmen, and having agreed for the Price, which for the most part does not exceed the Sum of Forty Crowns, he shews them the place which he thinks most fit, and they without further consideration dig a *Well* in a place mark'd out for them; and when they have come to a depth of about 63 Foot, they pierce the bottom with a great Auger, which when it has been driven down 5 Foot deep, immediately the Water gushes out with so great Force, throwing up Stones and Sand, that almost in a Moment all the Well is filled to the top, and the Water flows out thence constantly. — In some Places where the Situation of the City is lower, the Water arises above the Plain, from whence it runs easily down, but in higher Places it stops below the Surface of the Plain; so that it is necessary to make Conduits under the Ground, thro' which it falls into the public Canals, which afterwards meet into one Canal that is Navigable, and by which they sail conveniently even to *Venice*.   For this Canal falls into the *Scultenna*, and the *Scultenna* into the *Po*.

The number of these Fountains is very great, so that now almost every House has one: Having often understood by the Diggers of the *Wells*, that they heard a great noise of the water running under the bottom of the *Wells*, and that when it first begins to be heard they take it as a sign, that 'tis time to Bore. To be assur'd of this, I went down to the bottom of a *Well* in the beginning of *February*, holding a lighted Candle in my Hand, the *Well* being built in a place of no great light; having staid there a little, I perceived a manifest Murmur and Noise, yet not such as I expected. Then I stampt upon the Ground with all my force, upon which the Ground made a hideous Noise, so that I thought I had to do with Hell, and therefore quickly gave notice to those that were above, to pull me up with all possible speed, remembring that once the force of the Water throwing up the Earth prevented the boring. —

Seeing the Nature and Original of this hidden Source deserves to be as much enquir'd into as that of the *Nile* did formerly, let us pass through these Subterraneous Waters with the sails of our Reason, seeing we cannot do it otherwise. First, we may freely affirm, That these Waters are not standing, as they are when shut up in a Hogshead, but are in continual motion, and that pretty quick: For the Noise of that Water which is heard before the Perforation in the bottom of the *Wells* does make it manifest enough. . . . And I think 'tis probable the matter is so in our Fountains, to wit, the Water flows out of some Cistern plac'd in the neighboring Mountains, by subterraneous Passages, where the Earth is firm and hard; but when it has come into the Plain, it expatiates far over the Sand, and in the way is lifted up to this height when a Hole is made with an Auger, according to the Laws of Hydrosticks. . . .

To give some Specimen how the flowing of the Water may be according to my Explication: Suppose, as in figure 73 there is a Cistern in the Bowels of the *Apennine*, drawing Water from the Sea, and that the Water iscarry'd by subterraneous Pipes from the same Cistern, the Water is forc'd to run down by a more narrow space than it had in the beginning, and to follow its Course till it come into the Sea, or some great Gulph. Therefore Wells *E F G H* being digg'd, without any Choice in all the Tract lying upon this Spring, and a Hole being made by the Auger, the Water of necessity must be lifted up on high, being forc't by another, which descending from a higher Ground, presses on that which goes before, and drives it up. By this means these Waters receive a plentiful Supply from their Father *Appenine*, as does the Well of Waters which flows from Lebanon, of which there is mention in the Sacred History.

But 'tis, by far, more probable, that the Water is sent from the sea into such Cistern, than from Showers, or melted Snows, seeing Rain and Snow-waters run away for the most part by Rivers above Ground; neither can they enter into the ground so deep; as Seneca also testifies.

As I have deduc'd from the Original of this Water from the Sea, so I do not deny, that many Fountains owe their Originals to Rains and melted Snow; yet with this difference, that the Fountains which have their Spring from the Sea by hidden Passages continue perpetual, but those which rise from Showers and temporary Springs at some time of the year, are diminished and quite dry up. . . . Though I derive the Original of our Fountains from the Sea first, then from some Cistern of Water plac'd in our Mountains, into which the Vapors, sent up by the enclos'd Heat, are returned in Form of Waters. I would not thence infer, that this Cistern is plac'd in the tops of the *Apennine* Mountains, but I believe rather that 'tis plac'd in the Foot of the Mountain, than in the top. But I cannot certainly conjecture in what part, whether near the foot of the Mountain, or in their inner parts, this Cistern of Waters is plac'd by the Divine Architect.   I have spar'd no Labour nor Experiences to find out the Head of this Spring, and therefore I diligently viewed not only the Plain towards the Mountains, but the Mountains themselves, and could find no Marks of it.   I observed indeed some small Lakes, but such as dry up in the Summer, and so become Pasture for Cattel; of the number of which is Lake *Paulinus*, 25 Miles distant from this.   I thought best therefore to fetch the Original of these Waters from another source, *viz.* From some secret Cistern of Water plac'd in the inner parts of the *Apennine* Mountains.   And it is certain, that the inner parts of the Mountains are cavernous, and that there are in them Cisterns of Water, from whence Fountains and Rivers drawn their Original.

The waters of these fountains of Modena find their way eventually into a canal which runs from Modena into the river Panaro, a tributary of the Po.   The waters in the lower levels of the city rise to the surface and flow out as permanent streams.   In the borings in the higher parts of the city they rise nearly to the surface but do not form flowing springs. Even at the time when Ramazzini wrote his book, the former level of the water had already been lowered owing to the large number of borings which had been made to obtain the water.   The arms of Modena, suggested by these springs, are two augers with the motto "Avia, pervia"[49] (a path for the wanderer).

It may here be mentioned that Ramazzini was led by his investigation of these springs of Modena to indulge in certain speculations concerning the condition of the earth in former times and in the course of these introduces a long quotation from a work of Franciscus Patricius on the Rhetorick of the Ancients, written in Italian and printed in Venice in 1562. In this work there is a "Pleasant story" which was handed down "from a certain Abyssine philosopher in Spain.   This wise Abyssinian did say that in the most ancient annals of Ethiopia there was a history of the destruction of Mankind and the breaking of the Earth," which Patricius then goes on to relate and which Ramazzini copies.   This story so aroused the indignation of a Robert St. Clair, M.D. of London, that he translated Ramazzini's book into English, prefacing it with a long, violent and vituperative attack upon him.[50]   The reason which aroused his indignation was that the views set forth "contradicted the first chapter of *Genesis* and were regardless likewise of the second."

In 1715, Antonio Vallisnieri, President of the University of Padua, a

[49] Amédée Burat, *Traité de Géognosie*, Tome III, pp. 645, 647, Paris, 1835.

[50] *The Abyssinian Philosophy confuted or Telluris Theoria, neither sacred nor agreeable to Reason*, London, 1697.   It is from this translation of Ramazzini's Italian text that the extracts given above were taken.

PLATE XIV
PORTRAIT OF VALLISNIERI
(From the edition of Francesco Redi's *Works* published at Venice in 1712–30)

man of noble family, marked ability and very wide attainments made a further and very valuable contribution to the true understanding of the origin of springs and rivers. It was given in an address delivered at the University of Padua.[51] Vallisnieri, having been impressed by the results obtained by the French investigators referred to above, determined to go up into the Alps where so many of the Italian streams take their rise and trace certain of these streams up to their actual sources and see what could be learned concerning their origin by a careful study in the field. In his description of what he saw on the mountain heights, there can be found something of the aversion to the harsher aspects of the high mountain regions of the globe, as contrasted with the smiling plains below, noted by Geikie as characterizing the peoples of the classical period and down through the ages until toward the close of the eighteenth century.

Above the tree line he found shattered rocks and overhanging cliffs affording a spectacle "misto di compiacimento e d'orrore"—grottos, abysses, declivities, cracks, great valleys, caverns, trenches, ponds, gorges, craters, precipices with many basins and receptacles in which water was collecting. The water he found, was not being forced up out of the ground but was always trickling or running down the slopes in little streams. Then came the snow line above which were those great accumulations of snow and ice which formed "eternal reservoirs" from which the waters produced by their melting issued as rivulets, saturating the earth and swelling the streams which carry off the rain falling on the lower mountain slopes. The origin of the streams and rivers was manifest—they were fed by rain and the melting of the snows.

In the S. Pellegrino Alps, however, he observed a perplexing phenomenon. Here the snow fields were very extensive but the streams running down the mountain sides from them toward Modena were few and small. He expressed his surprise concerning this to certain shepherds in the high Alps, who showed him many places where the waters from the melting snows were seen to be running down into fissures in the rocky crust of the earth through which they evidently took their way, as hidden subterranean streams, to lower levels—whereupon, he tells us, he stood to use the words of Dante, "Like a man who when in doubt is reassured, and whose fear changes into comfort because the truth was now revealed to him," for there came to his mind those miraculous wells or fountains of Modena concerning whose origin there had been so much discussion from

[51] Antonio Vallisnieri, *Lezione Accademica intorno all'Origine delle Fontane*, 1st Edition, Venice, 1715. A 2nd edition, enlarged by the inclusion of a number of contributions to this subject by other authors, was published at Venice in 1726.

ancient times, and concerning which philosophers had long sought for a satisfactory explanation. Some thought of subterranean fires producing vapors from which these waters were condensed, others of great alembics in the adjacent hills, still others conjectured the existence of some complicated machinery hitherto unknown in nature. But here was the solution of the riddle; the streams from high up on Monte S. Pellegrino flowed down through subterranean channels and passed beneath Modena and on toward Bologna. A further important fact was noted by Vallisnieri, namely that while these underground waters could be heard passing beneath the city they did not come to the surface until a shaft was sunk and a certain floor of hard chalk was reached and penetrated. Beneath this the water ran, evidently under great pressure, and when the floor was pierced by the boring tool, the water burst up to the surface, forming a perennial fountain.

So Vallisnieri and certain other contemporary observers were able to demonstrate that the water derived from rain and from the melting snows on the mountains not only runs off the surface as streams and rivers but may pass underground beneath impervious strata and again come to the surface at lower levels when the continuity of the overlying stratum is broken. Further studies showed them that such strata are often thrown into folds, in which the beds may even assume a vertical position. In this way, water falling on high mountains may enter certain strata and be carried through them down to sea level or below it and then be forced again by hydrostatic head high above the level of the sea and appear as springs in unexpected places. Vallisnieri's little book is illustrated by a plate which presents six geological sections showing the structure of certain mountains in Switzerland and Germany, which illustrate his discussion of the subject. These he tells us were drawn and given to him by the renowned naturalist Scheuchzer, an account of whose extended Alpine explorations carried out during the years 1702 to 1711 was published in Leyden in 1723. These sections, reproduced in the figure 73 are interesting not only as illustrating Vallisnieri's observations and discoveries but also because they present what seems to be the earliest attempt to depict geological structure graphically in the form of horizontal sections. The only earlier attempt that has been met with by the writer, is a little sketch which depicts the structure of a mountain composed of inclined strata, recorded by Giuseppe de Lorenzo,[52] as occurring in Folio 36r of the Leicester Manuscript of

---

[52] *Leonardo da Vinci et la Geologia*, Bologna, 1920, p. 111.

Leonardo da Vinci.   It is Leonardo da Vinci's only sketch which illustrates a geological subject.   In these sections of Vallisnieri's, reproduced

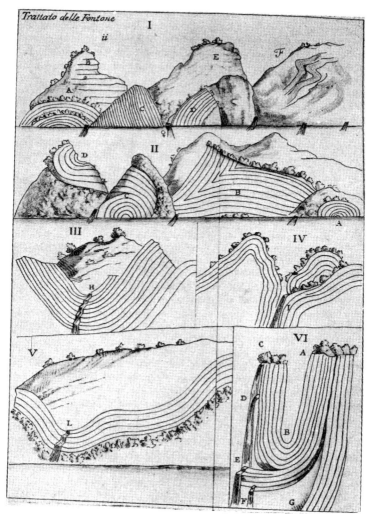

FIG. 73. Geological Sections of certain mountains in Switzerland and Germany.   (From Vallisnieri's *Lezione all'Origine delle Fontane*.)

in figure 73, the recognition of the exact line followed by the sections given in figures I and II presents some difficulties, as do also the names by which he designates certain localities.   The latter have been given as they occur in Vallisnieri's text.

I. This shows the shore of the Lake of Lucerne facing north.   *A* is the Geosberg or Goat Mountain.   The strata present a perfect arch so that there are no springs of water except at the side of the arch where it abuts against the adjacent mountain.

B. This is called Fronalp and presents a high wall composed of horizontal strata. No springs can be seen except where there are fissures passing down from inequalities on the surface.

C. Schibetemberg.   Formed of a series of highly inclined parallel strata with gentle undulations.   At the right hand of the series, at the point indicated by a sign, the strata are sharply bent and some of them are fractured.   From these fractures fountains issue into the valley between this and the adjacent mountain.

D. This is called Buggis-Grade.   It is formed of thicker and much fractured strata with a high dip.   Springs are not found because the rain runs down into the fissures and drains off by subterranean passages.

E. The mountain is called Gross-Axemberg.   It presents a rugged rocky wall without any distinct evidence of stratification which plunges down into the water of the lake.   It is wooded and springs are found in it.

F. Klein-Axemberg.   This mountain displays the most wonderfully contorted strata; at certain places about its foot Springs gush out.

II. This section is on the other side of the lake from that in Figure I and looks toward the south.   The two sections correspond to one another—the same sets of strata, on either side of the lake being indicated by the same letter.

A. Geelis-Berg, made of a series of wonderfully arched strata.

B. Teufess-Munster, called by the country people Munistero del Diavolo on account of its rough rocky and forbidding character.   As in the Geelis-Berg which corresponds to it in the other section, there are not fountains in the mountain itself but only at its foot on either side between it and the adjacent mountain.

C. This is called Auf-der-Woerche.   The upper portion is formed of strata dipping to the south but in the lower portion these form a perfect arch.   Here again there are springs only in the valley between this and the adjacent mountain.

D. Kolm.   This is a mountain situated back of *C*.   In it the strata are bent into a great horizontal fold.   This carries the rain water down into the interior of the mountain whence it passes away in subterranean channels.

III. A valley in Mount Schild in the Canton of Glarus, Switzerland.   The mountain is formed by a synclinal fold.   The waters derived from rain and melting snows pass down through porous strata within the mountain and issue at *H* as the stream called the Muhlebach.

IV. In the same part of Switzerland along the road called the Via Mala two sets of strata dip toward one another and assume a vertical attidude at *I*, at which point a spring issues.

V. In Switzerland also there is Mount Chattstoz on Lake Rivario.   The strata dip down from the summit of this mountain and rise again, the axis of the syncline thus formed lying above the town of Wallenstadt.   At *L* on this axis the waters collecting in this syncline, within the mountain, issue as a spring.

VI. Two mountains in Germany which are adjacent to one another. The strata forming the mountain A dip down and rise again forming the mountain C. The water from the rain and melting snow on the summit of the first mountain sinks into the earth and is carried down through previous strata and by its hydrostatic head rises to the top of C, where it issues as a spring. Some of the strata reach the surface at D, E and F and carry water which forms springs at these points. Still other waters pass off underground and disappear through the nearly perpendicular strata at G.

It may be of interest to compare with these very early and rough sections of Scheuchzer's, the following little sketch section dashed off for the writer by professor Albert Heim of Zurich, the greatest authority on the geology of this portion of the Alps, while on a geological excursion with him in this same region of the Lake of Lucerne in 1895. It was drawn for the purpose of illustrating exactly the same fact, the discovery

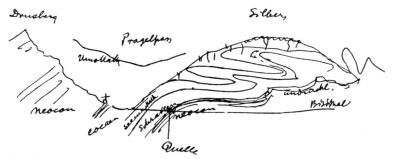

FIG. 74. Sketch by Professor Albert Heim of a section across the Muotathal in Switzerland, showing the origin of certain springs.

of which aroused such interest on the part of Vallisnieri. It is a section across the Muotathal from the Drusberg to the Silbern. The Muota River running in this valley presents an appearance which is quite different from that of the usual Swiss streams, in that its water is perfectly clear and free from sediment. This is due to the fact that its water is derived altogether from the many springs which issue from the limestone which is here the prevailing country rock. One of these springs has a discharge of no less than four cubic metres per second. On the heights of the Silbern there are hundreds of funnel shaped sink holes (indicated in the section) which are usually quite empty and dry, but down which in wet weather the rain water from the Karren Felder pours and fills the many fissures in the limestone feeding the springs, which even in dry weather show but little diminution in the amount of water which they discharge.   The mountain, as Aristotle says, hangs as a sponge over the lower lands.

As in the case of Ramazzini, so with Vallisnieri, there were many who criticized his conclusions.   Among these Niccolo Gualtieri[53] may be mentioned, who in his book expresses the opinion that Vallisnieri has almost brought himself into the company of the heretics in his treatment of the statements of Holy Writ, and further takes especial objection to two of his assertions namely, that sea water could not purify itself by passing through the earth and that rivers carry away less water than that which falls as rain.   A further discussion of Vallisnieri's work by contemporary Italian men of science will be found appended to the second edition of Vallisnieri's book.

The researches of Perrault, Marriott, La Hire and others had now definitely established that the rainfall on any area was more than sufficient to supply the waters carried off by its rivers and the observations of Vallisnieri and others had shown that the waters issuing as abundant springs from mountain slopes and the lower lands were, in many cases at least if not in all cases, derived from the rain and melting snows on the higher levels of the adjacent mountains.   It was no longer necessary to suppose that these must have come from great "abysses" or subterranean lakes supplied with water by the condensation of subterranean vapors.

This new conclusion received the most enthusiastic support of Thomas Bartholini[54] the renowned professor (of mathematics and later of anatomy) in the university of Copenhagen, who concludes his dissertation on the subject as follows:

> In drawing to the close of our dissertation we may sum up all our arguments in one brief statement; we see in the first place that floods, or the mighty inundations of entire rivers, manifestly arise from immediately preceding rains, or from the melting of snow and hail.   In the second place that there are very many fountains which subject to the heat of the sun dry up and that the flow of all is much diminished. Furthermore, that no fountains ever burst forth at the summit of a mountain, or near its head; but that always some portion of still higher land from which water may be supplied to them, overtops the fountains.   And what, indeed, is the most pertinent argument of all, the fact that fountains are but rarely observed in the stiffer, clayey soils, through which water with difficulty can permeate, while they are never wanting in the more sandy and stony parts of the soil, through which water readily percolates. Finally they never fail to yield approximately that fair and reasonable quantity of water which has been supplied to them by the rain.

[53] *Riflessioni sopra l'Origine delle Fontane*, Lucca, 1725.
[54] *De Fontium Fluviorumque Origine ex Pluviis*, Oxford, 1713.   (The work was evidently written before 1713 as the dedication bears the date 1692.)

The "Alembic" theory had become generally discredited but it is remarkable to note how this old conception set forth in detail by Kircher received the support of some writers until well on toward the middle of the eighteenth century.[55] Some of these writers while forced to recognize that mountains condensed watery vapor into rain upon their surface and not internally, still retained the old term of the alchemists and likened the mountains to *Alembics*.

Thus Edmund Halley[56] in a communication to the Royal Society of London in 1690 says that the vapor from the sea is swept by the winds against the high mountain tops, whose cold condenses it to water. Some of this "gleets down by the crannies" of the stone and part of the vapor enters into the caverns of the hills and the water thereof gathers as in an alembic and flows out in the form of springs but by far the greater part of the vapor condenses on the mountain sides and flows down on the surface slopes of the mountains, gathering into rills which flow together into larger streams of water and eventually form rivers. The mountain ranges act as *external alembics* to distil fresh water for the use of man and beast, water which finally returns to the ocean in streams "like so many veins in the microcosm."

Ray[57] a few years later writes:

Those long ridges and chains of lofty and topping mountains, which run thro' the whole continent East and West (as I have elsewhere observed) serve to stop the evagation of the vapours to the North and South in hot countries, condensing them like Alembic heads into water, and so by a kind of external distillation and likewise amassing, cooling, and constipating of them, turn them into rain by those means rendering the fervid Regions of the Torrid Zone habitable.

As an outcome of this long continued study of the problem of the Origin of Springs and Rivers it may now be considered as definitely established that these all have their origin in rain, snow, hail and other forms of watery precipitation falling upon the earth's surface. The rainfall in every country is now known to be disposed of in three ways: (1) A portion runs off the surface and forms brooks, streams and rivers. (2) A portion sinks into the earth, forms the ground water and may emerge again at lower levels in the form of springs. (3) A portion rises

---

[55] Belidor, *Architecture hydraulique*, Paris, 1737.
Fabricus, *Théologie de l'Eau*, Paris, 1743.
Kulm, *Indications sur l'origine des fontaines et l'eau des Puits*, Bordeaux, 1741.
[56] *Philosophical Transactions*.
[57] John Ray, *The Wisdom of God manifested in the Works of the Creation*, 6th Edition, 1714, p. 220.

into the air again in the form of vapor.   The relative proportion of the rainfall which is carried away in these respective ways depends on local conditions such as the porosity of the soil or rock on which it falls, and the temperature and dryness of the air.

In the United States about one third of the water that falls as rain or snow reaches the sea.   About one half in the eastern part of the country, only a small percentage on the Great Plains and virtually none in the Great Basin.   All the rest sinks into the earth or is evaporated.

4. CONCLUSION

Looking back over this long history of varying opinion the chief reason why the recognition of the true origin of springs and rivers was so long delayed can now be determined.   Those who devoted their attention to the study of this subject, seeing the great volume of water which the rivers bore away to the sea, could not believe that the rainfall of the country was sufficient to supply so great a quantity of water.   They did not know what vast expanses of the earth's surface were covered by the waters of the ocean, nor did they recognize how great a volume of water was raised from its surface by evaporation, and hence sought for some other source of water from which the springs and rivers might derive their supplies.   It was not until it had been demonstrated by actual measurements that the rainfall was more than sufficient to supply the rivers with their water and that the water derived from the ocean was freed from its saline contents by the process of evaporation, that the true origin of springs and rivers was finally established, although even yet there are some European hydrologists who maintain that the ground water is derived in part from underground condensation.[58]

[58] See O. E. Meisner, *The History and Development of Ground Water Hydrology*, Jour. Washington Acad. of Sciences, Vol. 24, No. 1, Jan. 15, 1934, p. 13.

Chapter XIII

QUAINT STORIES AND BELIEFS

Many quaint stories and old beliefs are met with in these ancient books, most of which probably came down from primitive times. Reference has already been made incidentally to some of them in former chapters, as for instance to that of the "Witterung" which had at times been seen by certain persons issuing from the outcrops of very rich ore deposits in the form of a faintly luminous mist and by which in mineral veins the *baser* were gradually "matured" into the *precious* metals. Also the fancy that pearls, so highly prized and of such signal beauty, were really dew drops which certain molluscs had taken up and by the occult powers which they possessed, had transmuted into the substance of the pearl, while the original form and beauty of the dew drop was preserved.

From the earliest times also asbestos attracted the attention of mankind in every country where it was found, owing to its remarkable resemblance to silk or cotton fiber, together with the fact that it was incombustible. Its strange deportment when submitted to fire gave rise to the still stranger fiction related by some writers, that it was a stone which can be set on fire and which when ignited will burn forever. One of the strangest names ever given to a mineral is that which Kentmann mentions as having been applied to this mineral by "Superstitious Germans" who call it "Plumae Spiritus Sancti" or Feathers of the Holy Spirit. He also says that the Italians call it the "Wood of the true Cross" because it is incombustible.[1]

The universal belief in the "virtues" of certain minerals and especially of gems has been referred to repeatedly, but De Boot actually gives a recipe for extracting the "virtue" of the emerald, in the preparation of a "Tinctura Smaragdi" which he says is a most efficacious medicine in certain diseases: "Triturate the emerald in an iron mortar, sift the powder through muslin, then cover it with spiritus urinae (sal volatile): the spirit is then distilled off, leaving the powder of a grey color, but which will communicate the colors of the emerald to spirits of wine!"[2]

Some of these legends and stories however would seem to merit some-

---

[1] *Dresdensis Medici Nomenclaturae Rerum fossilium, que in Misnia praecipue & in aliis quoque regionibus inveniuntur*, Zurich, 1561, leaf 27.

[2] *Le Parfaict Joaillier ou Histoire des Pierres*, Lyons, 1644, Book II, chap. 53.

thing more than a mere passing reference, and a few of these have been selected for further consideration in the present chapter.

1. "LAC LUNAE," THE MOON'S MILK OR MINERAL MOONSHINE

A group of the most curious stories concerning minerals which have come down to us from classical times are those which are associated with the mineral selenite. This mineral, as is well known, is a variety of ordinary gypsum, occurring as transparent crystals. In this form the mineral is colorless and can be easily cleaved along one plane, into transparent sheets bearing a certain resemblance to those of colorless mica or muscovite. These cleavage plates of selenite, especially when thick and translucent rather than transparent, have a pale silvery luster which is suggestive of moonlight. This fact is apparently the basis of the stories related concerning the mineral, as also the reason for its name selenite, derived from the Greek work *Selene*, signifying the moon. The mineral has indeed been called by many names at various times and in different countries. Among these may be mentioned: "Marien Eis," "Unser Lieben Frauen Eis," "Lac Lunae," "Lapis Lunaris," "Lapis Specularis," "Aphroselenon," "Spuma Lunae" (Froth of the Moon), "Star Jelly" and "Speccio d'Asino" (the Asses' Looking-glass). The stories concerning it must have originated in very early times, for Pliny, who for his information concerning minerals draws on the current opinion of his age, mentions them, as does also Dioscorides, both of whom wrote in the first century of our era.

One story is that looking into the smooth cleavage surface of a piece of selenite one sees an image of the moon, which increases and decreases in size as the moon waxes or wanes. The story varies somewhat in its details as set forth by many different writers. Pliny[3] himself says, "Selenites is a precious stone, white and transparent, yielding from it a yellow luster in manner of hony and representing within it a proportion (i.e. a form) of the moone, according as she groweth toward to the full, or decreaseth in the wane." Psellus[4] says much to the same effect: "It is so named because it displays as it were an eye within itself, which increases or diminishes according to the growth or decline of the moon." Marbodus, improving upon this, makes the *stone itself* to increase or decrease proportionately with the moon, which caused it to be called the

---

[3] *Natural History*, Book 37, Holland's translation, Pt. II, p. 629.
[4] Quoted in King's *Natural History of Gems & Decorative Stones*, London, 1867, p. 301.

Sacred Stone. Isidore[5] and St. Augustine[6] mention the same increasing and diminishing of the image in the crystal, accompanying the waxing and waning of the moon.

Agricola[7] however endeavors to explain the origin of this remarkable fable. The name "Lapis Specularis" he says is given to it because of its transparency. The Grecians call it selenite, either because it is found in the middle of the night when the moon is waxing, or because, by virtue of its transparency, the picture of the moon which it takes up at night is reproduced again during the day to match the waxing or waning of the moon.

The Emperor Julian mentioned that the ice of the river Seine looked like the "Phrygian Stone" which was dug up out of the earth in several places in Phrygia and which he calls the Lapis Specularis.

Gesner[8] repeats the usual story about the increasing and decreasing of the image of the moon, and refers to Agricola's observations. He also tries to explain away the story, stating that for the expression "Waxing and waning of the moon" can perhaps be better substituted the expression "In the time of the full moon," the idea to be conveyed being that at that time the mineral can best be discovered where it is being looked for. He goes on to say that no figure of the moon can be seen in the mineral, but that when the moon is brightest, the mineral presents a more brilliant appearance than at other times. An interesting old woodcut in the *Hortus Sanitatis* shows a man pointing to his collection of selenites lying on a table, each of which has in it a figure of the man in the moon. (see figure 75)

Baccius[9] repeats the same old story of the increase and decrease of the image of the moon and says that while the Germans call the mineral "Erdglass," the Romans call it "Lunaris," which indicates their opinion that the mineral derives its influence from the moon, as other minerals derive their power from certain stars.

De Boot,[10] in a somewhat extended reference to the mineral, mentions that Albertus Magnus says selenite is made by a certain kind of shell fish which has its habitat in India, Persia and Arabia. Here he is

[5] *Etymologiae*, Book XVI, 9 & XVI 7.
[6] *The City of God*, Book 21, chap. V.
[7] *De Natura Fossilium*, Book 5, p. 257, Basel, 1546.
[8] *De Rerum Fossilium*, 1565, p. 45 et seq.
[9] *De Gemmis et Lapidibus*, 1603, p. 155.
[10] *Le Parfaict Joillier*, 1644, p. 508.

evidently confounding selenite with pearls, which have a somewhat similar luster. He continues:

> Garcias ab Horto, Physician to the King of the Indies, we are informed states that perfect pearls are formed during the waxing of the moon; but that if the pearls are taken when the moon is waning, these become much smaller as time goes on, which is not the case with those which are collected when the moon is waxing.

This curious legend is evidently connected with the fable of the increase and decrease of the moon's image in selenite.

There is another and even quainter story with reference to selenite, one which is repreated by almost every writer on this mineral, down

FIG. 75. Specimens of Selenite showing the figure of the Man in the Moon. (From the *Hortus Sanitatis.*)

through the centuries. It was current at least as far back as the time of Dioscorides,[11] who says that some call selenite "aphroselenon" (the Grecian equivalent of "Lunae Spuma" or spume of the moon,) adding that it is a mineral which cures epilepsy and is thought if tied to the trunk of a tree will make the tree more fruitful. In the early Middle Ages this same idea is met with in an unexpected quarter, namely in *Philobiblon*,[12] a charming little work on the *Love of Books* (to which reference has already been made in Chapter V) written in 1345 by Richard de Bury, Bishop

[11] Book V, 159, p. 655.

[12] *The Love of Books*, the *Philobiblon* of Richard de Bury newly translated into English by E. C. Thomas (The King's Classics) London, 1902, pp. 3, 4, & 124.

of Durham, High Chancellor of England and a friend of Petrarch. In this, referring to the sad fact that many men of marked intellectual power and who promised to be scholars of great ability and accomplishments were unable to continue their studies, since, "With poverty only as their stepmother, they are repelled violently from the nectared cup of philosophy as soon as they have tasted of it," he goes on to say of them:

Alas, how is the sun eclipsed in the brightness of the dawn, and the planet in its course is hurled backwards, and while it bears the nature and likeness of a star, suddenly drops and becomes a meteor.

FIG. 76. "Spuma Lunae" dropping from the moon.   (From the *Hortus Sanitatis*.)

The following note to this passage is given on page 124 of the book:

A meteor: the word used in the original "assub" is derived from the translations of Aristotle made from the Arabic; it is found in Latin-English handbooks of the Middle Ages glossing "sterre-slyme" i.e. the star jelly which was supposed to be deposited by falling stars.

In the earliest edition of the *Hortus Sanitatis* in which minerals as well as plants and animals are included, there is a paragraph on "selenites" illustrated by a very interesting little woodcut showing tears of star jelly falling from the bright and shining face of the moon· (see figure 76).

Gesner[13] mentions the ancient story, that selenites came from the moon, ("The common crowd can be persuaded that the most absurd

[13] *De Rerum Fossilium*, leaf 45.

statements are really true"), and observes that *lunatics* when seized with an attack are said to *froth* at the mouth. He says that the mineral called Aphroselenum (moon froth) is found only in Egypt where dew from the sky is deposited in a time of bright moonlight in a form "quam specularem vocamus" and after deposition becomes hard and solid.

Henckel[14] says that falling stars are composed of limpid water, some of which he had caught and distilled, obtaining from it "Une terre phlogistique charbonneuse qui n'avoit rien de nitreux, qu'on pouvoit réduire en cendres, & même vitrifier, ou plutôt mêler à la frite du verre." He also states that Menzelius reports that on one bright moonlight night some of this "Gélée aerienne" actually fell at his very feet.

These stories, however, like so many other legends generally accepted in medieval times, gradually faded away when they were subjected to close observation and investigation. Already as far back as 1546, Agricola[15] with his wide practical experience of selenite, "Others call it Aphroselenon because in all seriousness they believe (and what absurdities can the ignorant mass of the people not be guilty of) that it is the foam of the moon."

Much light was thrown on the actual nature of Lac Lunae by the work of Longius,[16] who in his chapter on this mineral says that it occurs in Switzerland in a number of places, on the tops of the highest mountains, where it is found in caves. Among these is Mons Pilatus near Lucerne, toward the summit of which is the Moonloch or Caverna Lunaris of which Longius gives a picture. In this well known cave Lac Lunae occurs abundantly. It forms, he says, a white coating on the walls and also assumes imitative shapes resembling leaves and icicles. Longius makes the important statement that the Lac Lunae effervesces when treated with acetic acid, and goes on to say that the waters from the melting snows on the summit of Pilatus, sinking down through cracks and fissures in the rock which forms the roof of the cavern, dissolve this shattered and porous rock, converting it into Lac Lunae which it then deposits in the cave. In a succeeding section entitled *Selenites seu Lapis Specularis* he gives brief references to several varieties of "Selenites" and three figures of this mineral. One of these shows an excellent cleavage rhombohedron of calcite or Iceland spar. Another is a group of dog tooth crystals of this same mineral. There is nothing in

[14] *Origine des Pierres*, chap. I, p. 397, (Translated from the German, Paris, 1760.
[15] *De Natura Fossilium*, Basel, 1546, Book V, p. 257.
[16] *Historia Lapidum Figuratorum Helvetiae*, Venice, 1708, p. 6.

his description or figures to indicate that under the name selenite he includes any true selenite at all. His "selenites" is calcite.

Wagner[17] also records that Lac Lunae is found in the "Moonmilchloch" on Mount Pilatus and adds that it is light and "Fungosa" in character, being fluid at first and subsequently hardening.

Bruckman[18] writing a few years later says that the mineral has been called by various names. Among these *Marien Glas or Eis* "aus blinde Pabsthum" since there is no reference made to it in the bible in connection with the Virgin Mary. It has also been called *Argyrolithus* or silver stone on account of its silvery luster, and *Selenite* because it shines clearly in bright moonlight and shows the figure of the waxing and waning moon, also *moon spume*. The stories connected with these two latter names, he remarks, are untrue and ridiculous. It is evident that he confuses three minerals under the term selenite, namely, true selenite, which he says occurs as veins in alabaster quarries, calcite which occurs at Tideberg and cleaves into rhombs, and muscovite which comes from Moscow and is used for windows, a mineral which in later years was commonly known as *Muscovy glass*.

2. Amber or Sermons in Stones

Amber, like coral, was in classical and medieval times considered to be a mineral, its origin being unknown or at best uncertain. It is mentioned by the earliest writers. Homer refers to it, as do also Thales of Miletus, Aristotle and Theophrastus. Pliny whose descriptions of most minerals are brief, devotes several pages in his *Natural History* to amber, making reference not only to its properties, origin and history, but also giving the opinions of many former writers concerning it, and using some strong language in his denunciation of their fables and the excessive value placed upon amber in his time, some people paying more for a small figure carved in amber than the price of a living and healthy slave. Ovid refers to its origin, Martial has several epigrams with reference to the remains of insects and other living things which are sometimes enclosed in it, and a host of later writers mention it, many of them drawing their material from Pliny and often quoting him verbatim.[19]

---

[17] *Historia Naturalis Helvetiae*, Zurich, 1680, p. 340.

[18] *Thesaurus Subterraneus Ducatus Brunsvigii*, Brunswick, 1728, pp. 104 & 108.

[19] For an account of the literature of the subject see Gimma, *Della Storia Naturale delle Gemme*, Naples, 1730, Vol. I, p. 381 et seq. Also Bock, *Versuch einer kurtzen Naturgeschichte des Preussischen Bernsteins*, Königsberg, 1767, and Stoppani, *L'Ambra nella storia e nella Geologia*, Milan, 1886.

The unusual interest manifested in this mineral was due to three causes:

First.   The bright yellow or golden color which it generally displays;

Secondly.   The fact, which was regarded as a very remarkable one, that when rubbed it attracted to itself little scraps of paper, twigs and other similar objects, a property which led many writers to classify it with the magnet which drew to itself little pieces of iron.

Thirdly.   Because enclosed in amber small insects of many different species were found, while in the clouded varieties strange and curious forms were often seen, which in the imagination of the observer seemed to bear a resemblance to scenes, landscapes, human forms or strange figures which were believed to have mystical meanings or which conveyed strange suggestions.

Very diverse views concerning the origin of amber were current among the ancients and there was much discussion concerning them.   There were many fables on this subject.   One was that Phoebus, the charioteer of the sun, yielding to the entreaties of his son Phaëthon, allowed him to attempt to drive that orb for one day across the sky, and he not having sufficient strength to control the horses, these ran wild and carried the sun down so near the earth that the latter was set on fire and would have been destroyed had not Jove, seeing the danger, killed Phaëthon with a thunderbolt, who being hurled down, fell into the river Eridanus or Padus, and his sisters who had yoked the horses to the chariot were changed into poplar trees by the side of the river, and wept tears which changed to amber.   Ovid telling the sad story says, "Still their tears flow on and these hardened into amber by the sun, drop down from the new-made trees.   The clear river receives them and bears them onward, one day to be worn by the brides of Rome."[20]   Another story is told by *Sophocles* and is referred to by Pliny as follows:[21]

But I wonder most at *Sophocles* The Tragicall Poët (a man who wrote his Poësies, with so grave and lofty a stile, and lived besides in so good reputation; being otherwise borne at Athens, and descended from a noble house, emploied also in the managing of state affairs, as who had the charge & conduct of an army) that he should go beyond all others in fabulous reports, as touching Amber: for he sticketh not to avouch, That beyond India it proceedeth from the tears that fall from the eies of the birds Meleagrides, wailing and weeping for the death of Meleager.   Who would not marvell, that either himselfe should be of that beliefe, or hope to persuade others to his opinion?   For what child is there to be found so simple and ignorant, who will beleeve, that birds should keep their times to shed tears every yere so duly, and

---

[20] Ovid's *Metamorpheses*, Book II, v. 364–366.
[21] *Natural History*, Book 37 (Holland's translation.)

especially so great drops and in such quantitie, sufficient to engender Amber in that abundance? Besides, what congruitie is there, that birds should depart as far as to the Indians and beyond, for to mourn and lament the death of *Meleager* when he died in Greece? What should a man say to this? Are there not many more as goodly tales as these, which Poëts have sent abroad into the world? And their profession of Poëtry, that is to say, of faining and devising fables, may in some sort excuse them. But that any man should seriously and by way of history deliver such stuffe, as touching a thing so rife and common, brought in every day in abundance by merchants which were ynough to convince such impudent lies, is a meere mockerie of the world in the highest degree; a contempt offered unto all men, and argueth an habit of lying, and an impunitie of that vice intollerable. . . .

But to leave Poëts with their tales, and to speak resolutely and with knowledge of Amber, knowne it is for certain, That engendered it is in certaine Islands of the Ocean Septentrionall, where it beateth upon the coasts of Germany: and the Almanes call it Glessum. And in very truth, in that voyage by sea which Germanicus Caesar made into those parts, our countrymen named one of those Islands Glessaria, by reason of the Amber there found; which Island the Barbarians call Austravia. It is engendered then in certaine trees, resembling Pines in some sort, and issueth forth from the marrow of them, like as gum in Cherrie trees, and Rosin in Pines. And verily, these trees are so full of this liquor, that it swelleth and breaketh forth in abundance: which afterwards either congealeth with the cold, or thickeneth by the heat of Autumn. Now if at any time the sea rise by any extraordinary tide, and catch any of it away out of the Islands, then verily it is cast a shore upon the coast of Germany, where it is so apt to roule, that it seemeth (as it were) to hang and settle lightly upon the sands, whereby it is the more easily gotten.

Niceas expresses his opinion, which is fanciful as that of Ovid or Sophocles, and which Pliny sets forth as follows:

Niceas[22] would have us conceive that it should be a certain juice or humour proceeding (I wot not how) from the raies of the Sun, and yet he maketh a reason thereof, imagining that the said beames should be exceeding hot toward the Sun-setting, which rebounding from the earth, leave behind them a certain fatty sweat, in that part of the ocean and the same is afterwards cast up with tides into the sea-shore and sands of the Germans.

Others again held that amber is the consolidated urine of the lynx, the semen of elephants or of certain fishes, views which Aldrovandus sweeps aside as ridiculous.[23]

De Boodt[24] says:

Some tell us that there is a place, in the "Sudinorum Oceano," not more than 30 stadia from the shore, where if when the sky is clear one looks down into the water

---

[22] *Natural History*, Book 37 (Holland's translation).
[23] *Musaeum Metallicum*, Bologna, 1648, p. 405.
[24] *Gemmarum et Lapidum Historia*, Hanover, 1609, chap. CLIX, p. 163.

he will see something which shines & glistens at the bottom of the sea looking like transparent bitumen, and he will see the fishes playing about it.   They believe that this is a fountain of amber or a mountain mass of consolidated and hardened amber from which fragments were broken off by the waves of the sea in times of tempest, and cast up on the shore, but all these things are doubtful.

Amber is now known to be a resin derived from certain fossil trees which occur in early tertiary strata along the shores of the Baltic Sea, and which is found most abundantly on the coast of Samland and the Kurische Nehrung.   This is the district from which the chief supplies of amber have been obtained from classical times down to the present era.   Some of it is washed up by the sea from submarine extensions of the amber-bearing beds and some of it is obtained by digging pits or trenches on the land near, or in some cases at considerable distances from, the coast.   It is associated with lignite into which the wood of the trees has been transformed.   Hence the various writers who in ancient times ascribed to it certain widely different modes of origin were often in a measure right.   Some, Theophrus, Tacitus and others, said it came from the sea.   Some, Theophrastus, Agricola, Erasmus, Stella, said it was a product of the earth.   Others, Aristotle, Pliny, Dioscorides, the *Hortus Sanitatis*, said it was the product of a tree.

A fact which, as already mentioned, attracted especial attention to it all through the ages, was that there could frequently be seen enclosed in the clear transparent amber, insects of various kinds, flies, ants, beetles, spiders, etc. and it was reported by some writers that bees, wasps and even small reptiles and fishes were sometimes also found.   It seems highly probable that these latter existed only in the imagination of the writers who recorded them, or that they were introduced into the amber artificially, since amber can be softened and then hardened again.

Sendelius describes many of these and his descriptions are accompanied by thirteen quarto pages of illustrations.   He mentions a piece of amber enclosing three bees, which was presented to Pope Urban VIII (1623) and Stoppani[25] referring to this remarks, that whether these were three bees, or three gnats or three winged insects of any kind, it was a gift which would have pleased the Pontiff greatly, for he belonged to the Barberini family whose coat of arms displayed three bees: *"Azure three bees Or with wings folded, two and one."*[26]   He would have been impressed by the fact that his coat of arms had already been recorded by nature so

[25] *Historia Succinorum corpora aliena involventium*, Leipzig, 1742, p. 9.
[26] *L'Ambra nella Storia e nella Geologia*, Milan, 1886, p. 32.

many centuries before the Barberini family, not to say the family of Adam, had come into existence.

The figures, forms and pictures seen in clouded amber, undoubtedly elaborated by the imagination of those who studied them, led to the building of a body of mystical interpretation and moral instruction, which has made amber perhaps the outstanding example of a "fossil" in which, for those who had eyes to see and ears to hear, there were indeed to be found *"Sermons in stones."*

Aldrovandus[27] mentions several of these moral lessons to be drawn from amber:

Men who like amber are at first soft, changeable, fickle and prone to commit offences of every kind, nevertheless when chilled by the divine fear or having been exposed to the experiences of human suffering, become confirmed in a life of constancy & perseverance.

As Amber, when warmed by rubbing, attracts to it various scraps of rubbish which are brought near to it, so Man, glowing with a love of earthy things, draws to him and embraces all manner of temporal goods and good things, none of which have any true value.

Or again:

Amber which is at first soft but at the same time very tenacious, makes prisoners of many living things, which find it very difficult or impossible to again regain their freedom, so Heresies (or Heretics) with ease lay hold on many unwary souls and entangle them in evil ways.

But the most remarkable book on this subject is that of Severin Göbels, which bears the title *De Succino*, the first (quarto) edition of which appeared in 1558 and which was reproduced in small octavo form, in Gesner's collection of eight works by various authors, of which his own *De Rerum Fossilium* was one, and which bore the title *De Omni Rerum Fossilium Genere* published in Zurich in 1665.[28] The work is divided into two "books" the second of which contains a general account of the origin, character and uses of amber, as known at that time. The different varieties of clear and cloudy amber when intermingled in the same specimen often give rise to a confused structure in which strange resemblances to various objects, scenes or figures can be traced by the help of a vivid imagination, and the first book of Severin Göbels work

---

[27] *Musaeum Metallicum*, Bologna, 1648, p. 414.

[28] See reference to the four editions of this work in Bock: *Versuch einer kurtzen Naturgeschichte des Preussischen Bernsteins*, Königsberg, 1767, p. 13.

is devoted entirely to an exposition of the fact (or the author's conviction) that in the forms and pictures which they present is to be found a complete allegorical exposition or revelation of the teachings of Christianity:

This precious balsam passed down by subterranean passages, often defiled by clay and filth, into the sea whose waves took it hither and thither, until finally it is solidified by the coldness of the sea, and altered by its saltness.   The saltness and the bitterness of the sea represent the bonds and the sharpness of the law; and furnish a symbol of the anger of God, wherein our first parents and their off-spring, as it were, overwhelmed by the storms of the ocean, were long tossed about at random, until, by the admirable goodness of God, they were once again restored to the shore and harbour of a quiet conscience.   And that restitution takes place through the prophets and apostles, who are the poor little fishermen, who collect the amber thrown upon the shores of the ocean.[29]

The treatment of the subject is quaint and interesting, the book having its basis in that doctrine of signatures to which reference has already been made in Chapter III.

3. ROCK CRYSTAL AS A FORM OF ICE

The mineral quartz, which in certain of its common and more or less impure forms is known as flint, chert and jasper, and in some of its purer and more beautiful forms constitutes chalcedony, amethyst, carnelian and agate, when in perfectly pure, clear, transparent and colorless crystals is known as rock crystal, from the Greek word "Krustallos" meaning ice, because it was the common opinion of the Greeks from the earliest times that it was really ice which by the long continued and intense cold, in the high mountainous regions from which it was supposed to come, had become permanently frozen into this hard, and ice like form.   It is referred to as ice by Homer, Thucidides and Plutarch.[30]

The belief that rock crystal is a form of ice was the basis of many interesting opinions and fancies on the part of medieval theologians. This Peter Lombard in his *Book of Sentences* discussing the division of the "Waters which were above the Firmament" from the "Waters which were below the Firmament," mentions an opinion put forward by the Venerable Bede, to the effect that the former are solid and form the *crystalline heavens* of the Ptolemaic System, "For crystal which is so hard and transparent is made of water."[31]   There are two passages in

[29] *De Succino*, Zurich, 1665, leaves 1 & 2.

[30] Gimma, *Storia Naturale delle Gemme ovvero Fisica Sotterranea*, Vol. I, p. 297.

[31] Quoted in Whewell's *"History of the Inductive Sciences*, Vol. I, p. 242.

Holy Writ to which the theologians repeatedly refer, in this connection. The first is in Psalm CXLVII, V. 17: Mittet *crystallum* suam sicut buccellas"—"He casteth forth his *ice* like morsels." The other is from *Ecclesiasticus*, Chapter 43, V. 20: "Frigidus ventus aquilo flavit et gelavit crystallus ab aqua."—"When the cold north wind bloweth and the water is congealed into *ice.*" Their custom of trying to read into the plain text of scripture recondite and mystical meanings was abundantly exercised on these texts and much admonition extracted from them.

Among those who in classical and medieval times wrote on natural history subjects Pliny, Seneca, Isidore of Seville, Marbodus and Neckham say that rock crystal is ice and Marbodus in his *De Lapidibus Pretiosis* begins his statement concerning crystal with the following words:

Cristallus glacies multos durata per annos,
Ut placuit doctis qui sic scripsere quibusdam
Carminis antiqui, frigus tenet atque calorem.

It was, however, pointed out by other writers that there were grave objections to this view namely, that if rock crystal were ice it should be possible to change it into water by heating it, even the sun's rays should be sufficient to bring about this result and thaw out the rock crystal on the high mountains where it is found just as it does the ice which it so much resembles. Furthermore, if it be merely frozen ice, it should be far more abundant in high, snowclad mountains and in the cold portions of the earth, but it is found abundantly in hot countries such as Madagascar and the island of Cyprus, and finally while ice will float on water, rock crystal sinks.

For these reasons, the writers in question rejected the idea that rock crystal was a kind of very hard ice and were led to believe that it was composed in part at least of "earth" in the Aristotelian sense. Agricola classes it among his "succi concreti" to which salt, alum, sulphur and auripigment belong, and De Boodt[32] agrees with him. Gimma,[33] Rosnel,[34] Matthiole[35] and others consider it to be a "humour," by which they probably intended to convey the same idea.

Diodorous Siculus is said by Baccius[36] to bridge the gap between these two opinions when he states that "Crystal is a stone which is pure water

[32] *Gemmarum et Lapidum Historia*, Leyden, 1647, p. 222 (edited by Tollius).
[33] *Fisica Sotterranea*, Naples, 1730, Vol. L, p. 302.
[34] *Le Mercure Indien*, Paris, 1668, p. 64.
[35] *Commentary on Dioscorides*, Lyons, 1579, p. 780.
[36] *De Gemmis et Lapidibus pretiosis*, Frankfort, 1603, p. 102.

consolidated, not by cold, but by the power of a divine heat which gives to it its hardness.''

The belief that rock crystal was frozen water still survived in spite of the very valid reasons which had been urged against it, as is shown in a reference to it by Boyle in 1672 who, as he says:

Hydrostatically and with a tender balance examined the weight of crystal first in the air and then in water. I found its weight to be to that of water of equal bulk as two and almost two-thirds to one. (And he adds) Which, by the way, shews us, how groundlessly many learned men, as well ancient as *modern*, make crystal to be but ice extraordinarily hardened by a long & vehement cold; whereas ice is bulk for bulk lighter than water and therefore swims upon it.[37]

It is to be noted however that while Boyle probably in this passage announces for the first time the specific gravity of rock crystal, the fact that this mineral was heavier than water and would sink in it, while ice floated on water was known and was mentioned by several writers before Boyle's time. The first to draw attention to this fact seems to have been Baccius who wrote in 1603,[38] and who cites it as evidence that rock crystal is not ice.

While preparing this chapter the writer happened to attend a dinner of the Geological Society of America and found himself seated beside one of the most distinuished American palaeontologists. In the course of converstion this gentleman mentioned that one of his sons, although still a young boy, had developed a great interest in geology and had become an enthusiastic collector. Among the specimens which he showed to his father with great pride was one of "fossil ice" which he had found filling a narrow crack in a cliff of granite. His father went on to say that he had explained to the boy that it was not ice but a mineral called quartz, but added that he was very doubtful whether he had really convinced his son that such was the case. It was an interesting incident, as his father expressed surprise when it was pointed out to him that, in his reluctance to abandon his belief that the treasured specimen was a piece of fossil ice, his son was merely following in the footsteps of a long line of most distinguished men of science down through the past centuries.

4. Stones that Shine in the Dark

The gems or precious stones, owing to their beauty of color and especially owing to their brilliance due to a high refractive index which causes

---

[37] *An Essay about the Origine & Virtues of Gems*, London, 1672, p. 81.
[38] *De Gemmis et Lapidibus*, Frankfort, 1603, p. 101.

them, when suitably cut, to reflect a large part of the light which falls upon them, have always attracted especial attention. But in certain gems which combine with this brilliancy special colors, the quality seemed to be so enhanced that they were commonly regarded among the ancients as actually giving out light which originated within their own substance, that is to say, as shining with their own light. They were therefore believed to be visible in the dark and in fact many stories coming down from very early times tell of them as lighting up the blackest night with a radiance like that of the sun at noon. Among these gems the *carbuncle* and the *topaz* were preëminent, especially the former, although certain others have had the same power attributed to them, especially lychinites, orfano, chrysophrase and selenite, to these, however, no further reference need here be made.

In the Middle Ages no sharp distinction was or could be made between the bright red varieties of a number of different minerals, as for instance, ruby, garnet and spinel. Nothing was known of the composition of these minerals and their color was their outstanding physical character. Carbuncle was a name given to the purest and most brilliant specimens of any of these minerals having a deep red color. The name "carbunculus" signifies a small live coal, to which the mineral bears a close resemblance. In French the gem was known as the "Escarboucle" and passed into heraldic usage under this name.

Aldrovandus[39] says: "Among the gems the carbuncle shines most brightly, as does gold among the metals or the sun among the stars." A hundred authors might be cited who relate that carbuncle shines brilliantly in the dark, but a few of these stories will suffice.

Ludovicus Vartomanus[40] relates that "The King of Pege, a city of India, wears a carbuncle of such size and splendor that anyone looking at the king in the darkness sees him resplendent as the sun."

Bartolommeo Cassaneo[41] tells of a carbuncle which set aloft on a high pyramid lighted up a whole city. Francesco Spinola of Milan,[41] in a letter addressed to a priest, Francesco Pilo of Brescia, in 1561 speaks of carbuncles which exist in Pegu which had "from nature" the property of shining with their own light and it is stated in the Hortus Sanitatis that the gem shines more brilliantly by night than by day, that in fact it turns night into day.

[39] *Musaeum Metallicum*, Bologna, 1648, p. 958.
[40] See Cardanus, *De Subtilitate*, Nuremberg, 1550, p. 169.
[41] Gimma, *Della Fisica Sotteranea*, Naples, 1730, vol. I, pp. 142 & 146.

The authors of the *Histoire Prodigieuses*[42] state that the escarboucle is a ruby and that if thrown into a blazing fire, it outshines the fire itself in brilliancy. But Archelaus goes a step further in enlarging upon the marvellous properties of this wonderful mineral, and says that it will melt wax which is brought near to it, even in the dark.[43]

A striking tale in which a brightly shining carbuncle plays an important rôle was told by Sylvester the Second, who died in the year 1003. It is to be found in the *Gesta Romanorum*, the most popular story book of the Middle Ages, and runs as follows:[44]

There was an image in the city of Rome standing in the erect posture, with the dexter hand outstretched; and upon the middle finger was written, "STRIKE HERE." The image stood a long time in this manner, and no one understood what the inscription signified. It was much wondered at, and commented on; but this was all, for they invariably departed as wise as they came. At last, a certain subtle clerk, hearing of the image, felt anxious to see it; and when he had done so, he observed the superscription, "*Strike here.*" He noticed that when the sun shone upon the image, the outstretched finger was discernible in the lengthened shadow. After a little consideration he took a spade, and where the shadow ceased, dug to a depth of about three feet. This brought him to a number of steps, which led into a subterranean cavity. Not a little exhilarated with his discovery, the clerk prosecuted the adventure. Descending the steps, he entered the hall of a magnificent palace, in which he perceived a king and a queen and many nobles seated at table, and the hall itself filled with men. They were all habited in costly apparel, and kept the most rigid silence. Looking about, he beheld in one corner of the place a polished stone, called a carbuncle, by the single aid of which the hall was lighted. In the opposite corner stood a man armed with a bow and arrow, in the act of taking aim at the precious stone. Upon his brow was inscribed, "I am what I am: my shaft is inevitable; least of all can yon luminous carbuncle escape its stroke." The clerk, amazed at what he saw, entered the bedchamber, and found a multitude of beautiful women arrayed in purple garments, but not a sound escaped them. From thence he proceeded to the stables, and observed a number of horses and asses in their stalls. He touched them, but they were nothing but stone. He visited all the various buildings of the palace, and whatsoever his heart desired was to be found there. Returning to the hall, he thought of making good his retreat. "I have seen wonders to-day" said he to himself, "but nobody will credit the relation, unless I carry back with me some incontrovertible testimony." Casting his eyes upon the highest table, he beheld a quantity of golden cups and beautiful knives, which he approached, and laid his hands upon one of each, designing to carry them away. But no sooner had he placed them in his bosom, than the archer struck the carbuncle with the arrow, and

[42] Antwerp, 1594, p. 83.

[43] See Agricola, De Natura Fossilium, Basel, 1546, p. 299.

[44] *Gesta Romanorum* or *Entertaining Moral Stories*, translated from the Latin by the Rev. Charles Swan (Bohn's Antiquarian Library) London, 1877, p. 185.

shivered it into a thousand atoms. Instantly, the whole building was enveloped in thick darkness, and the clerk, in utter consternation, sought his way back. But being unable, in consequence of the darkness, to discover it, he perished in the greatest misery, amid the mysterious statues of the palace.

APPLICATION

My beloved, the image is the devil; the clerk is any covetous man, who sacrifices himself to the cupidity of his desires. The steps by which he descends are the passions. The archer is death, the carbuncle is human life, and the cup and the knife are worldly possessions.[45]

But the carbuncle was of interest for another reason, it was always associated with that mysterious animal, of great renown in these early times and mentioned repeatedly in Holy Writ—the dragon. The dragon as represented by contemporary painters was a horrible monster, somewhat serpent-like in form but with legs armed with claws and with wings and terrible teeth, a tongue coming to a spear-like point, vomiting forth fire and smoke. Several diverse views were held concerning its origin. One story which met with widespread acceptance was that when certain great serpents attained to a very old age, they developed feet and wings and turned into dragons; others held that the dragon was a separate species and reproduced its kind like other creatures. Still others, judging from its traditional form, believed the dragon to originate from the interbreeding of great snakes with certain lizard-like animals of enormous size, which were believed to exist in those countries of romance and magic, the Indies and Egypt, in which dragons had their homes. It is to be remarked, however, that the great dragons which are so notable in history as having laid waste whole provinces and terrified the peoples of many lands in various parts of the world and which, apparently, could only be overcome and destroyed by someone renowned for sanctity and leading a life of distinguished holiness, was evidence to the minds of many, that such dragons at least were brought into being by some diabolical agency which not only gave them such powers of destruction, but which at the same time rendered them invulnerable and invincible to all purely human forces. Such was the dragon St. George slew on the coast of Syria, that which St. Martha subdued at Tarascon and that which St. Francis drove away from Assisi by making the sign of the Cross.

[45] For additional stories concerning the carbuncle and other gems see William Jones, *History & Mystery of Precious Stones*, London, 1880.

Now every dragon was believed to carry about with it a carbuncle to give light in the caverns which it inhabited or at night time in the open countries where it lived.   This glowing gem it held between its teeth and never laid it down except to eat or drink.   To get this rare and much prized gem, the dragon had to be killed, hence one of the reasons for the high value set upon the stone.   A very interesting little treatise on this subject, written "Avec beaucoup de Politesse et de vérité" as one of the letters of recommendation says, is that of Jean B. Panthot, Doyen du College des Medecins at Lyons, published at Lyons in 1691 under the title, *Traité des Dragons et des Escarboucles*.   Scheuchzer also, who made an especial study of dragons, refers at some length to those mentioned in the book of Job.[46] Aldrovandus says that according to some, the carbuncle grows in the forehead of certain animals and gives them light.[47]

The topaz ranked next to the carbuncle among the gems as a source of light.   But among the ancients this name was given by different authors and at different times to other gems of different colors, as well as to the true yellow topaz.   Caesius says that held up to the sun it is seen to emit flames, and Baccius[48] states that some varieties of this gem actually give out light.   Strabo says the same, and a topaz was said to have been presented to a monastery by the noble Lady Hildegard, wife of Theodoric, Count of Holland, which at night emitted so brilliant a light that, in the chapel where it was kept, prayers were read without the aid of a lamp.[49]   It comes, according to Archelaus, from the country of the Troglodites in Arabia, who discovered it in the soil when grubbing for roots on which to feed.   Strabo[50] adds some interesting details concerning the method in which these gems are collected:

After the bay is the island Ophiodes, so called from the accidental circumstances of its having once been infested by serpents.   It was cleared of serpents by the king, on account of the destruction occasioned by those noxious animals to the persons who frequented the island and on account of the topazes found there.   The topaz is a transparent stone, sparkling with a golden lustre, which however is not easy to be distinguished in the day-time, on account of the brightness of the surrounding light, but at night the stones are visible to those who collect them.   The collectors place a vessel over the spot, where the topazes are seen, as a mark, and dig them up in the day.

[46] *Jobi Physica sacra*, Zurich, 1721, p. 257.

[47] Milton in describing the cobra says: ". . . his head crested aloof and carbuncles his eyes" (*Paradise Lost*, IX, 499).

[48] *De Gemmis et Lapidibus pretiosis*, Frankfort, 1603, p. 45.

[49] William Jones, *History & Mystery of Precious Stones*, London, 1880, p. 14.

[50] *Geography*, Book XVI, chap. iv, Section 6, (Bohn's Classical Library).

A body of men was appointed and maintained by the kings of Egypt to guard the place where these stones were found, and to superintend the collection of them.

But one of the most striking and interesting occurrences of topaz is that referred to by Seyfrid:[51]

In that tract of country which lies along the course of the mighty stream known as the Riodella Plata, there is found a kind of stone or rather concretion, to which the natives give the name *Coccos*. They grow under the earth and when fully mature, explode with a loud report throwing into the air amethysts, topazes and gems of many other kinds which they enclose. So soon as the inhabitants of the district hear the reports of these explosions, guided by the noise, they hasten to the place and search for the scattered gems, but in many cases return disappointed, having been able to find nothing but common crystals.

But as the years passed on, doubts were expressed concerning the truth of the stories of ancient time concerning these refulgent gems. Some said that it was impossible for nature to produce a stone which shone in the dark. De Boodt[52] doubts the truth of such a statement for, he says, in the vegetable kingdom certain substances as for instance rotten wood, give forth light, and in the animal kingdom the eyes of cats and other animals shine in the night, why not therefore may not some gems do likewise, but he confesses that whether this actually is or is not the case had not been as yet definitely decided. Giovanni Renodea[53] expresses a more definite opinion, and says that no stone shines by its own light "Ut idiotae putant."

When the various stories were investigated it was found that those who told them had not actually seen the occurrences themselves or when they said they had, there was reason to doubt their veracity. They had heard them from someone who had seen them, or they had read about the wonderful occurrences in a book. And so the great collection of splendid stories which had gathered around these gems that shone at night were relegated to the realms of romance, and the refulgence of the gems themselves faded into the light of common day.

5. THE LODESTONE

No mineral has, from the earliest times, attracted more attention and awakened more interest than the lodestone or natural magnet, on account

---

[51] *Medulla Mirabilium Naturae*, Sulzbach, 1679, p. 468.
See also J. L. Bausch, *De Lapide Hematite et Aetite*, Leipzig, 1665, p. 23.
[51] *Gemmarum et Lapidum Historia*, Hanover, 1609, p. 70.
[53] See Gimma, *Fisica Sotterranea*, Naples, Vol. I, p. 147.

of the remarkable property which it possessed of attracting fragments of iron to it. "Is there anything," asks Pliny[54] "more wonderful, anything in which nature had displayed her power in a more striking manner than in this mineral? What is there that seems to us more dull or inert that this hard stone and yet nature has given it sense, yes and hands also, and the ability to use them." And he quotes Nicander as saying that its name is derived from that of a shepherd who was keeping his sheep on Mount Ida and who discovered it there, owing to the fact that as he was going to and fro on the mountain the nails of his boots and the tips of his iron-shod staff stuck to a certain exposure of rock when he walked over it. And as far back as the time of Pliny the bipolar nature of lodestone was known. For having referred to the "True Ethiopian Magnet" he says that "There is another mountain in the same Ethiopia not far from Zimiris, which breedeth the stone Theamedes which will abide no iron, but rejecteth and driveth the same from it."[55] And in another chapter of the *Natural History* he records that there are two hills near the river Indus, one of which will attract to itself iron of any kind, while the other repels the metal. If a man with iron nails in his boots tries to walk of the first of these mountains he is held fast and cannot move his feet, while if he tries to ascend the other he cannot put his foot down anywhere. One hill is composed of lodestone and the other of theamedes. Agricola[56] says that Pliny mentions stones which are magnet on one side and theamedes on the other, that is to say, pieces of lodestone which display polarity. Albertus Magnus[57] says in his time magnets were found which at one corner attracted iron and at another repelled it.

The power of magnet to attract iron and especially to magnetize the attracted fragments so that they in their turn attracted others, was a source of never-ending wonder to those who lived in former times—even St. Augustine refers to it:[58]

We know that lodestone draws iron strangely and surely, when I observed it at first, it made me much aghast. For I beheld the stone draw up an iron ring, and then as if it had given its own power to the ring, the ring drew up another and made it hang fast by it, as it hung to the stone. So did a third by that, and so on until

[54] *Natural History*, Book XXXVI, Chapter 16.
[55] *Natural History*, Book XXXVI, Chapter 16.
[56] *Natural History*, Book II, chap. 96.
[57] *De Mineralibus*, Cologne, 1569, Book II, p. 158.
[58] *De Civitate Dei*, Book 21, chap. 4.

there was, as it were, a chain of rings only by touch of one another, without any interlinking. Who would not admire the power in this stone, not only inherent in it, but also extending itself through so many circles and such a distance? But daily trial ever takes off the edge of admiration.

As might be expected, many other and wholly fabulous properties in the course of time came to be attributed to this remarkable mineral.

Gesner[59] gives a little picture of a fragment of lodestone, probably the first that ever appeared in a printed book, with iron filings adhering to it and a needle placed beside it.

This power of attracting iron was in some cases the cause of great disasters, for Serapion states, and the story is repeated by successive writers down through the centuries, that on the coast of India there are certain rocky hills composed of lodestone and that if any ship whose planks are fastened together by iron nails, sails too near them, every nail and every other article in the ship which is made of iron flies to the mountain like a bird to its nest, the ship falls apart and the crew are cast into the sea. In the *Hortus Sanitatis* a striking picture is given of such a disaster. Similar mountains or reefs of lodestone in other parts of the world are mentioned by various writers, adjacent to which the remains or débris of many ships can be seen through the clear waters, lying on the bottom of the sea.

The cause or reason why the lodestone or magnet attracted iron was naturally a subject of widespread speculation. The consideration of these speculations would here be out of place, they lie rather within the domain of physics. The literature of the subject is immense and goes back to the time of Anaxagorus. "I shall not," says Porta in his book on *Natural Magic*, "pass by the opinion of Anaxagorus set down by Aristotle in his book *De Anima*, calling lodestone, by a similitide a 'living stone' and that therefore it draws iron." The opinion expressed by many of the ancient writers when treating of the mineral lodestone, is that its power was exerted in some manner through a peculiar "effluvium" which it gave forth.

Cardan[60] and others as has been mentioned in Chapter IV, held that minerals possess life and need nutriment. In his opinion the magnet draws iron to itself, in order that it may feed upon this metal, both being closely similar in their natures.

[59] *De Rerum Fossilium*, Zurich, 1565, leaf 84.
[60] *De Subtilitate*, Nuremberg, 1550, Book 7, p. 187.

Agricola[61] after setting forth the strange powers possessed by the lodestone makes the characteristic remark: "Theologians assert that these powers are supernatural, the Physicians (and Naturalists) believe them to be natural but inexplicable." And finally a rather amusing contribution to this subject, made a century later by Col. Thomas Blount, a Fellow of the Royal Society of London, may be quoted, being interesting not only on account of what this gentleman says but owing to the euphuistic style, fashionable at that time, in which his observations are conveyed:

FIG. 77. A ship sailing too near a cliff of lodestone has its iron nails drawn out causing it to fall to pieces and sink.   (From the *Hortus Sanitatis*.)

Dr. Highmore tells us that the Magnetical Expirations of the Lodestone may be discovered by the help of glasses and be seen in the form of a Mist, to flow from the Lodestone.   This indeed would be a most incomparable Eviction of the Corporeity of Magnetical Effluviums and sensibly decide the controversy.   But I am sure he hath either better Eyes, or else better glasses than I ever saw (tho' I have looked through as good as England affords) . . . Nay, I could never see the grosser Streams that continually transpire out of our own Bodies and are the fuliginous Eructations of that Internal Fire which continually burns within us.   Indeed if our Dioptricks could attain to that Curiosity, as to grind us such glasses as would present the Effluvium of the Magnet we might hope to discover all Epicurus's Atoms and all those insensible Corpuscles which daily produce such considerable effects in the Generation

[61] *De Natura Fossilium*, Basel, 1546, Book V, p. 250.

and Corruption of Bodies about us: Nay might not such Microscopes hazard the discovery of the Aerial Genii and present even Spiritualities themselves to our view?[62]

6. ANTS WHICH DIG FOR GOLD

It was a widespread belief among the ancients that the search for gold was attended not only by much labor but in many cases at least by great danger.   Thus the Golden Apples of the Herperides were guarded by the dragon Ladon, the Golden Fleece also, sought by Jason and the Argonauts was, according to one version of the story, guarded by a dragon, this beast being generally considered as the appropriate warden of treasures.   But certain other gold deposits in some distant lands were not only guarded but actually worked by an even stranger and more remarkable group of animals.   Before referring to these a word more may be said concerning Jason and the Argonautic Expedition.   Homer in his *Odyssey* describes the adventures of Odysseus when returning from the siege of Troy, but before Odysseus sailed Jason started to find the Golden Fleece.   In its original form the story of the Argonauts was the narrative of an actual voyage to the Euxine or as it is now called the Black Sea, made by some "Minyans" of Thessaly in the late fourteenth or early thirteenth century B.C.[63] Jason, their leader, caused Argus to build for him a ship of fifty oars, which he named the *Argo* after its builder, and from this ship in its turn the Argonautic Expedition derived its name.   This was in the Bronze Age.   The story in the course of time became embellished and adorned by many fairy tales and appeared in several different versions.   Many and various conjectures have been put forward as to what this Golden Fleece really was which Jason and his men went to find.   Of these the explanation of Strabo seems to be by far the most probable.   The Golden Fleece he says was the gold of Colchis which was washed down in the river Phasis and collected by the Colchians in fleeces.   Agricola[64] was also of the same opinion and says:

The Colchians placed the skins of animals in the pools of springs: and since many particles of gold had clung to them when they were removed, the poets invented the *Golden Fleece* of the Colchians.   In like manner it can be contrived by the methods of miners that skins should take up, not only particles of gold, but also of silver and gems.

[62] *Philosophical Transactions of the Royal Society of London.*
[63] Janet Ruth Bacon, *The Voyage of the Argonauts*, London, 1925, pp. 141 & 168
[64] *De Re Metallica*, Book VIII.

This is a well known method of separating gold from the sand or other material with which it may be associated. In the present time a piece of blanket is sometimes employed instead of a fleece, the gold bearing sand after passing down a sluice box being carried over the blanket in the stream of water. The heavier gold sinks to the bottom and becomes entangled in the wool of the blanket, while the lighter sandy material is washed away. When sufficient gold has been gathered the blanket is washed or burned and the gold thus secured. Agricola's statement is

FIG. 78. The Argonauts at Colchis examining the fleece on which the gold is being collected. (From Agricola: *De Re Metallica*.)

accompanied by an interesting woodcut of the Argonauts being welcomed upon their arrival at Colchis and shown the stream of water carrying the gold-bearing sand, passing over a ram's fleece to which the head and horns of the animal were still attached. This was the golden fleece, the dragon was the dangers which the navigators had already overcome. The Argonautic Expedition is the first "gold rush" which is recorded in history.

The story of the Gold Digging Ants is a very old one also. It first appears in Herodotus,[65] the Father of History, 484 (c)–425 B.C. He says:

Besides these, there are Indians of another tribe, who border the city of Caspatyrus, and the country of Pactyica; these people dwell northward of all the rest of the Indians, and follow nearly the same mode of life as the Bactrians. They are more warlike than any of the other tribes, and from them the men are sent forth who go to procure the gold. For it is in this part of India that the sandy desert lies. Here, in this desert, there live amid the sand great ants, in size somewhat less than dogs, but bigger than foxes. The Persian King has a number of them, which have been caught by the hunters in the land whereof we are speaking. Those ants make their dwellings under ground, and like the Greek ants, which they very much resemble in shape, throw up sand-heaps as they burrow. Now the sand which they throw up is full of gold. The Indians, when they go into the desert to collect this sand, take three camels and harness them together, a female in the middle and a male on either side, in a leading rein. The rider sits on the female, and they are particular to choose for the purpose one that but just dropped her young; for their female camels can run as fast as horses, while they bear burthens very much better. . . .

When the Indians therefore have thus equipped themselves they set off in quest of the gold, calculating the time so that they may be engaged in seizing it during the most sultry part of the day, when the ants hide themselves to escape the heat.

When the Indians reach the place where the gold is, they fill their bags with the sand, and ride away at their best speed: the ants, however, scenting them, as the Persians say, rush forth in pursuit. Now these animals are, they declare, so swift, that there is nothing in the world like them: if it were not, therefore, that the Indians get a start while the ants are mustering, not a single gold-gatherer could escape. During the flight the male camels, which are not so fleet as the females, grow tired and begin to drag, first one, and then the other; but the females recollect the young which they have left behind, and never give way or flag. Such, according to the Persians, is the manner in which the Indians get the greater part of their gold; some is dug out of the earth, but of this the supply is more scanty.

Herodotus did not, of course, originate this tale, it was one already generally current among people in his time. Indeed he says that he got it from the Persians, probably when he was in Susa gathering material for his history.

Strabo[66] (writing in 25 B.C.) says:

All the country on the other side of the Hyspanis is allowed to be very fertile, but we have no accurate knowledge of it. Either through ignorance or from its remote situation, everything relative to it is exaggerated or partakes of the wonderful. As, for example, the stories of myrmeces (or ants), which dig up gold. . . . Nearchus says

---

[65] *History*, Book III, 102, 104, 105. (Rawlinson's translation.)

[66] *Geography*, (Bohn's Classical Library) Book XV. chap. 1, Sections 37, 44 and 69.

that he saw skins of the myrmeces as large as the skins of leopards. Megasthenes, however, speaking of the myrmeces, says, among the Dardae, a populous nation of the Indians, living towards the east among the mountains, there was a certain plain of about 3000 stadia in circumference; that below this plain were mines containing gold, which the myrmeces, in size not less than foxes, dig up. They are excessively fleet, and subsist on what they catch. In winter they dig holes, and pile up the earth, like moles, at the mouths of the openings.

The gold-dust which they obtain requires little preparation by fire. The neighbouring people go after it by stealth, with beasts of burden; for if it is done openly, the myrmeces fight furiously, pursuing those that run away, and if they seize them, kill them and the beasts. In order to prevent discovery, they place in various parts pieces of wild beasts, and when the myrmeces are dispersed in various directions, they take away the gold-dust, and, not being acquainted with the mode of smelting it, dispose of it in its rude state at any price to the merchants. . . . They say that some of these gold-digging ants have wings.

It is not to be supposed that Pliny[67] (A.D. 76) would allow such an old and interesting story to escape his net. His version of it is as follows:

In the temple of Erythrae, there were to be seen the horns of a certain Indian Ant, which were there set up and fastened for a wonder to posteritie. In the countrey of the Northerne Indians, named Dardae, the Ants do cast up gold above ground from out of the holes and mines within the earth: these are in colour like to cats, and as big as the wolves of Aegypt. This gold beforesaid, which they worke up in the winter time, the Indians do steale from them in the extreme heate of Summer, waiting their opportunitie when the Pismires lie close within their caves under the ground, from the parching sun. Yet not without great danger: for if they happen to wind them and catch their sent, out they go, and follow after them in great hast, and with such fury they fly upon them, that oftentimes they teare them in pieces; let them make away as fast as they can upon their most swift camels, yet they are not able to save them. So fleet of pace, so fierce of courage are they, to recover gold that they love so well.

Solinus[68] (flourished about A.D. 250) says that "The Ants which guard this gold resemble great dogs in shape and dig up these golden sands with their feet which resemble those of small lions. They seize and kill any who try to carry it away.

This story is one or other of its various modification appears in the writings of a whole succession of writers, European and Arab, from the classical period through the middle ages and down into modern times. It would seem that there was something on which the story was originally based, but what this is remains a matter of conjecture. The locality where these "Gold-digging Ants" live, judging from hints to be gathered

[67] *Natural History*, Book 11, chapter 31.
[68] *De Memoralibus Mundi*, Venice, 1493.

from the various accounts, was some cold and distant country, probably on the border of the Gobi desert. The stories suggest extensive alluvial gold workings and an uncouth people intensely hostile to all intruders or visitors.

The Count von Veltheim[69] who in 1799 made a study of the various versions of this legend gives as the result of his investigations the following interpretation of the story. In this remote district in very early times the gold was obtained from the sand by washing it in the manner adopted in all places where alluvial gold is obtained. The ancient records show that a very considerable amount of gold was derived from this area as tribute to various rulers. Thus must have necessitated the presence of a large number of persons in this gold field, working the sand and throwing up great piles of the refuse from which the gold had been separated. The method adopted in such alluvial gold workings (as has been mentioned) is to throw the gold-bearing sand into long sluices through which water is running, and to pass the sand and water over a fleece, blanket or some such woolly or hairy surface on which the heavy particles of gold, sinking to the bottom of the sluice will become entangled while the lighter sand will be washed away. In the case of these Gobi washings Veltheim conjectures that the pelt of the natural Tartary fox was employed in lieu of a fleece as being suitable for the purpose and easily obtained. This fox itself is accustomed to dig into the sand and make shallow subterranean burrows for itself, throwing up irregular piles of sand, similar in a way to those made by the gold washers. It is customary for the inhabitants of the neighbouring portions of Tartary to keep large and very fierce dogs to drive away other animals or marauding strangers. As those in charge of the gold washing operations would naturally desire to prevent any passing caravans or travellers from approaching the workings, accounts of these would be given only by those who had seen them from a considerable distance, so that the cages of foxes trapped for their pelts, the tumultuous heaps of sand like those thrown up by foxes seen elsewhere, the uncouth appearance of the workers themselves and the fierce dogs would form the basis, which embellished by the oriental imagination, might give rise to the various elements of the wonderful stories passed down by successive generations of writers, concerning the "Gold-Digging Ants" of the Gobi desert.

[69] *Von den goldgrabenden Ameisen und Greiffen der Alten, eine Vermuthung*, Helmstädt 1799.

A more recent contribution has been made by T. A. Richard,[70] who after an extended discussion of the subject mentions a simpler explanation of these wonderful tales.   Professor Wilson, he says, has drawn attention to the fact that the word for ant-gold, *paippilika*, was a Sanscrit name for small fragments of alluvial gold.   This word is used in the Mahábárata.   It was the gold collected by the *pippilikis* or ants.   The adjectival form of this word is *paippilika*; it means, not the gold collected by the ants, but the ant-like gold, that is to say, rounded grains of gold that resembled ants in form.   If this be the true origin of the fable the ancient writers have certainly adorned a simple fact with a marvellous wealth of "circumstance and much embroidery."

[70] *The Story of the Gold-digging Ants*, The University of California *Chronicle*, Jan. 1930.

CHAPTER XIV

CONCLUSION

Our somewhat extended review of the evolution of the geological sciences down to the earlier part of the nineteenth century has now been brought to a close. In tracing out the gradual emergence of geological ideas it has been necessary to go back to the earliest literature in Europe—that of the ancient Grecian peoples.

The Greeks, as all agree, were a race of remarkable intellectual power and were guided by a disinterested spirit of free enquiry. As Sarton says, their scientific achievements were incomparably greater than those of all the centuries following until the seventeenth. While this is true of the sciences as a whole, it must be acknowledged that their contributions to the geological sciences were few and of small account.

When the remarkable contributions made to zoölogy by Aristotle in his *Historia Animalium* and the work of Theophrastus on botany are considered, it would seem that the failure of the Greeks to make similar advances in the geological sciences was due in part at least, to their having neglected the study of these sciences to devote their attention rather to mathematics, the physical sciences and to the study of animate nature. It may be however, since but little of the early literature of Greece has survived, that more important studies of inanimate nature were made, but have been lost.

There is no doubt however that many of the Greeks, as Schvarez has remarked, chose to speculate concerning the phenomena of nature, rather than to follow the path of extended observation with a view to arriving at a true understanding of the character and meaning of natural phenomena. It is interesting for us now, he says, looking back through the intervening centuries, to think what the results would have been, had the Greeks followed this latter path, since the conditions presented for research and investigation in classical times, free and untrammelled by any hostile influences from church or state, were in many respects ideal and not found again in Europe until the age of the Renaissance.

Seneca, who lived much nearer to those times than we do, in reviewing the opinions concerning the origin of earthquakes put forward by the old Greek writers says:

These old views were crude and inexact, men were groping their way round truth, it was a task demanding great courage to remove the veil that hid nature and not satisfied with a superficial view, to look below the surface and dive into the secrets of the Gods. A great contribution to discovery was made by the man who first conceived the hope of its possibility. *We must therefore listen indulgently to the ancients.*

In looking for the reason why the geological sciences failed to make any substantial progress during the long period of the Dark Ages and Middle Ages and indeed until the early part of the Renaissance, it has been seen that certain opposing influences prevailed during these times. Albertus Magnus and his pupil Thomas Aquinas attempted to weld together the teaching of Aristotle and the theology of the church fathers and thus founded the Scholastic Philosophy. The scholastic system gave rise to endless speculation and interminable discussion. Under it arose the quaint and even ridiculous explanations of natural processes referred to in the introductory chapter of this book. As a field for the display of intellectual adroitness it was unsurpassed, but for the development of the natural sciences and as a path to the true understanding of the phenomena of nature it was useless. It hampered progress instead of advancing it.

The rise of experimental philosophy ushered in a new day and was marked by two features; its opposition to authority and its appeal to experience. One of the earliest and most important exponents of the new movement was the Franciscan Roger Bacon (1214–1294), a contemporary of Thomas Aquinas and a thinker far in advance of his time.

Born in England near Ilchester, he studied with great distinction at the universities of Oxford and Paris, and thus acquired an intimate knowledge of the academic learning of the time. But he went further and opened up, by his experimental investigations, in spite of much opposition and discouragement, new fields of knowledge in the natural and physical sciences, writing upon the construction of optical lenses, telescopes, burning glasses and clocks, as well as many other subjects which soon after his death attracted widespread attention.

Roger Bacon showed from the testimony of the philosophers themselves, that the authority of antiquity, and especially of Aristotle, was not infallible. He urged the study of mathematics and the use of experiment, and he predicted by their aid a splendid advance for human knowledge. Centuries later, in the time of the Renaissance when all obstacles to freedom of thought and investigation had disappeared and men could pursue their studies without let or hindrance, as they had done in classical times, the path to a true understanding of the physical and natural sciences by experiment and observation, indicated by Roger Bacon, was followed and was developed into a great highway for learning by Leonardo da Vinci, Descartes and Francis Bacon, one in which all students of these sciences now pursue their way. The results obtained already have far exceeded the anticipations of its great herald, Roger Bacon.

In the more recent development of the sciences, the great *scientific societies* and *academies* in many lands have played and continue to play a very important part. Among the earliest of these was the *Accademia del Cimento* (meaning Trial or Experiment), founded in Italy in 1657, which has had as its motto the words, "Provando e Riprovando." The Royal Society of London was founded in 1662 "to encourage the study of experimental knowledge," and similar great national societies have been constituted in every land. Thomas Sprat, one of the original Fellows of the Royal Society of London and afterwards Bishop of Rochester concludes his *History of the Royal Society of London for the improving of Natural Knowledge, London, 1667*, with a quaint comparison of the Old and New Philosophies:

While the Old Philosophy could only at best pretend to the Portion of Napthali[1] *to give goodly words*, the New will have the Blessings of Joseph the younger and beloved son. *It shall be like a fruitful Bough, even a fruitful Bough by a Well, whose Branches run over the Wall. It shall have the blessings of the Heaven above and the blessings of the deep that lies under.* While the Old World only bestows on us some barren Terms and Notions, the New shall impart to us the uses of all the Creatures, and shall enrich us with the Benefits of Fruitfulness and Plenty.

Looking out into the future it is evident that this promise of fruitfulness will, in the geological sciences, be realized in an ever increasing measure as the years roll on. With each new discovery the field of the unknown expands displaying new vistas to be explored and conquered by further research.

Within the past few decades great problems, as for instance, those of the Constitution of Matter, the Age of the Earth and the Duration of Successive Periods of its History have been solved, or are in the process of solution, through the discovery and application of new methods of research which were unthought of a century ago. As an example of but one of the promising new fields in geological investigation which is

[1] *Genesis:* 49– vs. 21–25.

now opening up, the study of the character, depth and configuration of the ocean beds by sonic sounding may be instanced. By this method very recently discovered, not only may maps showing the topography of the deepest ocean beds be constructed, but even the character of the materials composing it, whether hard rock, sand or silt, can be ascertained[2] by observations made from a ship sailing on the surface. Furthermore by means of an ingenious apparatus designed by C. S. Piggot[3] (of the Geophysical Laboratory of the Carnegie Institute of Washington) cores 10 to 14 feet in length representing vertical sections through the deposits which form the ocean floor may now be secured, instead of merely surface scrapings by dredges drawn over the bottom, as formerly.

In manu Domini ſunt omnes ſines terræ.

FIG. 79. Illustration at end of *Joannis Velcurionis Commentarii in Universam Physicam Aristotelis*, Tubingen, 1547.

The new facts thus discovered are leading to a knowledge of the character and structure of that three quarters of the earth's surface which is mantled by the waters of the oceans, which will throw a new light on many of the outstanding problems of geology.[4]

[2] R. Raven-Hart. *Inaudible Sounds.* In *Discovery.* Jan. 1937.

[3] Charles S. Piggot, *Core Samples of the ocean bottom.* Smithsonian Report for 1936.

[4] See Richard M. Field, *Recent Developments in the Geophysical Study of Ocean Basins.* Trans. American Geophysical Union (National Research Council) 17th Annual Meeting, Washington, 1936, Part 1, p. 20.

Again, recent discoveries in seismology[5] are revealing many facts of the utmost importance concerning the nature of the interior of the earth, a field of research with reference to which speculation and conjecture, as has been seen, have busied themselves throughout all the ages. These together with the results of investigations into the elastic constants of rocks have made it certain that the earth does not consist of a crust floating over a thin liquid, but that it is as a whole "solid" and that below about 300 kilometers, the temperature is nearly the same as it was originally. It has also been found that many of the most violent earthquakes have no connection with volcanoes or volcanic action in the outer portion of the earth's crust, but are of deep seated or plutonic origin, with their source in some cases 1000 kilometers from the surface. "In the study of the earth's deeper structure seismology has no rival. In attempting to paint a picture of the deeper parts of the earth, the best we can do at present is to draw the outlines. Perhaps future developments may enable us to make a bolder drawing and even to fill in something of form and colour."

And so the "Splendid Advance" in the field of human knowledge, anticipated by Roger Bacon, continues and will continue, extending and widening out over the whole arena of geological science. It is now realized that this progress is based on the discovery of new facts and the acquisition of new knowledge. The continued and urgent need is for facts and more facts. These will be gathered year by year, through the diligent and careful work of many men in many lands and, as time advances, progress will be sure although it may be slow, for as the author of the Novum Organum has said:

Truth is the daughter of Time

[5] James B. Macelwane, *Modern Trends in Seismological Research. Trans. American Geophysical Union*, Washington, 1936, Part 1, p. 23. and
Leason H. Adams, *The Earth's Interior its Nature and Composition. The Scientific Monthly*, Vol. XLIV. March, 1937.

INDEX

Academia della Fama, 345.

Achates, 152.

Adamas, 153.

Aetites, (Etites or Eagle Stone) 29, 98, 151.

Agate, 50.

Agatharchides (or Agatharchus), 9, 21; holds that metals grow from their own "seeds," 289.

Agricola, Georg, 19; first clearly set forth the theory of the "Lapidifying Juice," 93; his *De Natura Fossilium*, the first handbook of Mineralogy in the modern sense of the term, 93; called the "Father of Mineralogy," 93, 175; his *De Ortu et Causis Subterraneorum*, 93; recognized the origin of the plastic rocks, 94; on "Donneräxte," 122; outline of his life and work, 183; his books, 188; his classification of "Inanimate subterranean bodies," 191; his belief that the heat of the earth's interior was due to the combustion of beds of coal or other inflammable materials, 282; his views on the source of subterranean vapors were very vague, 282; views of "common people" concerning the creation of the earth, according to Agricola, 305; leader of the "Freiberg School," 308; on the origin of mountains, 342; description of the Cycle of Erosion, 343; on the origin of rivers, 445.

Alabandina, 150.

Albertus Magnus, 56, 82; his *De Mineralibus*, 144, 160; "Plastic Force" at work in nature, 254; origin of mountains, 335.

Alchemists, believed that each planet influenced the development of a particular metal within the earth, 282; this view was supported by Roger Bacon and by Thomas Aquinas, 282; the seven metals and seven planets, 283; this opinion opposed by Becher, 285; sought in the laboratory to bring about rapidly, the change of the baser metals into gold which was going on more slowly within the earth, 298.

Aldrovandus, Ulysses, 165; his *Musaeum Metallicum* issued after his death by Bartholomeus Ambrosinus, 166.

Alectoria, 104; 151.

Alexandrine Lapidaries, the, 28.

Alstedius, 86.

Amber, 25, 44; stories with reference to, 467.

Amethystus, 152.

Anaxagoras of Glazomenae, 11; holds that metals grow from their own "seeds," 289; considered "fire" to be the cause of earthquakes, 400; on the origin of springs and rivers, 427.

Anaximander of Miletus, 10, 13.

Anaximenes, view governing the origin of earthquakes, 400.

Andreasberg, inception of mining at, 171.

Androdamas, 150.

Annaberg, inception of mining at, 171.

Ambrose, on the origin of Mountains, 332.

Ants which dig for gold, 483.

Aphroselenon (Selenite), 25.

Apistos, 150.

Aquamarine, 25.

Aquilinus, 104.

Arabian learning, age of, 55; introduction into Europe, 55; in middle ages, 169.

Arabs, their conquests, 54.

Archelaus, G., view concerning the origin of earthquakes, 400.

Arduino, Giovanni, 373; called by Ronconi the "Father of Modern Italian Geology," 373; recognizes a four-fold division in the rocks of the earth's crust, 373.

Argonauts, the, 484.

Aristotle, 11, 12; founder of the Lyceum, 15; his writings, 16; his four "elements" and two "exhalations," 17, 81; source of his *Die Mineralibus*, 18, 55; views concerning the origin of metals, minerals and stones, 277; his "Light Exhalation" and "Dark Exhalation," 277; origin of gems, 278; his belief that within the earth the baser metals gradually changed into gold, 297; on birth of the Aeolian Islands by the escape of winds from the interior of the earth, 331; on

latent in it may be changed into a metal, 293; informed by Cornish miners that the tin ore grew again in exhausted tailings, if these were allowed to remain undisturbed for long periods of time, 294; appointed by the Royal Society of London to make inquiries in various mining regions in Europe whether the growth of ores in veins had been observed, 295.

Breislak, Scipion, 320; origin of ore deposits through magmatic segregation, 320; a vulcanist, 320, 321.

Brongniart, Alex., 267.

Brontia, believed to have fallen from the skies, 118; called "Donnerkeile," 118.

Brunner, Joseph, on origin of ore bodies, 321.

Buch, Leopold von, 381; Craters of Elevation, 381; his study of the volcanoes of the Auvergne, 382; Geikie's reference to him, 385; his letter to Murchison, 385.

Buffon, his *Histoire Naturelle*, 209; 266; held that metals and their ores grew in a manner analogous to the growth of plants, 296.

Bufonites, from head of a toad, 110.

Burnet, Thomas, 209; his "Sacred Theory of the Earth," 209.

Bury, Richard de, 6; his Philobiblon in which the word "Geologia" first appeared being used to designate the Study of Law (the "earthly science" in antithesis to theology), 166.

Caesius, his work, 164.

Caius Suetonius Tranquillus, 48.

Calcedonius, 151, 152.

Calcophonus, 150.

Callistratus, 9.

Cantimpratensis, Thomas, 141.

Carbuncles and Dragons, 478.

Cardan, on the origin of earthquakes, 409.

Cato, ("The Censor"), his *De Re Rustica*, 48.

Catochites, 50.

Ceraunia, found only in places which had been struck by lightning, 118; 151.

Cesalpino, Andrea, 161; his recognition of the true nature of fossils, 261.

Chalazias, 151.

Chalcite, 49.

Charpentier, J. F. W., on the origin and nature of ore deposits, 313.

Chelidonius, 104.

Chelonites, 151.

Chemical classification of minerals introduced at opening of eighteenth century, 200.

Chrysolectrus, 153.

Chrysolite, 32, 153.

Chrysopasius, 151.

Chrysoprasus, 152.

Cimedia, 104.

Cinedios, 29.

Cinnabar, 35, 43.

Clarke, Edward D., 380; reference by him to Pallas, 381; remark concerning him by Sedgwick, 381.

Clave, Etienne de, 86.

Cleaveland, Parker, 276.

Collini, C., difficulties presented by the study of the nature and origin of mountain ranges, 398.

Colonne, Marie, Pompée, 363; some views concerning the origin of mountains; they "grow like plants," 363.

Conception of the Universe in the Middle Ages, 51.

Coral, 32, 130, 131, 132, 151.

Corneolus, 152.

Creation of the earth, 305; views of "common people" according to Agricola, 305; Franciscus Rueus on this question, 306; also Della Frata, 306; Barba, 306; Löhneyss, 307.

Crystal, (Rock Crystal) 50; considered to be ice, 44, 127, 133, 352, 472.

Crystallus, 152.

Cube, Johann von, his *Hortus Sanitatis* 5, 142.

Cuvier, Baron Leop. Georges, his account of Werner's Lectures, 214; his life and work, 263; his belief in a succession of great and sudden catastrophes, 266; the founder of Vertebrate Palaeontology, 268.

Cyranides, the, 28.

Damigeron, 32.

Dante Aligheri, 56; on the elevation of land, 341.

Featherstonhaugh and to William Smith, 275.

Murray, John, his *Comparative view of the Huttonian and Neptunian systems in Geology*, 227.

"Natural History" method of classifying minerals, 200.

Neckam, Alexander; foster-brother of King Richard I of England, 138; his works on mineralogy, 139, 140.

Nemesite, 29.

Neptunian Theory, its final overthrow, 245.

Niavis, (Paul Schneevogel) author of the *Jovis Judicium*, 171.

Nicanor, 9.

Nicias of Malea, 9.

Nicols, Thomas, 163; the three issues of his lapidary, 163.

Noachian Deluge, speculations on, by Medieval writers, 331.

Norton, Thomas, his *Ordinale of Alchemy*, 280.

Ochres, 43.

Ocytocios, 30.

Olaus Magnus, 303; thunder and lightning caused by poisonous exhalations issuing from rich mineral veins, 303.

Ombria, believed to have fallen from the skies, 118; called "Regensteine," 118.

Onyx, 30, 152.

Opal, 32.

Ophthalmius, 151.

Oppel, von, first recognized the difference between veins and bedded ore deposits, ("flotze") 311; his *Bericht von Bergbau*, 212.

Ore Deposits, in how far has modern science advanced toward a true understanding of their nature and origin, 322.

Orites, 151.

Orpheus, 9.

Oryktognosie, term coined by Werner for "Determinative" or "Practical" mineralogy, 216.

Pactolus, 31

Paeanites, 50

Palaeontology, birth of, 263.

Palissy, Bernard, the three kinds of water, 90, his recognition of the true nature of Fossils, 261; his life and work, 446; first to recognize that rain and melting snow were the source from which springs and rivers derived their waters, 446.

Pallas, Peter Simon, 378; his studies of mountain ranges, 379, especially in Russia, 379; Primitive granite mountains, 379; Secondary and Tertiary mountains, 379; reference to him by Clarke, 381.

Panchorus, 29.

Pantherus, 151.

Papin, holds that the surface of the ocean is higher than the land and hence the ocean waters rise through the earth and issue as rivers, 441.

Paracelsus, origin of pumice by consolidation of sea foam, 91; 284; refers to the "First Ens" of gold and antimony, which is in the form of a red liquor, 284.

Passeri, the Abbé, describing his visit to the caverns in Monte Cucco, 126.

Peanites, 151.

Period embraced by the present work, 3.

Peripatetics, 15.

Pernumia, Johannis, on Juvenile waters, 326.

Perrault, Pierre, his *L'Origine des Fontaines* first showed by actual measurement that, in the drainage area of the Seine, the rainfall was in excess of the "run off," 448.

Piggot, C. S., investigations on character of the ocean floor, 492.

Plato, founder of the Academy, 14.

Playfiar, 239; his *Illustrations of the Huttonian Theory*, 239.

Pliny, the Elder, 36; his *Natural History*, 37; his account of ancient methods of mining, 42; on rock crystal, 44; on amber, 44; on the origin of mountains, 331.

Pliny the Younger, 39; his account of the method of life of the Elder Pliny, 36; account of the death of the Elder Pliny at eruption of Vesuvius, 46.

Plot, Robert, views on the nature and origin of fossils, 258.

Porta, his *Phytognomonica*, 69.

Porus, 30.

A CATALOG OF SELECTED
DOVER BOOKS
IN ALL FIELDS OF INTEREST

A CATALOG OF SELECTED DOVER
BOOKS IN ALL FIELDS OF INTEREST

DRAWINGS OF REMBRANDT, edited by Seymour Slive. Updated Lippmann, Hofstede de Groot edition, with definitive scholarly apparatus. All portraits, biblical sketches, landscapes, nudes. Oriental figures, classical studies, together with selection of work by followers. 550 illustrations. Total of 630pp. 9⅛ × 12¼.
21485-0, 21486-9 Pa., Two-vol. set $25.00

GHOST AND HORROR STORIES OF AMBROSE BIERCE, Ambrose Bierce. 24 tales vividly imagined, strangely prophetic, and decades ahead of their time in technical skill: "The Damned Thing," "An Inhabitant of Carcosa," "The Eyes of the Panther," "Moxon's Master," and 20 more. 199pp. 5⅜ × 8½. 20767-6 Pa. $3.95

ETHICAL WRITINGS OF MAIMONIDES, Maimonides. Most significant ethical works of great medieval sage, newly translated for utmost precision, readability. Laws Concerning Character Traits, Eight Chapters, more. 192pp. 5⅜ × 8½.
24522-5 Pa. $4.50

THE EXPLORATION OF THE COLORADO RIVER AND ITS CANYONS, J. W. Powell. Full text of Powell's 1,000-mile expedition down the fabled Colorado in 1869. Superb account of terrain, geology, vegetation, Indians, famine, mutiny, treacherous rapids, mighty canyons, during exploration of last unknown part of continental U.S. 400pp. 5⅜ × 8½. 20094-9 Pa. $6.95

HISTORY OF PHILOSOPHY, Julián Marías. Clearest one-volume history on the market. Every major philosopher and dozens of others, to Existentialism and later. 505pp. 5⅜ × 8½. 21739-6 Pa. $8.50

ALL ABOUT LIGHTNING, Martin A. Uman. Highly readable non-technical survey of nature and causes of lightning, thunderstorms, ball lightning, St. Elmo's Fire, much more. Illustrated. 192pp. 5⅜ × 8½. 25237-X Pa. $5.95

SAILING ALONE AROUND THE WORLD, Captain Joshua Slocum. First man to sail around the world, alone, in small boat. One of great feats of seamanship told in delightful manner. 67 illustrations. 294pp. 5⅜ × 8½. 20326-3 Pa. $4.95

LETTERS AND NOTES ON THE MANNERS, CUSTOMS AND CONDITIONS OF THE NORTH AMERICAN INDIANS, George Catlin. Classic account of life among Plains Indians: ceremonies, hunt, warfare, etc. 312 plates. 572pp. of text. 6⅛ × 9¼. 22118-0, 22119-9 Pa. Two-vol. set $15.90

ALASKA: The Harriman Expedition, 1899, John Burroughs, John Muir, et al. Informative, engrossing accounts of two-month, 9,000-mile expedition. Native peoples, wildlife, forests, geography, salmon industry, glaciers, more. Profusely illustrated. 240 black-and-white line drawings. 124 black-and-white photographs. 3 maps. Index. 576pp. 5⅜ × 8½. 25109-8 Pa. $11.95

CATALOG OF DOVER BOOKS

AMERICAN CLIPPER SHIPS: 1833–1858, Octavius T. Howe & Frederick C. Matthews. Fully-illustrated, encyclopedic review of 352 clipper ships from the period of America's greatest maritime supremacy. Introduction. 109 halftones. 5 black-and-white line illustrations. Index. Total of 928pp. 5⅜ × 8½.
25115-2, 25116-0 Pa., Two-vol. set $17.90

TOWARDS A NEW ARCHITECTURE, Le Corbusier. Pioneering manifesto by great architect, near legendary founder of "International School." Technical and aesthetic theories, views on industry, economics, relation of form to function, "mass-production spirit," much more. Profusely illustrated. Unabridged translation of 13th French edition. Introduction by Frederick Etchells. 320pp. 6⅛ × 9¼. (Available in U.S. only)
25023-7 Pa. $8.95

THE BOOK OF KELLS, edited by Blanche Cirker. Inexpensive collection of 32 full-color, full-page plates from the greatest illuminated manuscript of the Middle Ages, painstakingly reproduced from rare facsimile edition. Publisher's Note. Captions. 32pp. 9⅜ × 12¼.
24345-1 Pa. $4.95

BEST SCIENCE FICTION STORIES OF H. G. WELLS, H. G. Wells. Full novel *The Invisible Man,* plus 17 short stories: "The Crystal Egg," "Aepyornis Island," "The Strange Orchid," etc. 303pp. 5⅜ × 8½. (Available in U.S. only)
21531-8 Pa. $4.95

AMERICAN SAILING SHIPS: Their Plans and History, Charles G. Davis. Photos, construction details of schooners, frigates, clippers, other sailcraft of 18th to early 20th centuries—plus entertaining discourse on design, rigging, nautical lore, much more. 137 black-and-white illustrations. 240pp. 6⅛ × 9¼.
24658-2 Pa. $5.95

ENTERTAINING MATHEMATICAL PUZZLES, Martin Gardner. Selection of author's favorite conundrums involving arithmetic, money, speed, etc., with lively commentary. Complete solutions. 112pp. 5⅜ × 8½.
25211-6 Pa. $2.95

THE WILL TO BELIEVE, HUMAN IMMORTALITY, William James. Two books bound together. Effect of irrational on logical, and arguments for human immortality. 402pp. 5⅜ × 8½.
20291-7 Pa. $7.50

THE HAUNTED MONASTERY and THE CHINESE MAZE MURDERS, Robert Van Gulik. 2 full novels by Van Gulik continue adventures of Judge Dee and his companions. An evil Taoist monastery, seemingly supernatural events; overgrown topiary maze that hides strange crimes. Set in 7th-century China. 27 illustrations. 328pp. 5⅜ × 8½.
23502-5 Pa. $5.95

CELEBRATED CASES OF JUDGE DEE (DEE GOONG AN), translated by Robert Van Gulik. Authentic 18th-century Chinese detective novel; Dee and associates solve three interlocked cases. Led to Van Gulik's own stories with same characters. Extensive introduction. 9 illustrations. 237pp. 5⅜ × 8½.
23337-5 Pa. $4.95

Prices subject to change without notice.
Available at your book dealer or write for free catalog to Dept. GI, Dover Publications, Inc., 31 East 2nd St., Mineola, N.Y. 11501. Dover publishes more than 175 books each year on science, elementary and advanced mathematics, biology, music, art, literary history, social sciences and other areas.